D0936336

# Springer Complexity

Springer Complexity is a publication program, cutting across all traditional disciplines of sciences as well as engineering, economics, medicine, psychology and computer sciences, which is aimed at researchers, students and practitioners working in the field of complex systems. Complex Systems are systems that comprise many interacting parts with the ability to generate a new quality of macroscopic collective behavior through self-organization, e.g., the spontaneous formation of temporal, spatial or functional structures. This recognition, that the collective behavior of the whole system cannot be simply inferred from the understanding of the behavior of the individual components, has led to various new concepts and sophisticated tools of complexity. The main concepts and tools – with sometimes overlapping contents and methodologies – are the theories of self-organization, complex systems, synergetics, dynamical systems, turbulence, catastrophes, instabilities, nonlinearity, stochastic processes, chaos, neural networks, cellular automata, adaptive systems, and genetic algorithms.

The topics treated within Springer Complexity are as diverse as lasers or fluids in physics, machine cutting phenomena of workpieces or electric circuits with feedback in engineering, growth of crystals or pattern formation in chemistry, morphogenesis in biology, brain function in neurology, behavior of stock exchange rates in economics, or the formation of public opinion in sociology. All these seemingly quite different kinds of structure formation have a number of important features and underlying structures in common. These deep structural similarities can be exploited to transfer analytical methods and understanding from one field to another. The Springer Complexity program therefore seeks to foster cross-fertilization between the disciplines and a dialogue between theoreticians and experimentalists for a deeper understanding of the general structure and behavior of complex systems.

The program consists of individual books, books series such as "Springer Series in Synergetics", "Institute of Nonlinear Science", "Physics of Neural Networks", and "Understanding Complex Systems", as well as various journals.

# Understanding Complex Systems



## Understanding Complex Systems

Future scientific and technological developments in many fields will necessarily depend upon coming to grips with complex systems. Such systems are complex in both their composition (typically many different kinds of components interacting with each other and their environments on multiple levels) and in the rich diversity of behavior of which they are capable. The Springer Series in Understanding Complex Systems series (UCS) promotes new strategies and paradigms for understanding and realizing applications of complex systems research in a wide variety of fields and endeavors. UCS is explicitly transdisciplinary. It has three main goals: First, to elaborate the concepts, methods and tools of self-organizing dynamical systems at all levels of description and in all scientific fields, especially newly emerging areas within the Life, Social, Behavioral, Economic, Neuro- and Cognitive Sciences (and derivatives thereof); second, to encourage novel applications of these ideas in various fields of Engineering and Computation such as robotics, nano-technology and informatics; third, to provide a single forum within which commonalities and differences in the workings of complex systems may be discerned, hence leading to deeper insight and understanding. UCS will publish monographs and selected edited contributions from specialized conferences and workshops aimed at communicating new findings to a large multidisciplinary audience.

K. Kaneko

# Life: An Introduction to Complex Systems Biology

With 127 Figures

Professor Dr. Kunihiko Kaneko
University of Tokyo
Department of Pure and Applied Sciences
Komaba, 153-8902
Tokyo, Japan
E-mail: kaneko@complex.c.u-tokyo.ac.jp

Library of Congress Control Number: 2006923819

ISSN 1860-0840
ISBN-10 3-540-32666-9 Springer Berlin Heidelberg New York
ISBN-13 978-3-540-32666-3 Springer Berlin Heidelberg New York



Springer is a part of Springer Science+Business Media
springer.com

Printed in The Netherlands



Typesetting: by the author and techbooks using a Springer LaTeX macro package
Cover design: *design & production* GmbH, Heidelberg

Printed on acid-free paper SPIN: 11501534 54/techbooks 5 4 3 2 1 0

# Preface

What is life? What type of system is life? How can we understand life? Or what does "understanding life" really mean? Sixty years since the publication of *What Is Life?* by Schrödinger and after the rise and success of molecular biology, have we reached the answer to these questions?

In the recent years, I have often been asked by young researchers and students in biology: "**I am afraid that such basic questions on a life system itself are not answered by the main-stream approach of current biology** that elucidates molecules and genes. We need some alternative approach. What am I to do?" They are satisfied neither with the current trend in bioinformatics nor with the detailed computer models, and are striving for a framework complementary to molecular biology, a one that does not rely on enumerative approach.

Responding to these voices, I have explained approaches my colleagues and I have been taking both theoretically and experimentally, in lectures and seminars. Although they show much interest, introduction of these rather interdisciplinary style of research is not easy, let alone discussing how we can understand life. Of course they ask for some books that describe a theoretical basis of our approach and the summaries of the recent studies. My desire to answer these requests from the students and researchers was the main force that had driven me to write the present book.

On the other hand, those working in nonlinear dynamics and theoretical physics have strived to set up a novel theoretical framework that is compatible with biological systems. They are both interested in understanding life as a part of universal dynamic processes in nature and in finding out novel concepts in physics, inspired by and suitable to biology. Indeed, when researchers embarked on the study of "complex systems" about two decades ago, their main motivation was to seek a theory to understand biological systems. However, studies in complex systems so far in most cases have remained as studies of dynamical systems with many elements or novel type of (nonequilibrium) phenomena. Although both dynamical systems and statistical physics may provide some theoretical framework for the study of biological systems, they by themselves are not sufficient. We need something beyond , in order to understand life system. In the present book, I also intend to answer such requests from physicists' side, as much as I can do.

To meet these two requests from rather different sides, I have made some effort in organizing the book.

The first three chapters form the "basic section" of the book. In Chap. 1, we critically review the current status of molecular biology. Of course, just making criticisms is not so difficult. Most people who have expressed some doubt on molecular biology to me share these criticisms, but what they need is a proposition of an alternative approach, what I here call "complex systems biology." It is based on constructive and dynamical systems approach, and intended to understand the basic features of a biological system – metabolism, heredity, development, evolution, and so forth. In Chap. 2, I outline the "constructive biology" approach, where one "constructs" such basic features in life both experimentally (in a laboratory) and theoretically (in a computer model) to "understand" them. In Chap. 3, some basic background of dynamical systems and statistical physics is described, as a basis for the studies described in the later chapters. Most part of this chapter (except the last few sections) is elementary, but here I explain the basic concepts rather intuitively in relationship with biology.

Following, and based on, the three introductory chapters, I treat basic problems in life systems in Chaps. 4–11. The problems include heredity (Chap. 4), reproduction and metabolism (Chaps. 5 and 6), cell differentiation, development, and morphogenesis (Chaps. 7–9), evolution in relationship with biological plasticity (Chap. 10), and speciation for diversification (Chap. 11). Here, to meet the requests from the two sides, I have organized each chapter as follows: (i) basic question to be addressed, (ii) outline of the logics to answer the question, (iii) theoretical model embodying the logics, (iv) numerical results of the model and theoretical consideration, (v) corresponding "constructive" experiment (e.g., laboratory construction of replication system, cell differentiation, morphogenesis, and evolution), and (vi) relevance of the results to biology.

With this organization, one can skip some parts depending on one's background and interest. For example, biology-oriented researchers and students (with little background in theoretical physics or dynamical systems) can skip the details in parts (iii) and (iv), while for theorists, part (iv) may be more interesting. Also it is possible for college students to read only (i), (ii), and (vi) and skim through the other parts. With this organization of (i)–(vi), however, I have to admit that there remains some overlaps between parts (ii) and (iv) that justifies such skipped reading.

In Chap. 12, I look back at the previous chapters by summarizing the basic concepts proposed in the book as universal statistics in steady growth system, isologous diversification, consolidation, itinerancy, and minority control principle. Then, I discuss how one can understand biological plasticity, recursiveness, and evolvability, while emphasizing the necessity and possibility of phenomenological theory at a system level for biology.

Although most topics in the present book come from the studies at the level of cell biology, the concepts presented here can be extended to other fields in biology. For example, I believe that those interested in neuro science can find sources for inspiration, as the picture 'dynamics first and consolidation to symbols later' will be relevant also to cognitive science.

The original version of the book was published in Japanese texts at the fall of 2003, from University of Tokyo Press. Most part in Chaps. 1–3 and Chap. 12 are translation of the Japanese version, while I have updated the contents to include the advances made since then. Chapters 5 and 10 are newly added here. As for translation of Chaps. 1–3, and Chap. 12, I am grateful to Dr. Glenn Paquette. I am indebted to Dr. Philippe Marcq for critical reading and kind comments on the manuscript.

Contents of the present book are based on collaborative studies and discussions with many researchers, including the members of my laboratory as well as those from the group of Professor Tetsuya Yomo who kindly provided their experimental results. Some of the experimental results in Chapters 5 and 9 are from two colleagues in my campus, Professor Tadashi Sugawara and Professor Makoto Asashima, while theoretical studies in Chaps. 6 and 8–10 are based on the collaboration with Professor Chikara Furusawa. I sincerely thank these people for their courtesy and collaboration.

Most researches presented in the book are supported by Center-of-Excellence grant by MEXT and ERATO 'Complex Systems Biology' project by JST, Japan, to which I am grateful.

**May the present volume be a source of inspiration to find appropriate answers to the question "What is Life?"**

May 2006 *Kunihiko Kaneko*

# Contents

# 1 How Should Living Systems Be Studied?

## 1.1 What Is Life?

What is life? Almost everyone has considered this question at some time, yet it has resisted any kind of definitive answer. For example, all life forms presently existing on Earth contain information-encoding DNA molecules. Considering this fact, a simple answer to this question might be that anything containing DNA is living. In this case, however, we would have to consider an omelette or a hamburger or vichyssoise to be "alive." Indeed there is no particular molecule, including DNA, just whose presence implies life. Further more, even if we were to gather all types of molecules possessed by present-day living organisms and mix them in some way, no one would believe that the result would be something that possesses life. The conclusion of these simple considerations would seem to be that the existence of certain molecules or some chemical structure is not a sufficient condition for life.

Now, let us suppose that we have been sent into outer space with the mission of determining the presence or absence of life. Then, let us consider the situation in which on this mission, we were somehow able to determine that there exists no DNA and, even, no protein molecules beyond Earth. What can be concluded from such a result? Despite its significance, most people would probably feel uneasy – on the basis of this evidence alone – to draw the conclusion that there exists no extra-terrestrial life, as it is certainly reasonable that the molecules used by living creatures on Earth are not the only possible molecules that could play such a role. In other words, it would seem that there are some conditions other than the existence of some particular chemical structure or the presence of certain molecules that defines life. In particular, use of the molecules or chemical structures of organisms existing now on Earth cannot be considered a necessary condition for life.

From the above elementary discussion, on the basis of very intuitive ideas, it is evident that the definition of conditions for *life* should not be based on some kind of molecular characteristics. However, identifying appropriate conditions is not a simple task. Certainly, different people would cite different properties as necessary conditions for "life." As just a few examples, the ability of reproduction (i.e., the spontaneous production of nearly identical individuals), the potentiality to undergo evolution, the existence of some kind of structure separating an individual's body from the external world, the ability to adapt to the external world through interaction with it, some kind of metabolic capacity through which an individual's body is maintained, and the existence of some degree of autonomy could all be considered reasonable necessary conditions for something to be considered "living." However, no matter what conditions we list in this way, there certainly would be no end to opinions that some are too weak or some are too strong.

Despite the problems inherent in attempting to specify the conditions for life, in most cases, we have an intuitive ability to distinguish between living and nonliving things. But if we are asked to identify the criteria on which we make such a determination, we cannot. Also, we are not able to easily identify the conditions that allow us to determine at what point a living entity becomes dead. This is in spite of the fact that as an empirical fact, we are all too aware that something once dead cannot come back to life.

Perhaps at the present time, it is too difficult to directly confront the problem of identifying the conditions for life. However, it is probably an indisputable fact that there are certain properties that are common to all living systems. Here we would first like to understand why such common properties (which we will refer to here as "the universal properties of living systems") arise and under what conditions they can arise. We consider this the first step to answering the question posed by this book.

Here we have to recall that the present mainstream approach in molecular biology is to count and map all genes, proteins, and other molecules. However, does accomplishing such a task necessarily provide us with an understanding of the nature of life itself? Instead of simply constructing such a map, it will also be necessary to determine some universal properties that are common to all life and construct some theory that provides an answer to the question of why these properties necessarily emerge. No matter how difficult this may be, it cannot be avoided, in order to realize such a goal.

Seeking to identify and understand Nature's "universal properties" is a way of thinking that has been developed among physicists. Thus, the approach we take requires the "spirit" of physics. However, this is not to say that we should seek a solution to this problem by simply employing the equations of physics or applying present fields of physics. Any attempt at such a simplistic approach would reflect only the arrogance of physicists and would fail. Rather, what we need in this pursuit is the spirit of striving to discover the universal logic that underlies natural phenomena. Indeed, this is, in its essence, the true scientific spirit.[1]

[1] In this sense, to understand the discussion in this book requires almost no physics "knowledge." What is necessary is only the intellectual desire to understand the logic underlying Nature, without being biased by "common sense," and a small

The field of molecular biology was indeed pioneered with such a truly scientific motivation and spirit in attempting to find a theory of life. From the rise of molecular biology, now half a century has passed. So, we ask again, what is life? Has molecular biology given us a satisfactory answer to this question? Are we at least making steady progress toward finding an answer? If, on the other hand, we do not yet have an answer, can we say why this is? And if this is the case, can we introduce some new methodology and with this set up an effective research program? In this book, we hope to answer these questions. With this goal, in Chap. 1, we carry out a critical examination of the field of molecular biology. On the basis this discussion, in Chaps. 2 and 3, we briefly describe a complementary methodology from both theoretical and experimental perspectives. These three chapters lay the foundation of the book. In the remaining chapters, we apply in a concrete manner the methodology laid out in the first three chapters to several specific problems. In each chapter we set up a general question associated with some universal property of life, and give an answer through a synergetic approach among theory, modelling, and experiments. The origin of reproductive cellular systems and the generation of genetic information are studied in Chap. 4. The universal properties of reaction networks allowing for self-sustained cellular reproduction are considered in Chaps. 5 and 6. In Chap. 7, we investigate the phenomena of cellular differentiation and robustness of development, while in Chap. 8 we present a treatment of stem cell systems and the irreversibility of differentiation through development. The generation of positional information and stable pattern formation are studied in Chap. 9. We consider evolution in relationship with plasticity of phenotype, measured by fluctuation in Chap. x10, while the problem of speciation (species differentiation) is studied as the genetic fixation of phenotype differentiation, by combining Darwinian process and dynamical systems with interaction, in Chap. 11. In Chap. 12, on the basis of the understanding gained from the preceding chapters, we reconsider the question of how living systems are to be comprehended, and formulate a phenomenological theory of biology. With this, we reconsider the question, what is life?

## 1.2 A Half Century of Molecular Biology

More than half a century ago, physicists introduced a major trend in biology. Putting aside the question of whether or not the individual studies made within this new approach were successful, it most certainly had a strong influence on the foundation of the field of molecular biology.

Life appears in a grand variety of forms. Categorization of this variety is the main topic of natural history that indeed opened the field of biology. Later,

---

amount of imagination (in other words, faculties possessed by any curious high school student).

Darwin proposed the theory of evolution to describe the universal process underlying the formation of these categories (Darwin, 1859). Since that time, one of the main goals of biology has been to connect this process of evolution with the taxonomical description of the species that has been developed.

It is significant that Darwin's thought is quite general and written in a mathematical form. Consider a system that consists of a set of units that exhibit reproduction accompanied by some degree of variation. With the increase in the number of units that results from reproduction, there emerges a competition between units. In such a system consisting of *variation*, *reproduction*, and *selection*, there exists a universal logic. In this sense, Darwin's manner of thinking provides an extremely important first step toward understanding the universal structure of living systems. For example, if there exists some property of an organism that bestows it with a particular faculty, even if this faculty exists at some very low level initially, through evolution, it can come to attain some higher level of functioning. Thus, high-level functions, whose formation would at first sight seem unfeasible, can emerge through evolution. However, it must be kept in mind here that Darwin's theory of evolution alone does not allow us to determine what kinds of properties or functions of organisms are possible, nor does it allow us to determine whether any particular property can be realized. This leads us to the following fundamentally important question: Does there exist no universally applicable logic that would allow us to understand the emergence of the properties fundamental to living organisms?

Molecular biology attempted to describe the universal properties of the phenomena exhibited by living systems in terms of molecules. Establishing the approach of searching for such universal properties itself was a role played by physicists at the birth of the field of molecular biology. The goal of molecular biology is to trace down from the level of cells to the molecules that compose them and to understand the functioning of each molecule in such biological processes such as heredity, metabolism, and motility. It takes the following viewpoint to attain the goal. With the assumption that these molecular functions are shared by most living organisms, it should be possible to determine the structure and organization of the universal properties displayed by living organisms through studying combinations of the roles played by individual molecules. Of course, behind this assumption of universality is the assumption that the life forms under consideration all evolved from a common ancestor. This would imply that these life forms employ the same (or, at least, similar) molecules and thus the same molecular processes.

This approach of molecular biology was developed by physicists (most notably, Delbrück and collaborators) who were strongly influenced by the lectures of Neils Bohr (Bohr, 1958) concerning the application of the idea of complementarity to the study of life. Schrödinger's *What Is Life?* (1946) had a particularly strong influence on the establishment of the direction to be taken in the study of molecular biology, and particularly with regard to the search for a universal molecule responsible for the process of heredity. Later,

Delbrück laid out an effective research program for the advancement of molecular biology. One apex in the half century of the study that succeeded these efforts was the discovery by Watson and Crick of the double helix of DNA. With the revelation that this molecule is the universal agent controlling the process of heredity, the mechanism of reproduction as governed by its double helix structure became clear. Later, Watson established the methodology in the textbook *The Molecular Biology of Gene* (Watson, 1965). Even today, this book exists in an updated form entitled *The Molecular Biology of the Cell*, which is one of the most influential books in cell biology. We can understand the developments described above as the attempt through molecular biology, by describing biological phenomena in the universal terms of molecules, to make a break from the 'enumerationism" that preceded it.

Let us consider a specific example of the manner in which the approach of molecular biology is applied. Suppose that we are interested in the motor activity of an animal. Then, we would, of course, focus on muscles. This would lead us to study fibers from which muscles are composed, and then eventually to the underlying proteins. Doing so, we would find two types of proteins, actin and myosin. The actin forms a rail, while the myosin sometime attaches this rail protein, and moves along it, to generate a force, by consuming the chemical energy (supplied by ATP). Furthermore, such type of motor proteins then have turned out to be used in a variety of organisms, including unicellular organisms, suggesting the universality at a molecular level. Then, we could investigate the nature of the chemical reaction process with these proteins that takes place to cause the motor activity and the manner in which chemical energy is converted into mechanical energy. Having elucidated such processes, we would then – following the established program of molecular biology – naturally turn to the task of determining to what extent this type of protein structure is common among species, and doing so, we would come to understand the commonality of the mechanism of motor activity.

Now that half a century has passed, molecular biology has become an essential part of our current description of the phenomena of living systems. Roughly stated, the present approach of molecular biology can be summarized as follows (see, e.g., Alberts et al., 1983).

1) Consider a system at the microscopic level. Then, at that level, identify the molecules or genes that are important in some function in question and the role played by each.
2) Of course, such molecules do not act as independent entities but, rather, through their reactions and interactions with other molecules. Thus, we are to determine the manner in which the various microscopic elements combine and explicate the rules governing such behavior. For example, let us consider the case of genes. Here, as the result of a certain gene being expressed, that is, a particular mRNA is synthesized, which results in the production of a corresponding protein. Then it acts as an enzyme to activate and control some other gene. The rules according to which these

various interactions are carried out and such activation and control are exercised represent the target of such study.

3) Through such cooperative activity of the microscopic components, the macroscopic behavior of an organism is realized. Therefore, we must determine the rules according to which such components combine and how, as a result, macroscopic functions are carried out. For example, we are to discover how all of the genes within a given genome are wonderfully combined to call forth some life function.
4) In most cases, the number and variety of molecules or genes that act together are large. Therefore, the number of possible ways in which combinations of such microscopic elements can be realized is enormous. We thus face the task of picking out of such combinatorial complexity the particular sets of relationships among the elements that are actually selected and realized in living organisms.

For example, within present-day cellular biology, the following is a scenario for cell differentiation. When the concentration of some signal molecule impinging on a cell membrane is increased beyond some threshold value, a chemical reaction takes place through which some gene is caused to express. As a result, some protein is produced, and the state of the cell thereby undergoes a change. Thus, according to this description, the concentration of some molecule controls an on–off type switching of a gene, and this, in turn, causes the execution and cessation of some function; that is, within this scenario, this gene plays the role of the on–off switch of this function. In fact, the search for such one-to-one correspondence between individual genes and individual functions is a mainstream research in the field of cellular biology.

Of course, in some cases the type of one-to-one correspondence described above does not exist. For example, through the action of one gene some protein is produced, while through the action of a second gene some other protein is produced, and as a result of a reaction involving these two types of proteins some function comes to be carried out. Indeed, in myriad cases the exercise of some faculty results from such interactive effects of multiple genes. For this reason, research within the so-called "genome project" is now proceeding to determine how the expressions of multiple genes combine to yield the functions found within organisms.

For example, from this point of view, the cell differentiation that occurs during development, simply stated, consists of a combination of processes in which a gene is turned on (or off), when the concentration of some signal molecule exceeds a threshold value. More explicitly, a typical process would proceed something like the following: Through the turning on and off of one gene, as controlled by one signal molecule, the properties of a cell come to change, and as a result, the concentration of some other signal molecule is altered, with the effect of turning on or off some other gene. Such a process could be described as a sequence of *if...then* program (i.e., a combination of logical operations), and therefore, this point of view leads to the understanding that the process of differentiation can, through a step-by-step analysis, be broken down and written entirely by logical expressions.

The descriptive capability of present-day molecular biology and the success of its application, in particular to medicine, obviously cannot be denied. However, for the purpose of understanding what life is and, specifically, to elucidate the universal properties possessed by living systems, a description based on molecular biology encounters some difficulties. Below we give a brief outline of some of these problems.

Before proceeding, we note here that the point of view described above is not peculiar to the field of molecular biology but, indeed, represents the mainstream in all fields of biology. For example, in the field of brain research, a method of investigation commonly employed consists of (i) elucidating the function of individual neurons or area of the brain, (ii) investigating synaptic connections or interarea connections with regard to the effects of inhibition and excitation, and (iii) obtaining a global description by combining the individual neuronal activities deduced in (i) according to the results of (ii) (the neural network approach).

## 1.3 The Reemergence of *Diversity* and the Enumerative "-ome" Doctrine

With the advance of molecular biology, discovery by discovery, the number of molecules known to participate in biological processes has drastically increased, and it has come to be realized that the molecular diversity possessed by living organisms far exceeds that imagined by the pioneers in the field. Of course, among the ever-growing number of molecular species found to be connected with biological processes, some can be considered relatively important (e.g., genes) and some relatively unimportant. However, such a distinction is of a comparative nature, and despite the fact that such a distinction can be made, it is undeniable that the number of molecules that must be taken into account to obtain a molecular description of living systems is truly vast.

Now, suppose that we wish to construct some biological faculty. Then suppose that, to this end, we set out to create some kind of molecular system whose function can be described by the simple set of rules that if particular conditions are met, certain actions result. To create a system of this kind, a large variety of molecules would not be necessary. Instead, it would seem much more convenient to realize such a process using a small number of molecule species, with the simple prescription that the presence or absence of each determines whether a given action is to be carried out. Such a simple system would be sufficient to carry out a function described above. Why then are, for example, genetic networks, metabolic networks, and signal transduction systems as complicated as they are? It would seem evident that simple combinations of molecules with its own role and the combinations of them to form some optimal faculty is all that we need. Let us pursue this point.

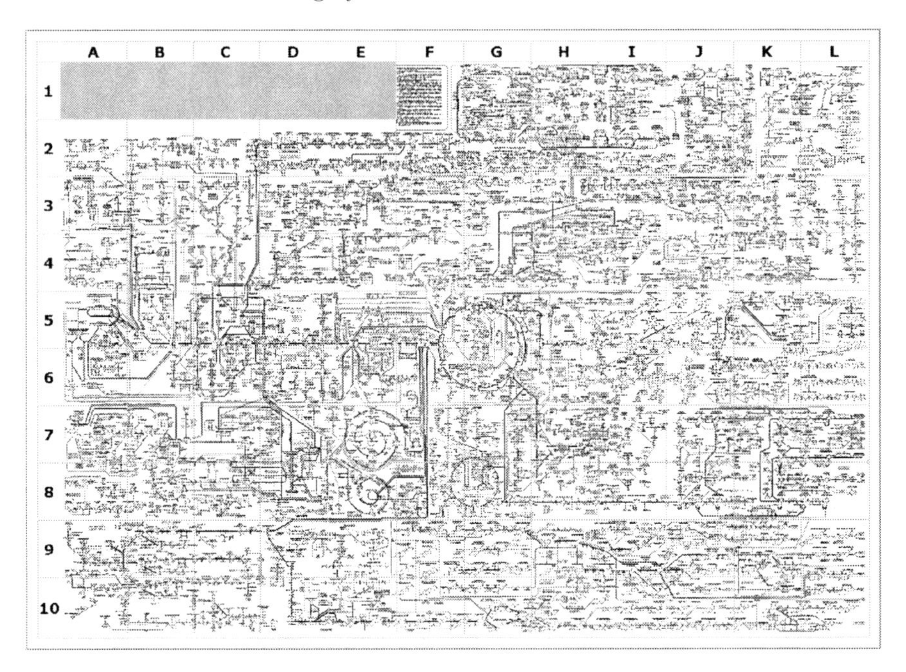


**Fig. 1.1.** The complexity of a metabolic system. Pictured is a part of the reaction network of a metabolic process. Although the network shown here is very detailed, this represents only a small part of the entire reaction (This figure is a reproduction of that posted at *http://www.expasy.org/cgi-bin/show_thumbnails.pl*. It is presented here by permission of Roche Applied Science)

As one example demonstrating the complexity of biological processes, let us consider Fig. 1.1, which depicts part of the metabolic reaction network existing within *Escherichia coli*, a type of bacteria. It is seen that this network includes many reaction cycles and possesses a great variety of distinct structures. Obviously, this network is far from simple. And that which is displayed here represents only a small part of entire metabolic reaction.

Similar complexity is found in the basic biological reproduction process. However, note that this would not necessarily have to be the case. Indeed, if we are interested simply in the multiplication of number of a unit, then, as an extreme situation, in analogy to the "reproduction" of crystals, a collection of very few types of molecules would be sufficient to exhibit such phenomenon. However, actual biological reproduction consists of the replication of systems consisting of molecules not of a few but of myriad types.

Now, let us briefly consider signal transduction. After a particular type of molecule has adhered to a receptor attached to a cell membrane, as a result, a signal is transmitted within the cell that leads to the cell responding in a certain manner. The nature of the signal transmission exhibited in this kind of process has been investigated in detail. In general, in this transmission, there will be one enzyme kinase that promotes one type of reaction, and another kinase that promotes the reaction involving the first kinase, and another kinase acting as a catalyst for the second kinase. In this way, the signal transmission is realized from a reaction composed of several levels. However, for the purpose of simply responding to the external environment, a simpler reaction consisting of a single level would be sufficient. Furthermore, there are enzyme phosphatases that act to terminate such reactions through dephosphorylation. In addition, a single such phosphatase can interact not with a single enzyme but with multiple enzymes, and for this reason, the path of the reaction will separate into distinct branches. Such branches can then themselves split into branches or even recombine. The question that interests us here is the following: Despite the fact that there exist feasible reaction systems consisting of a single component and a single level that could respond optimally to the external environment, why do the actual reaction networks found in organisms consist of complicated networks exhibiting many-to-many relations among numerous components? What is the origin of this complexity? How can we understand the complexity seen in biological processes, mysterious in its degree of diversity and, from the point of view of functionality, in its redundancy? Also, does the methodology of present-day molecular biology offer the means Sto gain such an understanding?

Actually, to the present time, the question of why there exists such diversity and complexity has been left largely unaddressed. Instead, the conventional approach in molecular biology has been to study the molecules themselves. Indeed the so-called "–ome" projects of identifying and enumerating all molecules, such as genome (the complete set of genes of an organism), proteome (the complete set of proteins), metabolome (the complete set of molecules involved in metabolic processes) are the present trend. Thus, while molecular biology originally stared as a pursuit of universality, rejecting the "enumerationism" that preceded it, it appears that it has now lost its original spirit and itself has become a kind of enumerative science. We will revisit the problem of this enumerationism in a subsequent section.

As one possible explanation of the mysterious degree of diversity exhibited by organic processes, the following simple idea comes immediately to mind. Because, through the ages, living organisms have experienced many types of environments and circumstances, it is reasonable that in the process of evolution, such experiences have in some sense been accumulated within them. It would follow, then, that great diversity or complexity we observe is a result of the history of organisms that experienced a great variety of circumstances. While the idea that through the evolution, the past experiences have come to be embedded in the structure and function of organisms seems appealing, the situation may not be so simple. Following this idea, as organic systems became more and more complex, the relations among them would have become extremely complicated. Then, the question becomes whether within such a complicated system it would be feasible that an advantageous state could be selected through the evolution. Furthermore, is it feasible that stable and proper functioning could emerge from such an intricate system? Or, could it be that in fact there is some inherent and inevitable relation between stability and this intricacy? Such questions regarding diversity and complexity are yet unanswered.

Perhaps we need to change the point of view from which we pose such questions. Rather than inquiring whether it is somehow advantageous for some faculty of an organism to be built upon a complex and diverse process, it is better to inquire why such complexity and diversity appears and, indeed, why it appears universally.

## 1.4 Diversity and Dependence on Environmental Circumstances

In studying the roles played by various molecules in organic systems, obviously, the clearest picture would emerge if each molecule plays a single role and each role is carried out chiefly by one molecule. In such a situation, the important task would be to identify the gene responsible for a certain faculty or the molecule that exhibits a certain type of behavior. However, in fact such one-to-one relationships among molecules and functions are exceptional in living systems. Below we consider a number of examples that demonstrate this point.

In recent years, it has been found that there are a number of molecules, discovered as certain functional units, whose function changes in response to environmental conditions. A typical example is signal transmission channels. As discussed above, in general, the relation between the input to a cell and the resulting output (i.e., the manner in which the state of the cell changes in response to this input) is not one-to-one. Rather, owing to the presence of certain proteins in these channels, many paths can emerge. Of particular interest here is the protein Ras, as transmission channels that include it are involved in many types of functions. For example, so-called Ras kinase, Raf kinase, and Map kinase channels have a number of roles. In nematodes[2] these channels play many important roles, including those in the formation of the excretory tube, in the formation of the vulva (in hermaphrodites), in the control of meiosis, and in the olfactory system (Sternberg & Han, 1998).

The so-called gene *p53* acts to suppress the growth of cancer and, for this reason, has lately been the subject of much interest. It has been found that this gene possesses abilities to control, among other things, cellular cycles, the regeneration of blood vessels, apoptosis, and DNA repair, and the particular role that it plays in any given case is determined by the environmental circumstances.

In immune systems there exist diverse responses to the invasion of various kinds of antigens. In such responses, the cytokine interleukin can play a number of roles, depending on the particular situation.

---

[2] This example was brought to my attention by Isao Katsura.

In many cases it is known that there exist parallel pathways from genes to functions. For example, there are situations in which removal of a single gene has no effect on function, but the removal of this and a second gene leads to mutated functionality (for discussion of the case of nematodes, see, e.g., Solari & Ahringer, 2000.) Then, it is also known that a change in environmental conditions often leads to the expression of different genes. Thus, when new enzymes are synthesized, pathways that previously existed only potentially can come into use, and molecules that previously possessed no such capacity can come to play a principal role in some newly appearing process.

In nervous systems, it is often the case that when the neural circuit responsible for some specific function is destroyed, a new circuit that eventually comes to play the same role is formed. Such phenomena have been investigated in relation to the reversibility of neural networks. In the past, it was thought that new neurons were no longer generated in adults. However, in recent years it has been discovered that neurons can be regenerated even in adults. At this time, such regeneration is known to exist only for certain local neural circuits, but it is also known that when certain areas of the brain are damaged, other areas can take over the role once played by the damaged area. Indeed, such cases of the recovery of brain function are described in many textbooks on the nervous system.

The approach of investigating one-to-one correspondences is not limited to molecular biology. Indeed, it is commonly employed in many areas of biological research. Such an approach is typically used in studies aimed at the construction of "maps." For example, there exists research whose goal is to develop a functional map of the brain. Such research is based on the premise that certain regions of the brain correspond to specific faculties – for example, sight, movement, high-level information processing, and language. However, it is known that even for such high-level functions, when one area of the brain that has been identified as that which controls a particular function becomes damaged, it is sometimes possible for a different area to take over in this role.

Let us now return to where this discussion began. As discussed with regard to the various examples given above, the correspondences between molecules and functions depend strongly on the environmental circumstances. These examples demonstrate that such correspondences are ambiguous. More specifically, even when we can identify such correspondences, they are generally not one-to-one, as often several molecules possess the ability to play a particular role and, conversely, a single molecule can play many roles. In addition, the types of correspondences that exist in any given case depend on the environmental conditions, and therefore change in time. In this sense, it is appropriate to refer to these as "dynamic many-to-many" correspondences.

From the above discussion, we have come to see that to understand the functioning of living systems, it is not sufficient to simply investigate the roles of molecules under certain specific conditions. In the background of the diversity discussed in the previous section is the possibility that given conditions differing from those with which it presently faces, an organism may be capable, due to the diversity it possesses, of functioning in a different manner. If we consider the case in which an organism is faced with very different conditions, it may be insufficient to adapt to this change through the gradual "knockout" of genes. Here, it may be the case that something existing within the diversity that once appeared useless comes to play an important role.

Of course, however, it is not possible to enumerate all kinds of behavior that could appear under all feasible types of conditions. (While such a task may be possible if there were a finite number of such cases, obviously it is not possible in the actual situation of an infinite – indeed uncountably infinite – number of cases.) Given this reality, it is clearly necessary to find some methodology to replace that of simple enumeration.

Here, we are reminded of the following story about the Chinese philosopher Zhuangzi (2–3 century B.C.)[3] In the story, Zhuangzi is told by his friend, "The things you say are of no use." In response to this, Zhuangzi says "When you understand what is useless, for the first time you also understand what is useful. For example, the earth is vast, but that part of it that any person uses is just that upon which he steps. So it follows that the rest is of no use. Now, if that beyond which we are able to set foot is useless to us, one can make a steep cliff from the part one uses to make a deep valley. Then, can one live at the part left over? To this, his friend replied, "No! Of course, one cannot live upon the left over." In this way, Zhuangzi was able to elucidate the "use" existing *in potentia* in "useless" things. This story points to the idea that it is possible for humans to live stable lives only when the things existing around them are not limited to simply those that happen to be of use at the present time – when there are also things whose potential use is realized only in some broader set of experiences.

The above story tells that a stable existence is not possible if that which surrounds us is no more than the land that we will use in our lifetime. There is a clear analogy between this and the discussion given in this chapter. The lesson to be learned here is that if we confine our investigation of life to only the behavior seen under environmental conditions close to those that exist in the present world, we will be able to obtain only a very limited picture of living systems. If this study is not extended to include the investigation of such systems within a broader range of possibility, we will not be able to truly understand life. In response to this assertion, some may wonder that, although the set of conditions that exist in the world today is inadequate as a background for the study of life, it is certainly sufficient if we expand this to include all the conditions that all livings on this planet have experienced through the course of evolution. However, considering only the history of the actual experiences of organisms on Earth corresponds simply to add to the path taken by all of its ancestors in Zhuangzi's anecdote. But, returning to the lesson learned from the story above, addition of only this limited set of paths would certainly not be sufficient for security, as indeed there may arise

[3] Chuang Tzu is another pronunciation.

a situation in which a novel path is required. Indeed, the range of possible experiences that a creature could face is far greater than the cumulative experiences of its ancestors. For this reason, rather than looking to the past to solve the problem we consider, we propose to search for universal characteristics of life that holds under a broad range of conditions. A point of view that provides a global understanding is necessary. In other words, even if it means giving up a detailed picture of local topography, we wish to take a point of view to see the distant mountains and valleys. It is with such a point of view that we set out to construct a theory of life.

## 1.5 Systems of Strongly Interacting Elements

Let us take a step back here and reconsider the type of methodology that we have criticized, that is, understanding living systems by isolating individual elements and then determining the role played by each. Of course, we do not wish to claim that such a methodology is meritless, and indeed it is worthwhile discussing the extent of its usefulness. First, in order for it to be possible to determine some one-to-one relation between elements and roles and then to investigate its small change by interaction with other elements, it is necessary that the properties that these elements possess in isolation are nearly unchanged when they are combined with the others. In other words, such an approach is based on the premise that the properties of individual elements change little in response to their interaction with other elements. Of course, there are situations in which such a premise is valid and the description that it leads to is in fact useful.

However, as will be discussed below, often interactions are not small in a biological system. Hence, we need an alternative approach that considers the system globally and then studies the roles played by its elements, as determined dynamically through the interactions among them and the manner how these roles can change.

When a person builds a machine, she or he does so in such a way that the interactions among parts are limited according to some design. The system is constructed according to a plan in which quantities (e.g., energy) are exchanged among parts in some predetermined manner. This is done so that the designated role of each part does not change through the interactions among parts. Actually, with the present design of computers, as the physical extent of their circuits becomes smaller, eventually we will face the situation in which the interactions among parts communicated through the magnetic fields they generate will not be negligible. Indeed, this is a serious problem that will need to be addressed in the design of computers in the future. By contrast, living systems apparently do not possess a design in which the interactions among parts are limited. Rather, they constitute systems of strongly interacting elements. (See Sect. 1.4 for related discussion.) For example, proteins are packed into a cell like passengers on a rush hour train (Fig. 1.2)

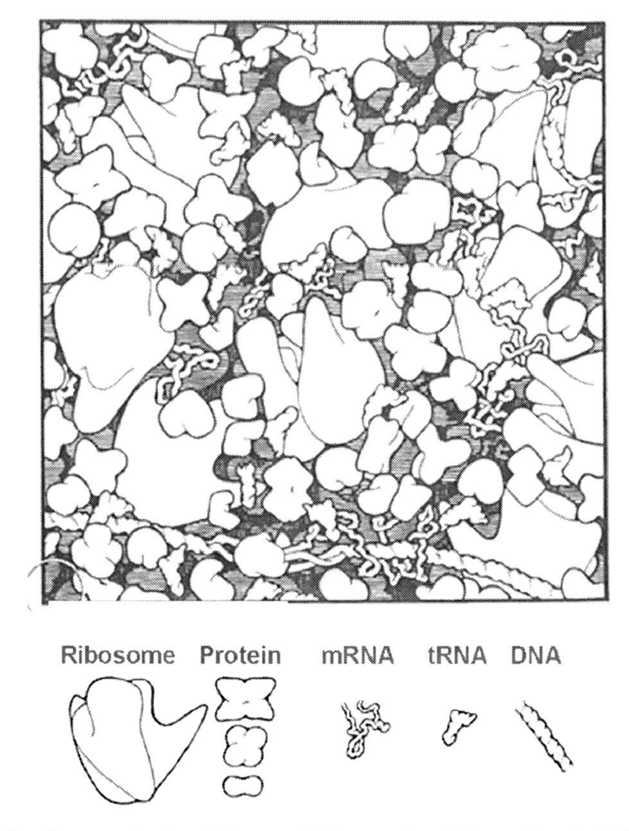


**Fig. 1.2.** The "congestion" within a cell (Reproduced from Goodsell (1991) under the courtesy of the author). Often the distances between protein molecules and the size of the molecules are of the same order

(Goodsell, 1998). In fact, in some cases, the distance between two neighboring atoms on a single protein molecule is greater than that between either of these atoms and the closest atoms of other protein molecules. For this reason, in general it is not the case that interactions among protein molecules are negligible in comparison with the interactions existing within a given protein molecule. It is thus seen that if we attempt to treat such a system by first regarding each protein molecule as an independent unit and then adding some weak interactions among such units, the description we obtain will be dubious at best. The proteins within a cell function are under so strong interaction that it is not possible to treat them separately. Therefore, the protein molecules in a cell are "elements" of a very different nature than the elements of a machine. In multicellular organisms, in the same way, cells are packed, and the effects of the communication among cells cannot be treated as a small perturbation of the processes going on within cells. In such a situation too, the methodology of first determining the roles of individual parts and then adding to this some small interaction among them cannot yield a faithful description.

It is thus clear that what we seek is to model organisms as systems in which dynamic and condition-dependent functions appear by virtue of the interaction among constituent elements. Such an endeavor is discussed further in Chap. 3.

Before moving on, we should make a brief comment on the use of the term "interaction." Usually, when molecular biologists make a statement that cells interact, probably what they have in mind is, for example, some effect communicated between cells through receptors or the exchange of certain molecules. Ecologists would most likely consider interaction to be the relation of individuals in an ecological system such as prey–predator relationship. Physicists and chemists would consider interaction to be the influences communicated by forces – intermolecular, electromagnetic, etc. In this book, we use the term *interaction* with a very broad meaning that includes all of the above senses and more. Simply stated, we use it in reference to any mutual effect, direct or indirect, that exists among elements. For example, we regard as an interaction the competition between two cells that seek the same nutrient.

## 1.6 Are Living Organisms "Computing Machines"?

Presently, in the field of molecular biology, great emphasis is placed on the study of the *information* stored within cells and the *programs* that allow for its expression. The terms *information* and *program* are, of course, used in the context of computers. The fact that they are also used with regard to biological systems, however, is not accidental. Indeed, there is a trend to describe the life phenomena as computational processes, and evidently it is this manner of thinking that has led to the deliberate choice of such terminology. For example, let us recall the threshold mechanism involved in the process of development, touched upon in Sect. 1.2. In general, the state of a cell is determined in accordance with the concentrations of certain molecules existing in its neighborhood, as detected by the cell membrane. The process of cell differentiation is an example of a transition in the state of a cell. The mechanism responsible for this transition is the following: If the concentration of a particular molecule exceeds a certain level, a certain gene will be expressed, and as a result, the chemical composition of a cell will change to a given direction. In other words, the process of cell differentiation can be expressed in terms of a simple *if–then* logical rule relating the concentration of the signal molecule and the change of the cell state, triggered by the expression of a particular gene. Then, because the development of a multicellular organism takes place through repeated cell differentiation, it can be described in terms of such *if–then* logical rules. This leads to the conclusion that the process of development exhibited by living creatures can be represented as some kind of computer program – a logical expression in the form of a chain of *if–then* statements:

if (the concentration of molecule 1 exceeds a threshold value) then (gene 1 is expressed)
if (gene 1 is expressed) then (protein 2 is synthesized)
if (the concentration of protein 2 exceeds a threshold value) then (the concentration of molecule 3 increases)
if (the concentration of molecule 3 exceeds a threshold value) then (gene 2 is expressed)...

Discoveries have been made of a number of such chain of successive reactions, in which the expression of one gene leads to the expression of another. For this reason, the developmental process is often referred to as a "program."

However, all processes occurring within a cell are reactions taking place among molecules, and therefore it cannot be the case that they proceed as a succession of errorfree logical operations like a computer program. For example, let us once again consider the logical *if–then* representation of successive gene expressions. Now, whether or not the condition for the expression of a particular gene (that the corresponding signal molecule has surpassed a certain concentration) is met is determined by the chemical reactions taking place within the cell. Of course, the concentration of a particular molecule at the position of a cell is given by the number of molecules in this region divided by its volume. But this number is subject to random fluctuations. For example, let us consider the case of gene expression. The expression, that is, if a protein corresponding to the gene is synthesized or not depends on if the concentration of some signal molecule is larger than some threshold or not. Suppose that the threshold, is given by some concentration.[4] The concentration is given by the number of molecules divided by the volume. However, for the number of molecules $N$, we would expect fluctuations in a number of order $\sqrt{N}$, as a natural consequence of probability theory (recall the law of large numbers and the central limit theorem). Thus, in terms of the concentration, there would be a blurring of the threshold value of relative magnitude $\frac{1}{\sqrt{N}}$. In the case of gene expression, typically, $N$ is of order 1000, and thus there will be an error of order 3% in the "computation" involved in gene expression (Fig. 1.3).

While it may be thought that a "mistake" rate of 3% is not large, it must be remembered that this is the error corresponding to a single gene expression. With such a level of error for each incidence of gene expression, when we consider the entire chain of reactions involved in some developmental process, this level of error is easily large enough that the resulting behavior will differ greatly from that of a computer program. To understand this, let us consider a computer in which, on average, for every 100 operations there are three errors. Then, using this computer, we could not expect a program that calls, say, on the order of 10 operations to give a correct result.

One might think that the problem caused by such a high rate of error could be solved by some kind of error correction program, and that through

[4] Detailed discussion of this point is given in Chap. 7.

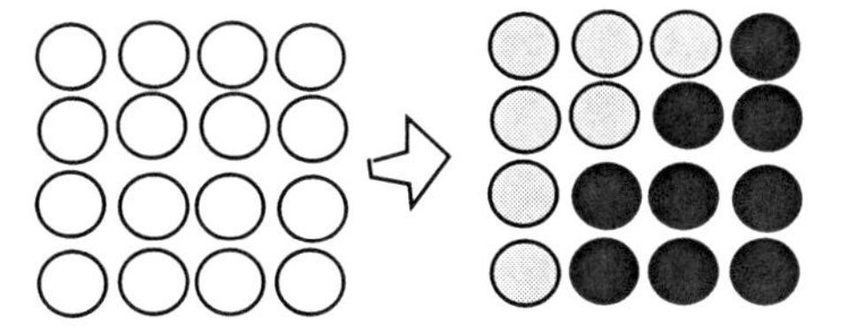


**Fig. 1.3.** Schematic depiction of an assembly of equivalent cells. Initially the cells are all of the same type, but through the interactions among cells, they differentiate into two types

such a mechanism the developmental process could indeed proceed as a logical *if–then* computation. This might be true to some extent. However, such an error correction program too would have to be based on chemical reactions, and, here again, the number of molecules involved in these reactions could not be large. Thus the error elimination to be carried out through these chemical reactions would itself be subject to significantly large fluctuations, and hence, we would simply have a new source of the same kind of error. Of course, it may be possible to realize some kind of error reduction with an intricately designed chain of multiple-level reactions, but in order to correct errors effectively, it would be necessary to have a design that can recognize whether each molecular process is correct or incorrect and act accordingly. In the case of a computer program, such error correction can be carried out, because the programmer is capable of making a judgment of correctness. However, in the problem we consider, everything is carried out through (molecular) chemical reactions, and therefore the situation is very different, as molecules do not know which the correct process is. Even if we were to assume that such an exquisite design does exist, in order for it to have come into existence, there must be some underlying principle ensuring its stable functioning.

The points we have raised above concerning fluctuations are universal questions encountered in the study of physical phenomena, and it is quite unlikely that they could be resolved through the introduction of some specialized "stabilizing genes" or "stabilizing molecules." As will be discussed later in the present chapter, the fluctuations are neither negligible nor eliminated. Indeed, we report a quantitative measurement of fluctuations in Chap. 5. To our surprise, the fluctuations are quite large. For example, for bacteria having same genes (i.e., clones), the number of proteins are different to the factor 10 or more by each individual. Then, we need to understand how a biological system is stable under such large fluctuations, by establishing a novel viewpoint.

One possibility that comes to mind here is that in a developmental process, the interactions between cells lead to stability. In fact, according to the theory to be presented in Chap. 7, in an assembly of cells, the interactions between cells do indeed introduce a degree of stabilization. In experiments also we can see such stabilization effects. First, for a cell assembly consisting of cells possessing identical potentialities, it has been observed that small fluctuations can lead to differentiation in which some subset of this assembly undergoes a transition to a new cell type. This differentiation results from only the interactions among the cells making up this assembly. Because these cells initially possess the same properties, they are referred to as "equivalent cells" (Greenwald & Rubin, 1992). In this case, as depicted in Fig. 1.3, if the cells on one side of the assembly are of one type, due to the interactions among cells, the transition of neighboring cells to this same type will be suppressed. In such a situation, it is not clear from the outset which cells will end up as one or the other type, but in any case, the system will come to exist in a state in which cells of the two types coexist. Here, the state of each cell is stabilized by the interactions among cells. This type of stabilization could perhaps be realized in the situation that, for example, the state of the cell assembly as a whole in some way reflects the state of each cell, and as a result, some appropriate distribution of cell types is realized. However, it is not clear how the condition of the assembly as a whole would be capable of reflecting the state of each individual cell. This problem is treated in detail in Chap. 7. There, we present a theory describing how fluctuations between cells become amplified, and as a result, cells undergo transitions to new types, and through the cell–cell interactions, the type of each cell becomes stabilized.

Summarizing the above discussion, we have seen that attempting to understanding a living system as a type of computer, we inevitably encounter difficulty regarding the stability of the system due to the error introduced by the presence of fluctuations. Given this situation, however, we then have to ask why it has become genetic information be regarded as constituting a program. This problem will be discussed in Chap. 4.

## 1.7 Problems with the "Program" Point of View

The premise behind the point of view that regards genetic information as constituting a program is that a specific input leads to a specific output and that the chain of events that connects the two provides a description of the actually observed phenomena. Research in biology is often aimed at detangling such cause-and-effect chains. Indeed, in many cases this approach has clearly elucidated *causal relations*. However, what we must keep in mind here is that when considering complex phenomena, usually all that can be determined experimentally is whether or not some phenomenon $A$ and some phenomenon $B$ exhibit a *correlation*. In particular, it is generally not easy to establish some kind of $A \to B$ causal relation.

Even if this point is recognized and experiments are carried out very carefully, keeping in mind the distinction between correlation and causation, it is quite difficult in studying the kinds of systems considered here to demonstrate cause–effect relations. There are two reasons for this. First, for most processes of interest, there are many causes, and even if for each of these in isolation there may be some clear *if A then B* causation, because they, in fact, do not act in isolation but in some collective manner, it is quite difficult to consider only one part of the whole and establish a cause–effect relation involving it. Suppose we are considering a process for which there are $N$ causes. Then, if there are two effects, the number of possible cause–effect relations is $N(N-1)/2$, if there are three effects, this number is $N(N-1)(N-2)/6\ldots$. Thus it is seen that if we must consider all such possibilities, the number of cause–effect combinations grows very rapidly, and if $N$ is large, it is essentially impossible to identify with any degree of certainty-specific causal relations. To see this more clearly, let us consider a set of $N$ variables that can each take the value "true" or the value "false." The value taken by each of these variables at any time is determined through a set of relations that themselves can each be either "true" or "false." In such a situation, if our only means of investigating the system is to observe the values of the $N$ "effect" variables, for large $N$ it would be extremely difficult to determine the truth values of the "cause" relations and the function that assigns the values to the $N$ "effect" variables.

The second problem that we must consider here is that there are cases in which the variables in question cannot be treated as discrete but continuous, and in such cases, it is not possible to describe the system in terms of program-like logical relations. For example, let us consider the situation in which there are three quantities, $X$, $Y$, and $Z$, and these vary in time in accordance with some reaction rate equation. As explained in Chap. 3, the evolution of these quantities can be expressed as a dynamical system in which the rate of variation in time of each of the quantities $X$, $Y$, and $Z$ is given by some function of $X$, $Y$, and $Z$. For example, let us suppose that the variation in time (the time derivative) of $Y$ is given by something of the form $C \times (X-\text{threshold})$, where $C > 0$. In this case, there is a simple causal relation that can be expressed in words as "if $X$ is greater than some constant threshold value, then $Y$ increases." If $Y$ does not affect the evolution of $X$, this expression, to a certain extent, describes the actual behavior. However, if the variation of $X$ depends on $Y$, then in general this is not the case. In such a situation, because $X$ and $Y$ evolve together, the expression properly describing their dynamics and the expression above describing a causal sequence of events will not necessarily be consistent. Next, consider a dynamical process in which there are several threshold values characterized by such a condition that, for example, if $X$ is greater than one of these and $Z$ is less than another, then $Y$ increases. In general, such a process cannot be expressed as a combination of some logical formulas. It is thus seen that in the context of describing the types of processes in which we are interested, the distinction between the dynamical and logical points of view must be kept in mind (Fig. 1.4).

As the discussion given here has made clear, it is important not to confuse correlation and causation in the study of living systems. In general, the question of whether a particular gene is the cause of some particular behavior

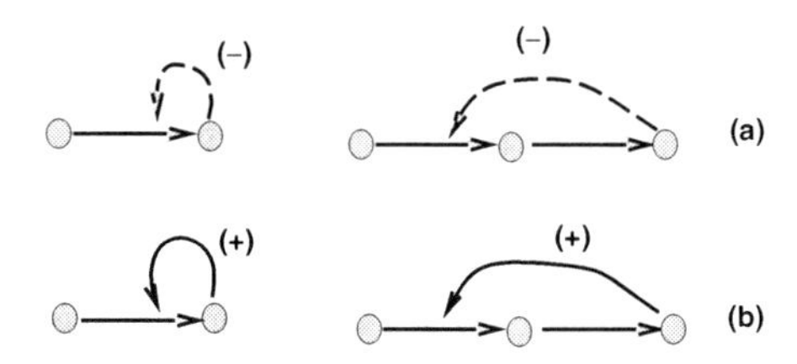


**Fig. 1.4.** (**a**) A reaction circuit system exhibiting negative feedback. (**b**) A reaction circuit system exhibiting positive feedback (an autocatalytic reaction system)

cannot be answered by simply removing this gene and observing the result. Indeed, it is often the case that, even if each gene on DNA codes the synthesis of some protein, the expression of each gene is determined by the concentration of some other enzyme (i.e., a protein). Thus the situation is one of mutual influence, not unidirectional causation. Hence, although the genes can be thought of as in some sense controlling such processes, in fact it is not true that an understanding of the genes alone is sufficient for their complete description. For example, even if we were somehow able to obtain the DNA of a dinosaur, unless we also knew the initial conditions of the cellular composition that allow their proper expression of genes, we would not be able to create a Jurassic Park. The conclusion we reach from these considerations is that, rather than seeking some sequences of logical formulas, in the modeling of living systems, we should be studying models of interactive dynamics. Then, we should inquire whether, within such dynamics, the asymmetric relation between two molecules is generated so that one plays a more controlling role and therefore can be regarded as the bearer of genetic information. (See Chap. 4 for more detailed discussion on this topic.)

## 1.8 The Problem of Stability

As discussed above, life processes involve an enormous variety of molecules. However, often the number of individual molecules of any given type is not large. For this reason, these systems are subject to a very large degree of "noise" (i.e., random fluctuations). Despite this fact, however, under normal conditions these systems are very stable. As an example, let us consider the process of development of a multicellular organism. It seems miraculous that systems with such a vast variety of molecules participating in such a large number of processes can result again and again in almost identical macroscopic patterns. The situation here is analogous to a person attempting to stack many irregularly shaped blocks into a particular form during an intense earthquake.

In studying this type of stability, a commonly employed approach is to consider the system slightly perturbed from the stationary state in question and to observe whether the evolution of the system then brings it back toward the stationary state. The mechanism how such displacement by perturbation decays is provided by "negative feedback," and the approach employing this concept, due to Wiener, is indeed quite effective in a wide range of contexts (Wiener, 1948). One of the principal approaches in the stability analysis of reaction networks is to seek "circuits" displaying such negative feedback (see Fig. 1.8).

For the case of living systems, however, the problem of stability cannot be fully understood in terms of negative feedback alone. First, there exists a mechanism acting at every point through which small disturbances are amplified. Indeed, this mechanism exercises a strong influence on the evolution of the system. For example, let us consider the vision by the eye. The retina contains a structure that amplifies the excitation imparted by a few photons. It is for this reason that we can perceive the very faint light coming from distant stars in the night sky. As another example, it is known that there is a mechanism through which the presence of just a few molecules at a cell receptor can be amplified into a recognizable signal.

It is also important here to note that living systems have the characteristic ability of reproduction. In other words, they have the ability to increase the number of molecules and cells. Such increase implies the existence of an amplification mechanism. In order to make cell division possible, it is first necessary that within the cell there take place chemical reactions through which the number of each of the various types of molecules essentially doubles. For this purpose, amplification is necessary. The typical type of mechanism through which this occurs is a type of autocatalytic process in which the presence of a certain type of molecule leads to the increase in number of the same type. In fact, this kind of process is such that the number of molecules of a given type that are created increases with the number that existed originally. Thus it is an example of positive feedback. For example, the following describes a simple reaction through which molecules of some species $X$ are reproduced: $X + A \rightarrow 2X$.

In this case, the change in the number of molecules of this species, $N_x$, is described by the equation

$$dN_x/dt = aN_x \ . \tag{1.1}$$

It is therefore easy to see that if the number of molecules is increased, then the rate of its synthesis also increases. Thus, some random fluctuation in $N_x$ becomes magnified in time. In other words, increase leads to further increase. Of course, with an equation of the above type, the amount of molecule $X$ increases without bound, but in general, molecules of this species are consumed in a reaction through which some other constituent is synthesized, and in this way, their number is prevented from exploding. In this kind of system, which includes some positive feedback mechanism and a mechanism that counteracts it, if there exists a time lag between the two, then there can appear a stable oscillatory behavior. The number of molecules in question starts to decrease only after it goes above the value for which the two mechanisms are balanced, and this decrease turns to the increase after it passes the balance point. Therefore, the stability of a system that includes positive feedback cannot be described simply in terms of the stabilization of a stationary state provided by negative feedback (see Fig. 1.4).

Living creatures also exhibit transitions over several states, as exemplified by the process of cell differentiation. Such behavior, in which the system makes a transition from one state to another, clearly cannot be understood entirely in terms of the relaxation to a stable state through the mechanism of negative feedback. Therefore, the concept of negative feedback does not provide an adequate understanding of differentiation and the resulting diversity of cell types displayed by multicellular organisms.

Furthermore, it is often the case that the type of diversity discussed in Sect. 1.3 prevents the stabilizing relaxation toward stationary states facilitated by negative feedback. The reason for this is that in a system consisting of a large number of interacting constituent elements, as the number of interacting elements increases it becomes increasingly difficult for the system as a whole to exhibit negative feedback, Such systems will generally include positive feedback, and small fluctuations become amplified. For this reason, to understand such systems, it is necessary to think in terms of a type of stability that preserves the existing diversity (Fig. 1.5).

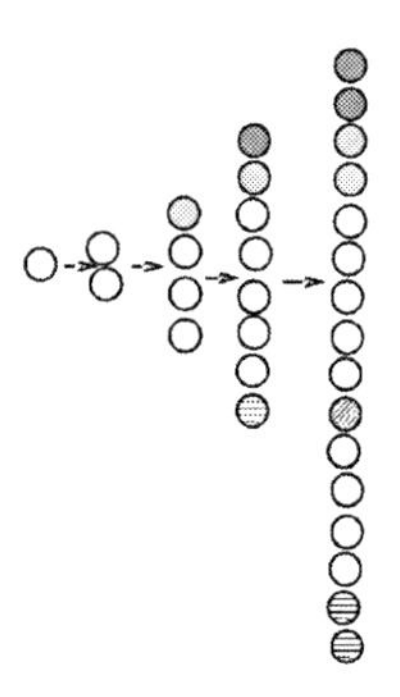

**Fig. 1.5.** In a system exhibiting reproduction, if some fluctuation occurs, it is amplified through the process of reproduction itself

## 1.9 Systems Evolving Amongst Fluctuations

Above we considered the question of whether a living system can be regarded as some kind of computer. There also exists a point of view that, while not going as far as actually using the term "computer" in their description, does employ a mechanical analogy in an attempt to understand living systems. A machine is a device constructed according to some plan that, in accordance with some commands, converts some input (or material) into output (or a product). According to this analogy, cell acts under the plan constituted by the genes it possesses and converts some raw material into an output. In this case, if the output consists of the construction of a new cell of the same kind, then the original can be thought of as a reproduction machine (a self-replicating automaton), or if the output is motion, then the cell can be thought of as a molecular motor. However, despite the prevalence of this analogy, it would seem that normally, living organisms are not considered to be "machine-like" and that nobody actually regards them as machines in the ordinary sense. What, then, is the difference between machines and living creatures that we perceive intuitively? Also, if in fact we could understand living organisms as machines, what kind of machines would these be?

First, we note that probably many people have the image that, in contrast to machines, living systems possess some kind of flexibility and adaptability. That is to say, a living organism does not always behave in strict accordance with the rules applied to it, and depending on the situation, the rules it perceives will change. There are two points to consider here. First, the relation between input and output depends on the circumstances (the environmental conditions), and, second, this dependence itself is not completely determined, as it exhibits fluctuations.

To proceed with an investigation along this line, we should not focus solely on elucidating the behavior or average values of characteristic quantities that are observed under given conditions within a deterministic framework. Instead, we must consider the role of fluctuations. However, in the study of molecular biology, the idea of quantitatively measuring fluctuations has received little attention. Indeed, the mainstream approach in all fields of biological research has been to seek deterministic on-off type conditions – for example, "if the concentration of molecule $A$ exceeds the value $\rho$, then behavior $X$ results." Experiments have been designed and carried out in such manners that minimize stochastic noise and thereby yield results for which the types of behavior observed under different sets of conditions are as distinct as possible.

However, to properly investigate the "softness" of living systems, it is also necessary to consider the question of how the nature of the fluctuations themselves depends on the environmental conditions. For that purpose, we must determine how the behavior of a cell can vary under fixed conditions, through the influence of the fluctuations. An example of such an investigation is the work of Fumio Oosawa and colleagues on the motion of the *paramecium* (Oosawa, 2001). It is known that a paramecium in a liquid displays Brownian motion in which it moves in a straight line for some time, then changes direction essentially randomly, and then again proceeds in a straight line. It is interesting to consider the question of the degree to which this motion is truly Brownian and how this degree depends on such parameters as the temperature of the liquid, the population density of paramecia, and other environmental conditions. Then, it is also of great interest to study how the paramecium uses fluctuations to adapt to the environment. The fluctuations of interest are those within the paramecium cell, and these perhaps originate as random behavior of the molecules inside the cell. However, as discussed above, within the cell, there are positive and negative feedback mechanisms through which various quantities existing in the cell are amplified and suppressed. For this reason, some such fluctuations come to be enhanced and others damped. Thus, the form of these fluctuations is not simply determined by the temperature. Rather, the macroscopic fluctuations we consider are involved directly in the behavior of the cell as a whole and depend on its state. Hence, to obtain an understanding of the *paramecium's* behavior, it is necessary to study how the fluctuations within it depend on the state of the cell.

Of course, the effect of fluctuations at the molecular level is also very important, and technologically, it is becoming possible to make precise measurements of these. In the study of molecular machines which was originated in attempting to elucidate the generation of force by muscle tissue, the importance of fluctuations has just begun to be recognized. The mechanism through which power is produced by muscle fiber has been studied in great detail. In a muscle there are the protein molecules actin and myosin. Myosin is bonded to the "rails" formed by the actin molecules. Then, it is believed that power is generated when myosin molecules excited by ATP move on these actin rails. If this mechanism moved like a gear, then the distance of the movement exhibited by the myosin molecule should be determined uniquely by the energy of hydrolysis of the ATP. However, carefully considering the fluctuations inherent in the motion of molecules, Fumio Oosawa hypothesized that in fact such "molecular machines" in living organisms do not behave like gears, suggesting, rather, that due to the influence of fluctuations, they behave in less restricted manners. As a characterization of this type of motion, he coined the term "loose coupling mechanism" (Oosawa & Hayashi, 1986; Oosawa, 2000). It has been found that indeed the actin molecule is not "stiff." As a result of thermal fluctuations, a myosin molecule is subject to violent shaking as if it were in a terrible earthquake, and it moves while repeatedly and randomly colliding with obstacles that possess energies on the same order as its own. For this reason, it is reasonable to conjecture that there would be no fixed relation between the dissociation energy of one ATP molecule and the distance moved by a myosin molecule, and that within actual muscle tissue, the distribution of such distances should vary from case to case. This conjecture has been confirmed experimentally by the Yanagida group (Ishijima et al., 1998).

## 1.10 Spontaneity

Another reason that we consider living creatures to be different from machines is that generally they are thought to possess "spontaneity." (This point is in fact related to the fluctuations considered in the previous section.) In other words, an organism may exhibit different types of behavior in two instances characterized by identical external conditions. Such variability is apparently due to some differences in states that we cannot observe directly from outside. In general, organisms possess internal states that cannot be controlled entirely by external conditions alone. For this reason, they appear to behave spontaneously (Fig. 1.6).

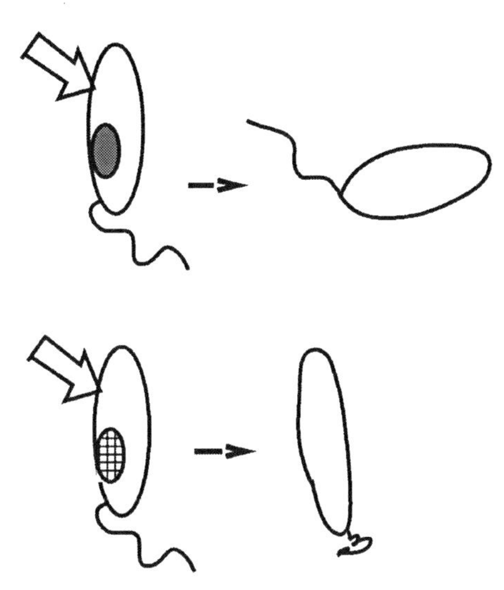

**Fig. 1.6.** Although the external stimuli are identical, the behavior of an organism depends on the internal state (represented by the grid pattern), which cannot be seen directly by an observer

In reflection, the same input results in the same output, while in a spontaneous behavior, different outputs are obtained from the same input, depending on the internal state of an organism. Figure 1.6 depicts the behavior of a paramecium in two instances in subject to identical external conditions. Even under the same environmental circumstances, in different cases, the *paramecium* may flee from, approach, or show no reaction to the source of the stimulation. Here we see spontaneity in its most rudimentary form.

In general, the behavior of a living creature is not uniquely determined by its genes and the environmental conditions to which it is subject. Each individual also carries with it the effect of its personal history. At any time, its internal state depends not only on its genes and environment but also on this history. Then, the behavior that it displays at any given time depends on this internal state.

Even the behavior of a single cell is not determined completely by its genes and environment but also depends on its physiological condition. That is, the manner in which it behaves depends on its internal "state," which includes, for example, the concentrations of various metabolic products. Now, if we were somehow able to obtain complete information concerning the internal state of a cell, knowing the number and position of each type of molecule present within it, and if these could be controlled externally, it may be possible to describe its behavior as that of a deterministic machine. However, in reality, the degree to which we are able to observe the state of a cell is limited, and it is certainly not possible for us to completely control this state. Even if we assume it is possible to control the concentration of every constituent molecule in the cell's environment by placing it in a carefully prepared culture, we would not be able to dictate the extent to which the various chemicals pass through the cell's membrane, and thus we could not control the internal chemical concentrations on which the state depends.

For example, suppose that we are performing experiments on a number of individual bacteria *E. coli*. Then, even if we were able to make the initial conditions of all the experiments, say the time at which they divided and their sizes, identical, we would not be able to make everything affecting the cells' behavior the same. Again, this illustrates how there is an internal state that cannot be observed externally (by humans or even by other cells), and because it cannot be observed, it cannot be prepared identically for each of the cells under investigation. Actually, experimentalists know that bacteria (*E. coli*) possessing identical genes and placed in identical environments can exhibit significantly different behavior if their past cultivation conditions differ.

From the above discussion, we have seen how living systems (including individual cells) carry their histories with them and that the internal states in which they exist at any time are partly determined by these histories. In addition, because we cannot observe all aspects of these internal states, as long as we rely on external observation to gather information, it appears to us that the behavior of a living organism has a dependence on history. Furthermore, we cannot know an organism's entire history and how the conditions that it has experienced are embedded within its internal state. Then, because the externally observable behavior exhibited by an organism depends on this internal state, as illustrated in Fig. 1.6, it is seen to have a "soft" input–output relation, in which a particular input can result in a range of different outputs. In this sense, this organism appears to possess "spontaneity."

If we hope to study the spontaneity of living organisms, we will have to find appropriate ways to express their internal states, and it will be necessary to determine how these internal states change according to environmental conditions and how the relation between input and output is made "soft" (i.e., not one-to-one) due to the influence of these internal states. It will not be possible to accomplish this by seeking the role played by each individual molecule. Rather, we must carry out research that attempts to understand the overall behavior resulting from the many processes occurring inside organisms.

## 1.11 How the Parts Composing the Whole Are Determined by the Whole

In living systems, it is often the case that each constituent behaves as if it "knows" the state of the whole. For example, in the process of development, when the heart attains a certain size, cells stop dividing. As a result, the heart realizes an appropriate size. As another example, each cell in the liver possesses the ability to divide, but when the liver reaches a certain size, division ceases, and from that point, the organ maintains a fixed size. Interestingly, if part of the liver is removed (e.g., for the purpose of a transplant), its cells will again begin dividing until the appropriate size is again realized. As a similar example, consider the planarian, which is an animal with the ability to regenerate itself from a piece of it that is cut off. However, if this is done, when the organism reaches a certain size, the cell multiplication responsible for the process of regeneration stops. This type of behavior occurs despite the fact that the growth of each cell is purely a local process. Therefore, unless the size of the group of cells in question can properly influence such local processes, it would not be possible for the appropriate size to be realized in this way. Thus, each cell behaves as if information regarding the properties of the group to which it belongs is somehow contained within it and, in this sense, as if it "knows" the whole (Fig. 1.7).

Among newts, there exists a diploid (i.e., a mutant with double the normal number of genes). In such an individual, each cell too is twice the normal size. However, in the growth process, the number of cell divisions is half that of a normal newt, and as a result, the sizes of the normal and mutant forms are essentially the same. This is an example of a case in which the size of the entire organism exercises control over the local property of the number of divisions undergone by each cell. A discovery that clearly demonstrates the existence of this kind of group effect in the developmental process was made by Gurdon and collaborators (Gurdon et al., 1993), termed as "community effect." In

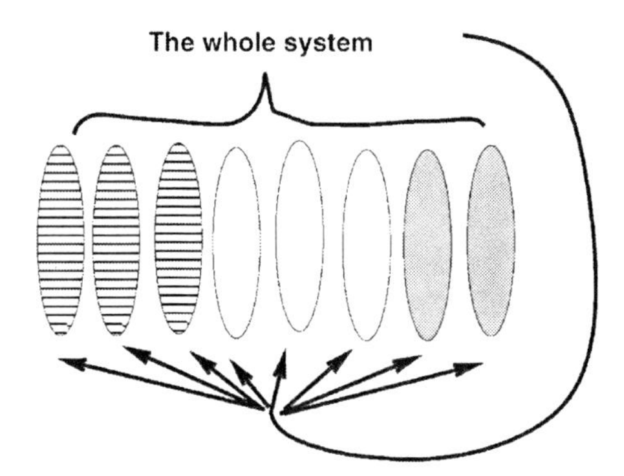


**Fig. 1.7.** The properties of each constituent comes to be determined through its relation to the whole

the developmental process, from the undifferentiated cell of the egg through repeated cell divisions, an organism undergoes the process of differentiation and comes to possess many types of cells. In this process, when the cells reach a certain stage in the differentiation process, they lose the ability to differentiate, and after that point they can produce only cells of their own type. When this stage is reached, a cell is referred to as "committed." Gurdon et al. began their experiment from a collection of undifferentiated cells, to find whether these cells would differentiate into certain cell types and eventually become committed. The purpose of the experiment was to determine whether this "committed" state depends on the number of cells present. The rationale here is that if a cell of a given type is surrounded by a sufficient number of cells of the same type, then perhaps that cell will be able to produce only cells of its own type. Thus this experiment was to test the hypothesis that a cell becomes committed through the influence of the surrounding cells. The result of the experiment was that this hypothesis is correct: Whether a cell retains the ability to differentiate depends strongly on the number of cells of its type that surround it. If this number is sufficiently small, the cell will continue to differentiate, but if it is large, it will become committed. This kind of importance of cell number was also demonstrated by the experiments of Asashima, as will be discussed in Chap. 9 (Ariizumi & Asashima, 2001).

How are individual cells apparently able to "know" the properties of the entire structure or entire organism to which they belong? It seems clear that this is due to the mutual influence imparted by the cells on each other. To elucidate the actual mechanism at work here, it is therefore necessary for us to study the interactions among cells and thereby seek to determine how the properties of each constituent and the properties of the whole are related. In other words, it is not sufficient to simply analyze the components and attempt to elucidate their properties alone. Rather, we must investigate how the properties of the individual components are related to the whole and the manner in which these properties are determined through these relations. To capture the spirit of this approach in a single expression, what we seek is a methodology to understand the *complementarity relation between the whole and the parts* (see Fig. 1.7).

## 1.12 Universal Properties That Cannot Be Traced Back to Molecules

As we have discussed, it is evident that living systems possess universal properties that cannot be expressed simply in terms of combinations of properties of individual genes or molecules. The most important and distinctive phenomena exhibited by living systems – the stability of the development process, the irreversibility in developmental process to lose the multipotency from embryonic stem cells to stem cells, and to committed cells capable of reproducing only their own kind, the compatibility of two faces of the reproduction, that is, the ability to produce the nearly identical offspring and capacity to generate variation leading to diversity through evolution, the autonomy possessed by living creatures to determine their own rules, and so forth – certainly cannot be attributed to the properties of individual molecules, and the very general questions that the existence of such phenomena pose cannot be answered by simply enumerating individual molecules and their roles and combining them to obtain some overall picture. In the research discussed in this book, we seek a fundamental theory of life based on complex systems, attempting to construct a mode of understanding through which such questions can be answered. Specifically, it is hoped that with this "mode of understanding," we can elucidate the structures of the dynamic circulation between microscopic and macroscopic levels of behavior. The problem addressed in this book is that of obtaining an understanding of the universal properties of living systems that does not rely on simple enumeration of molecular components and processes.

To address the problem raised above, perhaps it is first necessary to define what we regard as constituting an "understanding" in this context. For example, suppose that we are able to describe the motion of all of the molecules involved in some phenomenon and that, with this knowledge, in fact we have the ability to reproduce the phenomena. In general, this would certainly require an extremely large-scale numerical computation employing a vast amount of data and an enormous number of equations, but let us suppose hypothetically that despite such technical obstacles, we actually succeeded in such a project. For example, suppose that in this way we were able to simulate the metabolic process of some bacteria (say *E. coli*). However, if we then wished to construct some other type of bacteria, we would probably be required to again carry out a similarly large-scale simulation. Indeed, for each organism that we wished to construct in this manner, it would be necessary to perform an enormous computation. Of course, the accumulation of information regarding, for example, how individual cells behave under certain conditions could be of use in medical applications. However, for the purpose of obtaining an actual understanding, extracting, and elucidating the general properties of living systems, a more simplified, condensed description by removing nonessential details through abstraction, is necessary. Even if such a description is insufficient for the purpose of reproducing the phenomena in an organism, it is the type needed to answer the questions we consider. Certainly, a theory that simply reduces to a description of components would not be capable of extracting idealized, universal properties possessed by living creatures. For this reason, we seek an approach that is not based on the reduction to component elements.

Many people may doubt that the construction of the kind of theory that we seek is possible. However, it should be noted that this would not be the first theory of its kind. Thermodynamics, which is almost miraculous in its effectiveness, is a macroscopic phenomenological theory very similar in spirit to that which we seek. Of course, the description offered by thermodynamics is limited to equilibrium states and transitions between them, but this theoretical framework succeeds in describing the macroscopic state of a system and provides a general formulation of stability and irreversibility.

Here it is interesting to consider some hypothetical situations regarding the historical development of science. Suppose that before thermodynamics was developed, some technical ability that we now possess to study individual molecules existed. Then, suppose that with this ability, in an attempt to describe the behavior of a large number of molecules, people had taken the approach to measure the dynamics of each individual molecule and then construct a description of an ensemble of molecules through enumeration of its constituent dynamics. Assume that such an approach were successful, then could the concept of thermodynamics such as entropy be developed accordingly? Some people might believe that this would have been possible, noting that today we possess a methodology providing a means of deriving macroscopic representations from microscopic descriptions, namely, statistical mechanics. However, this is a serious misconception. It must be remembered that the theory of statistical mechanics was developed after (macroscopic) thermodynamics, and it was constructed by studying consistency of microscopic dynamics with thermodynamics. If thermodynamics had not existed previously, statistical mechanics could not have been developed.[5] In general, a knowledge of microscopic behavior does not necessarily allow us to automatically derive macroscopic behavior. One clear example of this is that despite the fact that thermodynamics provides an understanding and quantitative description of the irreversibility for a transition between equilibrium states, within statistical mechanics, such theory of irreversibility has not yet been completely established. From this consideration, it would seem that no matter how carefully and thoroughly it had been carried out, without thermodynamics, a study of microscopic dynamics of molecules would not have led to the concept of entropy. In the actual historical development, Carnot, Clausius, and others were able to construct the theory of thermodynamics with no consideration of molecules, focusing only on macroscopic properties.

At the present time, the state of molecular biology, with our increasing ability to observe individual molecules, is something like the hypothetical situation described above. Unfortunately, as we see from this case, it is not always true that a field of study benefits from an increased ability to observe microscopic phenomena. In the case of thermodynamics, inability to observe microscopic phenomena led necessarily the success of a theory of a strictly macroscopic scale The contrast between the situation with which molecular biology now faces and that which led to the development of thermodynamics, in this sense, is striking. Of course, we do not wish to claim that no consideration of the molecular level is needed to gain an understanding of life. However, it is not necessary to account precisely for each molecular process. On the contrary, it is often important to employ a coarse-grained description

---

[5] In statistical mechanics, which probability measure one should choose is constrained from a macroscopic condition provided by thermodynamics.

in which insignificant details are removed. In fact, one of the approaches we propound here is that of seeking universal properties by using a coarse-grained description of the systems under investigation.

The argument that enumeration of all of the molecular processes does not yield an "understanding" of the system can be made also in the context of computer simulations. To see this, let us suppose that we carried out an enormous molecular dynamics simulation incorporating the equations of motion for all the molecules in some system and in this way we were able to track the trajectory of each molecule. For example, assume that this was done for a gas or liquid system and in this way we were able to precisely mimic the behavior of an actual system at the molecular level. However, such a description would not provide us with an understanding of this behavior. In particular, without a knowledge of thermodynamics, the concept of entropy would bot be deduced. Only after we have become aware of the concepts of thermodynamics and we have attempted to imagine how these concepts are connected to microscopic behavior does it become possible to extract useful information from such a simulation. Presently, there are plans for several projects to carry out large-scale simulations employing models that incorporate all of the individual processes taking place in a cell. However, as made clear from the preceding discussion, even if such a project were successful and a precise description accounting for all of the molecular-level processes taking place in a cell were realized, this would not provide us with an understanding of living systems. This point is treated further in the next section.

## 1.13 Transcending Enumeration

The current trend in biology is to push ahead projects of enumeration, as if such a bookkeeping description would be sufficient. First, there was the "genome project," with the goal of determining the sequence of the four constituent compounds making up the DNA molecule. Then, there are the similar projects to determine the proteome (the set of all proteins) and the metabolome (the set of all metabolic elements). In fact, we have even come to hear the term "physiolome," referring to the specification of an organism's entire physiological state.[6] However, such a project is hopeless in some sense. Because the number of molecular species existing in an organism is itself extremely large, determining the number of individual molecules of each species would be essentially impossible.

Perhaps a comment regarding the field of systems biology is in order here. If this field were actually involved in an attempt to understand organisms as systems, then its aim would be the same as the field we hope to establish. However, that which is referred to as "systems biology" today has often degenerated into enumerationism, dedicated to projects of enumerating biological processes. In fact, it has gone so far that even the term "systeome" has

[6] Probably, we need "omome" soon, to enumerate all "–ome" projects!

become part of its vocabulary. For this reason, here we would like to make a clear distinction between the approach we advocate and that of so-called systems biology.

Of course, a set of data obtained in accordance with a fixed standard can be very useful, and in particular it is helpful in the development of medicines. However, such an approach amounts to nothing more than the compilation of data bases, and it cannot be regarded as science by itself. More precisely, an attempt to form a science of living systems on the basis of such an approach is inherently problematic. Below, revisiting the points made above, we explicitly discuss each of the problems that this would entail.

### 1.13.1 The Absolute Limitation of Enumeration

Obviously, it is not possible to enumerate every possible situation with regard to a living system or systems, as this would include an infinite (indeed, in many cases uncountably infinite) number of cases. First, there is the problem of the vast number of possible combinations of the ways in which genes can act, as discussed previously (although, strictly speaking, this is still a finite number). However, there are cases in which the expression of each gene is not a matter of simply "off" or "on." In the situation that the degree of gene expression is a continuous variable, enumeration of the number of possible types of behavior is impossible even in principle. Also, as discussed in Sect. 1.3, the number of possible situations that an organism could face is clearly uncountable.

### 1.13.2 Standard for Enumeration

It may be thought that, even if a total enumeration is impossible, it should be possible to define some finite bounds within which a complete enumeration can be carried out and it would be worthwhile to pursue something of this kind. However, it must be realized that there is a problem of ambiguity involved in the establishment of a standard that could be used to define appropriate bounds. For example, consider the metabolic map presented in Fig. 1.2. As mentioned above, this does not represent the entire reaction, but whether or not this describes the "important" part is open to debate. In general, the question of what standard to use in representing a reaction will be answered differently by each investigator, because in the study of a given system, the data that one is capable of obtaining or that one chooses to retain will depend on the particular experimental techniques that she or he can apply and the interests that (s)he may have.

### 1.13.3 The Essence of Life Does Not Lie in Combinatorial Complexity

Let us consider a typical protein molecule forming a chain of, say, 100 amino acids. In combinatorial terms, there are $20^{100} \approx 10^{130}$ possible realizations of a protein of this length. It is impossible that through the course of evolution, each one of these possibilities was tried and that through this process the realization that is actually found in biological systems was selected. In other words, it is not necessarily the case that the existence of a certain protein implies that its precise form was selected. Of course, there are certainly some sequences of amino acids that would be biologically worthless. However, this said, there is also no reason to believe that there are only a few special sequences that had to be selected. Rather, it is natural to believe that, to some extent, whatever protein molecules actually came into use, the nature of life would have been essentially the same. Realizing this point, we see that there is no reason to search within the vast realm of possibility for the particular sequences that happen to be employed by living systems, and taking great pains to construct models that accurately reflect these sequences will probably bring us no closer to an understanding of the essence of life.

Actually, in order for some degree of life function to emerge, no delicate selection of amino acid sequence is necessary. This point has been clearly demonstrated experimentally. Recently proteins have been synthesized by constructing random sequences of amino acids, and despite this randomness, these proteins possess some functions found in the proteins occurring naturally (Yamauchi et al., 2002). Although the sequence of amino acids is just given by a random sequence, the synthesized proteins possess a certain level of function, and if these are then allowed to undergo mutation, high degrees of functionality can be realized. In other words, in attempting to create protein molecules with levels of function sufficiently high to allow for life, it is not necessary to pinpoint some very special initial conditions in the space of possible sequences. Rather, although there are differences of degree, the faculties necessary for living systems are possessed quite universally among protein polymers.

Now, one might conclude from the foregoing discussion that enumeration may not require such an enormous task, as it would be necessary to consider only the "important" molecules. However, neither the methodology of enumerationism itself nor the field of systems biology in its present form provides the grounds to judge what molecules are "important." To make such a determination, some other theory is needed. In other words, until we can construct a theory that provides the basis to determine what is important, the information gathered in the enumeration procedure yields no definite conclusion.

### 1.13.4 The Importance of Fluctuations

Let us suppose that the rate equations of chemical reactions for all metabolic reactions and all other processes that take place within an organism could be written down, by assuming that we would know the concentration of each species of molecule (i.e., the numbers of individual molecules of all kinds). However, the rate equations of chemical reaction can be considered accurate descriptions of actual processes only if the number of molecules participating is very large. In the case of the reactions taking place within cells, the actual numbers of molecules are not large, and for this reason, fluctuations in molecule number play an important role. It is then in vain, for example, to attempt to fit as accurately as possible, say, 1,000 coupled differential equations describing processes that each contains some significant percentage of fluctuations.

### 1.13.5 The Essential Difficulty Involved in Constructing That Precisely Describe Biological Phenomena

When studying a simple system that can be described with just a few variables, it is often possible to precisely determine a unique fundamental relation between these variables that can account for experimental data. However, as the number of variables needed to describe a system becomes large, given a finite amount of data, it becomes impossible to carry out a fit that precisely specifies a particular relation among them. Of course, in such a situation, it is not difficult to find some relation (indeed, many relations) that provide a satisfactory fit to the data. (In the extreme case, we could define one parameter for each experimental datum.) However, doing this would be of little worth, as a relation obtained in this way would likely fail to describe the behavior observed in some other experiment. It thus becomes clear that there is a fundamental problem involving the question of how to limit the number of variables used to fit data. At the present time, there exists no guiding principle in this regard.

### 1.13.6 The Lack of Reliable Fundamental Equations

Now, having realized the problems inherent in fitting experimental data, one might be tempted to think that, without relying on experimental data, it should be possible to derive some basic equations describing the living process from the rate equations of chemical reactions, and to carry out simulations using these. However, there are several points that must be noted here. First, the reaction rate formulas used to describe the changes in concentration of various molecule species undergone in chemical reactions are usually only applicable for systems in (local) equilibrium or very close to equilibrium, and it is not necessarily the case that the reactions taking place within a cell actually satisfy such conditions (Blumenfeld & Tikhov, 1994). Also, in many cases, reaction rate equations are derived with the assumption of dilute solutions, and therefore it is not clear whether they can be used to describe the conditions of high concentration that exist within cells, as shown in Fig. 1.2. Of course, despite these complications, there are certainly many cases in which such equations yield useful approximate accounts of the phenomena found in cells. However, they cannot be regarded as providing precise quantitative descriptions. Indeed, equations capable of producing such descriptions cannot be derived within the present framework of thermodynamics. Employing equations of this kind in an attempt to construct a very quantitatively faithful model of such phenomena is almost meaningless, as this is simply beyond their realm of applicability.

### 1.13.7 A Description of Phenomena is Not Equivalent to an Understanding

As discussed in the previous section, an understanding of some phenomenon is not obtained by constructing and adjusting a set of equations in such a manner that it provides an accurate model. For example, suppose that we are studying a system that, under certain conditions, exhibits concentration oscillations of its various chemical species. Then, let us assume that we were able to write down all of the equations describing the reactions taking place in this system and that, upon investigation, we find that they indeed display concentration oscillations. (The Belousov–Zhabotinsky reaction is a well-known example of this kind. Even though this is a relatively simple inorganic chemical reaction, it consists of many intermediate reactions, of which nearly 100 have been identified, and it has not yet been fully elucidated.) Such a result may be considered "success," but we believe that, for the purpose of understanding the mechanism of the oscillation phenomena seen in chemical reactions, instead of such a complicated model, it is better to use a simple model describing just two variables, and elucidate how the oscillation appears through instability of a stationary state. Such an approach can demonstrate in general terms that under certain conditions chemical reactions exhibit concentration oscillations, and it can elucidate the mechanism at work in the original reaction in a variety of chemical species. Furthermore, with such an approach, the model we obtain is not specific to a particular reaction, and thus when considering some other system that displays similar behavior, it is likely that the same model can be applied to it as well. For these reasons, it is much more meaningful scientifically to seek the construction of simple models and endeavor to derive an understanding from these than to attempt to mimic every detail of each specific system we encounter.

### 1.13.8 The Necessity for a New Framework

Regarding a living creature as simply a mosaic of chemical processes precludes the realization of a deeper understanding of life. Of course, we could just concede that the problem here is too difficult and give up trying to obtain such an understanding. However, the fact that there exist properties common to all forms of life that we know – diversity, stability, the irreversibility of the process leading from birth to death – strongly suggests that the construction of a science capable of providing an understanding of these universal properties is possible.

In the task we face, the difficulty does not lie in the problem of accounting for unfathomably many molecules. Even if we are not able to enumerate all of the constituent elements and processes making up a living creature and construct a precise molecule by molecule account (in fact, we hold that this is probably impossible), it need not be concluded that living systems cannot be understood. The understanding of universal properties that do not depend on the details of individual systems is what we seek, and these should exist in some realm that is completely separate from that of enumeration and the mimicry of fine details. The study of combinatorial complexity (in other words, the so-called "complicated systems" in the enumerationism sense) represents an approach to the problem of understanding life that differs from that which we advocate, that of complex systems (Kaneko & Tsuda, 2000; Kaneko (editor), 2001; Kaneko, 1998b). What we hope to construct is a "mode of understanding" for the problems that cannot be understood by simply treating systems as a collection of microscopic *if–then* logical processes. Furthermore, we wish to construct a scientific method of treating a system with the circulation between micro and macro scales, a system in which the understanding of the parts requires understanding the whole system.

In this chapter, we have critically reviewed present-day molecular biology and the "-ome doctrine." However, with criticism alone and by simply stating that with the methods used to this time living systems cannot be understood, we cannot contribute to the advance of science. With this book, we hope to offer a system of study to replace these previously existing methods. Then, to promote the development of this system of study as a science, we propose a new point of view from which to develop experimental methodology and a concrete research program. What we describe in this book is the collaboration of theory and experiment. The importance of the book by Schrödinger mentioned in Sect. 1.1, which appeared when molecular biology was in its infancy, and the book by Watson, which was largely responsible for the rise in prominence of molecular biology, lies in the fact that they provided for biology a new point of view and a new system of study. In this book, we hope to present a point of view and system of study that differ from those which came to form the field of molecular biology. With this goal, in Chaps. 2 and 3, we summarize the two points of view that we take.

# 2 Constructive Biology

## 2.1 The Understanding Obtained Through Construction

As discussed in Chap. 1, we do not wish to investigate the detailed processes of present-day living creatures and to construct some kind of precise organism "blueprints" for them. Rather, our goal is to elucidate the universal properties possessed by living systems. With this goal, there is no reason to focus on the organisms that happen to exist on Earth at this time. However, as a minimal requirement, we set out to construct systems incorporating structure and processes common to the organisms of which we have direct knowledge. Thus, we attempt to uncover universal properties. We call this approach "constructive biology." The thinking that underlies this approach is to construct a world that reflects and captures the essence of biological systems and to find the phenomena that appear universally and inevitably in that world.

If there exists a universal structure of living systems that is potentially amenable to some general theory, it should be possible to make progress toward developing such a theory with the constructive approach that we propose. However, if such structure in fact does not exist, and if living systems are complicated and intricate in their very essence, then the problem of understanding life may be hopelessly difficult. If the reality is that living organisms represent extremely exquisitely formed designs precisely selected from unfathomably many possibilities, then there would be no way to elucidate the characteristics of living systems other than creating very intricate imitations. But, of course, producing models that could describe all of these details would be extremely difficult. By contrast, our basic philosophical point of view is that there is some universal structure for fundamental properties of living systems. Following this manner of thinking, we believe that if we consider systems that possess the proper conditions, the fundamental properties of biological systems should appear naturally. To carry through with such an approach, obviously it is necessary to develop appropriate experiments.

Now, before proceeding, we should first give some discussion of what we mean by "fundamental properties." What we are interested in is a universal theory of life, and therefore we do not wish to limit our attention to only the organisms that exist now or have existed in the past on Earth. We do not confine our investigation to the characteristics that this set of organisms happen to possess – the molecules they employ, the forms they take, or the chemical structures they exhibit. Rather, we seek the universal properties that all possible organisms must possess, without being constrained by the particular paths that by chance evolution took on this planet. So, what might these properties be?

The question we pose here is obviously not an easy one. The following are some properties that immediately come to mind: a distinct boundary separating the organism from the external world; some degree of autonomy with which the organism is able to maintain its internal state; some degree of complexity characterizing the metabolic activity that carries out this maintenance; the ability of production through which similar but not identical individuals come into being; the existence of a structure that carries the information allowing for such reproduction; for multicellular organisms, the existence of a mechanism of differentiation through which cells possessing a variety of roles are formed and a developmental process arises; the ability to pass the developmental process to the next generation through germ cells. In addition, one may include the capacity for diversification through evolution over generations.

Present living systems are extremely complicated products of history, and for this reason, it is probably not possible to judge what is essential by considering only these. In other words, studying these organisms alone, we cannot determine what properties are simply the accidental results of evolution and what properties are indeed inevitable. This point elucidates the potential usefulness of the constructive method in identifying truly universal properties.

There have been research programs in biology whose goal is to identify molecules essential for living systems – those molecules whose presence is absolutely necessary for such a system to function. For example, studies of the following type have been carried out. Suppose that we removed a particular gene from some organism, and that as a result, some important element $X$ became lost. Then, suppose that because of its absence, some function $A$ could no longer be carried out. Thus, because the absence of $X$ resulted in the loss of $A$, we are led to conclude that the former must be necessary for the latter. This simple description represents the manner in which such studies seek *necessary conditions* for the functions of living organisms.

This approach does not attempt to determine *sufficient conditions* for something to be considered a living creature because, clearly, the result described above would not establish sufficiency in any sense, as it could be the case that even if $X$ is present, the lack of some other element $Y$ would result in $A$ being lost. However, despite this fact, in actual cases when some element $X$ of this type is discovered, without looking for some other elements $Y$, $Z, \dots$ whose absence could yield a similar result, it is often asserted that $X$ is the element that causes $A$. In the case that $X$ is a gene, such a result is often (misinterpreted and erroneously) represented by such assertions as "$X$

is the $A$-gene." Such a tendency reveals the apparent motivation for seeking sufficient conditions.

Now, if we indeed are interested in determining a sufficient condition for a particular characteristic of a living system, how should we proceed? First, let us suppose that we have prepared a certain set of elemental processes and with these alone have constructed some system that possesses the characteristic in question. In this case, because we have constructed this system, we know what elements it possesses, and thus we are able to say that these elements are sufficient for the appearance of the characteristic of interest. Of course, for the purpose we consider, it is best if we include the fewest possible elemental processes.

To reiterate, what we propose here is to construct a system according to the conditions we have chosen, and with such a system, to recreate the fundamental properties and functions of living systems. This is the approach of constructive biology. Thus, we do not wish to study those forms of life that happen to exist here and now, after a certain set of "rules" has been selected by evolution. Rather, we hope that with the conditions we have established that we can elucidate how life's fundamental processes of, for example, reproduction and development come into existence.

## 2.2 The "Way" of Construction

The approach we advocate in constructing model living systems is not to aim for the description of systems that exist in the present world or systems that differ only slightly from these, but, rather, to investigate a very wide range of systems that can differ significantly from the present living systems. With such a broad scope, we hope to discover the properties common to all systems.

Here, small perturbations of presently existing systems are not sufficient, as we wish to confront situations far beyond the realm of common experience. As suggested by the writings of Zhuangzi presented in Chap. 1, we strive to behold the mountain peaks far from the road we have spent our whole life traveling. In this way, we believe that the large-scale properties of living systems will become clear.

Reiterating our stance in different terms, we have no desire to carefully add complicating details to our models, adjusting and appending in order to fit the phenomena exhibited by actual organisms. Our goal is not to construct theoretical models or experimental systems that copy the systems found on Earth. Instead, we seek a more universal theory of life, one that would apply even to creatures found on other planets.

At this point, allow us to relate another conversation between Zhuangzi and his friend. Zhuangzi often expounded the "Way" of things. (According to his thoughts, the term "Way" represented universal truth or a universal law.) On one occasion, his friend asked him where this "Way" could be found, where he could look to discover it. Naturally, Zhuangzi responded that this

"Way" is something universal that exists everywhere. But the friend was not satisfied and insisted that unless Zhuangzi explained to him where it is, he would not know. While Zhuangzi was at a loss to specify where universal laws can be found, he tried to offer his friend some understanding, saying that they are in crickets and ants, in millet, in roof tile, and in excrement. His friend replied that he had imagined that the "Way" is in noble and lofty things and that he was shocked to hear that it could be found in things of such a base nature. Actually, Zhuangzi mentioned such examples for just this purpose, emphasizing the idea that the "Way" is truly everywhere, and that it may in fact be better to search for it in the common rather than in the sublime. This manner of thinking expressed by Zhuangzi applies to our constructive approach as well. For example, suppose that we are interested in formulating a universal theory of multicellular organisms. Then, if we consider some highly developed creatures in whom the multicellular structure has been well established, the process of evolution must have already bestowed these creatures with some highly refined, detailed mechanisms. Because we are not interested in such detailed refinement, it would be better for our purposes to construct systems that possess only the minimal properties that allow for the functioning of a multicellular structure. Systems of this kind would be much more useful for formulating a theory to describe the emergence of multicellular organisms.

As the above discussion indicates, in the constructive approach that we employ, we aim to include in the systems we study only the basic processes necessary to realize and understand life. We incorporate only such basic properties as that, for example, elements possess internal degrees of freedom, elements can display a variety of properties, elements can reproduce, and elements interact with one another and therefore cannot be treated in complete isolation. Under such general conditions, we search for classes of phenomena that appear inevitably. Then, having identified such classes, we propose to reinterpret the world in which we live. This is the essence of the constructive method.

## 2.3 Examples of Studies in Constructive Biology

Our "constructive biology" consists of the following steps of studies.

(i) construct a model system by combining procedures;
(ii) clarify universal class of phenomena through the constructed model(s);
(iii) reveal the universal logic underlying the class of phenomena and extract logic that the life process should obey;
(iv) provide a new look at data on the present organisms from our discovered logic.

There are three possible levels to perform these steps. Three levels are (1) gedanken experiment (logic), (2) computer model, and (3) real experiment.

The first one is theoretical study, revealing a logic underlying universal features in life processes, essential to understand the logic of "what is life."

Still, life system is not so easily understood. In particular, such system has dynamic circulation between each part and the whole system. We have not gained sufficient theoretical intuition to such system. Then it is relevant to make computer experiments and heuristically find some logic that cannot be easily reached by logical reasoning only. This is the second approach mentioned above, that is, construction of artificial world in a computer. Here we combine well-defined simple procedures, to extract a general logic therein.

Still, in a system with potentially huge degrees of freedom like life, the construction in a computer may miss some essential factors. Hence, we need the third experimental approach, that is, construction in a laboratory. In this case again, one constructs a possible biology world in laboratory, by combining several procedures. This is nothing but experimental constructive biology that will be pursued in the present book at the biochemical reaction level, the organism level, and the level of ensembles of cells.

In the present book, we discuss the problems listed in the table. The first two items in the table are related with the construction of a replicating system with compartment. When we consider the origin of a cellular life, this self-replication is essential. However, we do not intend to reproduce what has occurred in the earth. We do not try to guess the environmental condition of the past earth. Rather we try to construct such replication system from complex reaction network under a condition set up by us. For example, by constructing a protocell, we ask the condition for the heredity, or search for universal features of the reaction process to allow for the recursive production of cells.

The third to fifth items are related with the construction of multicellular organisms with developmental process. When cells are aggregated, they start to form differentiation of roles, and then from a single cell, robust developmental process to form organized structure of differentiated cells is generated. This developmental process to form a cell aggregate is transferred to the next generation. Experimentally we try to construct such multicellular organisms from bacteria. Here again, we do not try to imitate the process of the present multicellular organisms. For example, by putting bacteria cells into some artificial condition, we study if the cells can differentiate into distinct types or form some robust distribution of cells. Also, construction of morphogenesis is discussed by putting undifferentiated cells of frog into some given artificial conditions. With these studies, we can establish a viewpoint of universal dynamics underlying development rather than finely tuned-up process in development.

The sixth and seventh items are with regards to the construction of evolution. The first topic concerns with the phenotypic fluctuation, where its relevance to genetic evolution is studied in relationship with artificial selection experiments. The second topic is about the speciation process, that is how a species splits into two distinct groups different both in phenotype and in genotype. We also discuss the problem of construction of symbiosis of organisms and other related problems.

To carry out this plan experimentally, we need a system to design a life system controlled as we like. As standard experimental tools, we will adopt recent techniques in molecular cell biology, such as flow-cytometry (cell sorter), imaging techniques, microarray, and so forth (see Sect. 3.1), while microfabrication techniques will be powerful in constructing a system regulating the behavior a single cell or multiple cells.

Here this construction is interesting by itself, but our goal is not the construction itself. Rather we try to extract general features that a life system should satisfy, and set up general questions. For example, we set up a question if there are some "information molecules" that control the replication system. Then we answer the question by setting up a theory. For each item, we set up general questions, make model simulations, and set up a general theory to answer the question. This theoretical part is carried out in tight collaboration with the experiment (Table 2.1).

**Table 2.1.** Examples of constructive biology under current investigation

| Construction of | Experiment | Theory | Question to be addressed |
|---|---|---|---|
| Replicating system | In vitro replicating system with several enzymes | Minority control | Origin of information |
| Cell system | Replicating liposome with internal reaction network | Dynamic bottleneck in autocatalytic reaction system | Evolvability and recursiveness for growth |
| Multicellular system | Interaction-induced differentiation of an ensemble of cells | Isologous diversification in inter-intra dynamics | Robustness in development |
| Developmental process (I) | Controlled differentiation from undifferentiated cells | Emergence of differentiation rule | Irreversibility in development |
| Developmental process (II) | Activin-controlled construction of tissues formation | Self-consistency between pattern and dynamics | Origin of positional information |
| Evolution | Artificial selection of bacteria | Fluctuation response relationship | Genetic assimilation of phenotypic fluctuation |
| Speciation | Interaction-dependent evolution of bacteria | Symbiotic sympatric speciation | Genetic fixation of phenotypic differentiation |
| Symbiotic system | Symbiosis between different organisms | Plasticity in coupled systems | Joinability of systems |

Three points are important here.

1) It should be stressed that we are not trying to be a "magician" to synthesize a life system. Rather, we intend to understand the logic of life through constructing a life system. By setting up a minimal condition from our side, we understand logic of life. Hence we pose questions common to a large class of life, as listed at the third column of the table.
2) What type of a model is best suited for a cell for the present purpose? With all the current biochemical knowledge, we can say that one could write down several types of intended models. Because of the complexity of a cell, there is a tendency of building a complicated model in trying to capture the essence of a cell. However, doing so only makes one difficult to extract new concepts, although simulation of the model may produce similar phenomena as those in living cells. Therefore, to avoid such failures, it may be more appropriate to start with a simple model that encompasses only the essential factors of living cells. Simple models may not produce all the observed natural phenomena, but are comprehensive enough to bring us new thoughts on the course of events taken in nature.

   In setting up a theoretical model here, we do not put many conditions to imitate the life process. Rather we impose the postulates as minimum as possible and study universal properties in such system. For example, as a minimal condition for a cell, we consider a system consisting of chemicals separated by a membrane. The chemicals are synthesized through catalytic reactions, and accordingly the amount of chemicals increases, including the membrane component. As the volume of this system is larger, the surface tension for the membrane can no longer sustain the system, and it will divide. After the division of this protocell systems, they should interact with each other, since they share resource chemicals. Under such minimum setup as will be discussed later, we study the condition for the recursive growth of a cell, as well as differentiation of the cell.

   Here we note the role of model again. It is a tool to reach some universal properties, and once we understand how such universal features appear through model simulations, then the features we found have to be described independently of the details of the models. We have to clarify the logic why such universality holds. At this stage, the specific model is not necessary. The model is a kind of scaffold to reach the universality structure; once the universal feature is understood, one can get away with the model, as scaffold. (For example, recall Carnot cycle in thermodynamics. It is a quite important model for gedanken experiment to construct thermodynamics, but once it is established, we can formulate it even without it).
3) We are not intending to imitate the real life. Often we take a situation far from in real-life phenomena, and then we try to understand the universal logic that the life system should obey (e.g., construction of a logic of multicellular organism by putting *E. coli* in a culture with very high density, that is far from in real life; construction of developmental process

by putting undifferentiated cells in the environment that does not exist in real developmental process [e.g., in a very high concentration of activin]).

## 2.4 On the Mode of Understanding

As stated above, the theoretical models considered here are solely for the purpose of acquiring understanding. In this sense, although these models may seem to be similar to the typical type of model used in the study of biological systems, as both are investigated through numerical simulation, they are in fact of very different natures, as the purpose of the latter is to precisely mimic phenomena observed in actual systems. The so-called systems biology approach of carrying out simulations of complicated models intended to closely fit the behavior of real systems is critically reviewed in Chap. 1. Now, what kind of models should we consider to realize an understanding of living systems? As previously asserted, we believe that carefully incorporating every process in a model and precisely describing some phenomenon does not in itself provide the understanding. So, if for our purposes, the standard by which we judge the worth of a model is not the precision of the description it provides, then what is this standard? We hold that to *understand* something is in some sense to be able to reduce it. One way to think of "reduction" is, for example, the reduction to a few specific molecules or combinations of molecules or the reduction to some fundamental system of equations. Such reduction is easily understood. However, "reduction" in science is not necessarily of such simple types. If we think of organizing natural phenomena into classes, it is often the case that phenomena that at some level appear to be quite different are seen to in fact belong to the same class when unimportant "details" are removed. In this way, it becomes clear why these different phenomena exhibit the same kind of behavior, and we come to understand why these phenomena appear.[1] When we observe phenomena with "half-closed eyes," ignoring details that may obscure the underlying essence, it comes to be realized that many phenomena that we previously regarded as different are actually of the same kind. In the case that the commonality of such phenomena can be characterized mathematically, we say that they belong to the same **universality class**.

We should point out here that what it means for two phenomena to belong to the same universality class is not that there is some quantity characterizing them that is equal in the two cases but that they exhibit behavior that is

[1] For example, consider the phenomena of fluid turbulence and the irregular fluctuations in a chemical reaction of chemical concentrations and molecule numbers. While physically these are different phenomena, when we consider the mathematical nature of the time evolution that each exhibits, with a knowledge of chaos theory, we realize that at this level they are of the same type.

qualitatively the same.[2] Therefore, we do not seek quantitative agreement but, rather, qualitative agreement regarding such questions as the existence of certain properties.

Later in this book, we consider a number of models. These models serve as representatives of certain qualitative universality classes. The purpose of this modeling is to obtain an understanding of why these types of universality exist and the range of phenomena that belong to the various classes. With this purpose in mind, the models we employ are chosen for their simplicity. Such models should possess only the most basic properties, because this will allow for their effectiveness in demonstrating universality. For example, we consider as a qualitative universality class the phenomenon of hierarchical differentiation, consisting of the process in which, with multiplication, cells differentiate into a number of distinct types. In this case, in modeling the properties of the cells, it is best not to attempt to faithfully mimic the complicated structure of the cells that exist in the actual organisms with which we are familiar. Instead, it is advantageous to choose models that exhibit the universal properties of interest while including the fewest number of conditions possible. For example, in the case of cell differentiation, the minimal set of conditions is that there exist chemical reactions within cells, that there are interactions among cells that are facilitated by the diffusion of the constituent chemical species, and that these properties allow for the multiplication of cells. Because we employ very simple models, more elaborate models based on the same basic structure should possess the same basic properties. Therefore, these simple models can be regarded as the basic forms of a wide range of more "realistic" models, and they can thus elucidate the types of phenomena common to many systems. In this way, studying models with very simple forms, we come to see the universality they possess. Then, if we can construct a theory that describes the manner in which the universal properties come into being, we should be able to approach an understanding of this universality itself.

### 2.4.1 Remark: Synthetic Biology

Constructive approach to complex systems was proposed in early 1990s (Kaneko & Tsuda, 1994, 2000), while the term "constructive biology" was coined in mid-1990s (Kaneko, 1998c). Nowadays the term "synthetic biology" is rather popular (Benner & Sismour, 2005; Sprinzak & Elowitz, 2005). In

[2] The term "universality class" originates from renormalization group theory, where two systems are said to belong to the same universality class if under the renormalization group transformation, which has the effect of smoothing out details, some of their properties come to be quantitatively identical. The concept of a qualitative universality class has not yet been precisely defined in a theoretical manner, nor is there a theoretical method to determine universality classes of this kind. Here we purposely use this term in a general sense, anticipating theoretical development through which we will obtain a more precise characterization of this concept.

both the approaches, intended is the artificial construction of a system that did not exist in Nature but has some biological features, while there is some difference in the emphasis in the direction of research. In the synthetic biology, studies oriented to design and engineering are stressed. In contrast, constructive biology does not aim at goal-oriented project to design some function. Rather, by setting minimal conditions for basic property of life, we try to unveil universal logic that a biological system has to satisfy. In this sense, construction is for the analysis or understanding what life is (Kaneko, 2003b). Of course, some studies in synthetic biology also aim at unveiling universal logic of life (Sprinzak & Elowitz, 2005) and the difference between the two could be in the emphasis in the study.

### 2.4.2 Remark: Artificial Life

Finally we give a few remarks on the study of the so-called Artificial Life (AL). Indeed, our approach may have something in common with AL (Langton, 1989, 1992, 1994). In AL study, people intended to construct life-as-it-could-be, not restricted to the present organisms. Originally, in the study of AL, they have been interested in logic of life that all possible biological system should obey, be it on this earth or in other conditions in the universe.

Indeed, there are some important studies on the origin of replicating structure from the side of computation (e.g., Fotnana & Buss, 1994). However, they often tried to imitate life, and the study often falls on superficial imitation. Even though they sometimes succeeded in making something similar to life, the success did not contribute in understanding the logic of life. As for the evolution, they usually adopt the genetic algorithm as a simplified version of Darwinian evolution, but the AL study has not contributed in proposing novel concepts in evolution.

Next, the study of "artificial life" is often biased into the study in a computer. Similarly with artificial intelligence study, the artificial life is mostly symbol based, and adopts a complicated combination of logical processes.

Our approach is clearly different from the conventional artificial life study in the two points. First, we do not take such symbol-based approach, but rather we use dynamical systems approach. Second, tight collaboration between experiment and theory is essential. Note, however, this collaboration is not of the type to "fit the data" by some theoretical expression, but rather, at much more fundamental and conceptual level (e.g., to understand what type of developmental process is irreversible, and what kind of operation is required to reverse it).

# 3

# Basic Concepts in Dynamical Systems and Statistical Physics for Biological System

## 3.1 Basic Picture in Dynamical Systems

This section is intended for biologists, who are not familiar with dynamical systems. Those who know dynamical systems theory and phenomena can skip this section, although explanation in connection with cell biology and recent experimental tool may be useful also to those. Note that this section is written rather loosely or intuitively (picturesque) without any mathematical rigor. Please refer to standard textbooks on dynamical systems or chaos.[1]

**State Space:** Dynamical system consists of time, a set of state variables, an evolution rule, an initial condition of the states, and a boundary condition. The state is represented by a set of $k$ variables. This $k$ is the number of "degrees of freedom." Thus the state at an instant is represented by a point in the $k$-dimensional space, called the state space (or phase space).

For example, consider a list of health check, represented by body temperature, pulse, pressure. Even if this set of numbers may be insufficient to define the condition for the health, it could be possible to characterize it by several variables, as long as one limits the characters in concern. Or characterize a state of a cell by a set of chemical concentrations $x_1, x_2, \ldots x_k$ in a cell. This set can be the gene expressions that give the numbers of mRNAs. Of course, it is a crucial question how many degrees of freedom $k$ are sufficient to characterize a cell state, but here we start by such approximate description that can capture some basic characteristics of a cell.

An example of state representation of a cell can be abundances of metabolites or proteins in a cell. Recently, there are techniques to measure the degree of gene expression, that is, the abundances of mRNA corresponding from each gene to a protein. With the use of microarray techniques, one can measure

[1] There is a classic book by Rosen (1970), which is pioneering and important, but unfortunately it is a bit out of date. For 'picturesque introduction, see Abrahama and Shaw (1988). For recent introductory textbooks, for example, Kaplan and Glass (1995), Alligood, et al. (1997), Strogatz (2001).

the abundances of a variety of mRNAs simultaneously. These roughly give a measure of abundances of corresponding proteins. Hence, a representation of $\{x_1, x_2, \ldots x_k\}$ with a huge-dimensional $k$ is available. Under the current (2006) standard technique, one could measure only the abundances from many cells, as the average abundances, but the data can provide a useful (statistical) information on a cell state (Fig. 3.1)

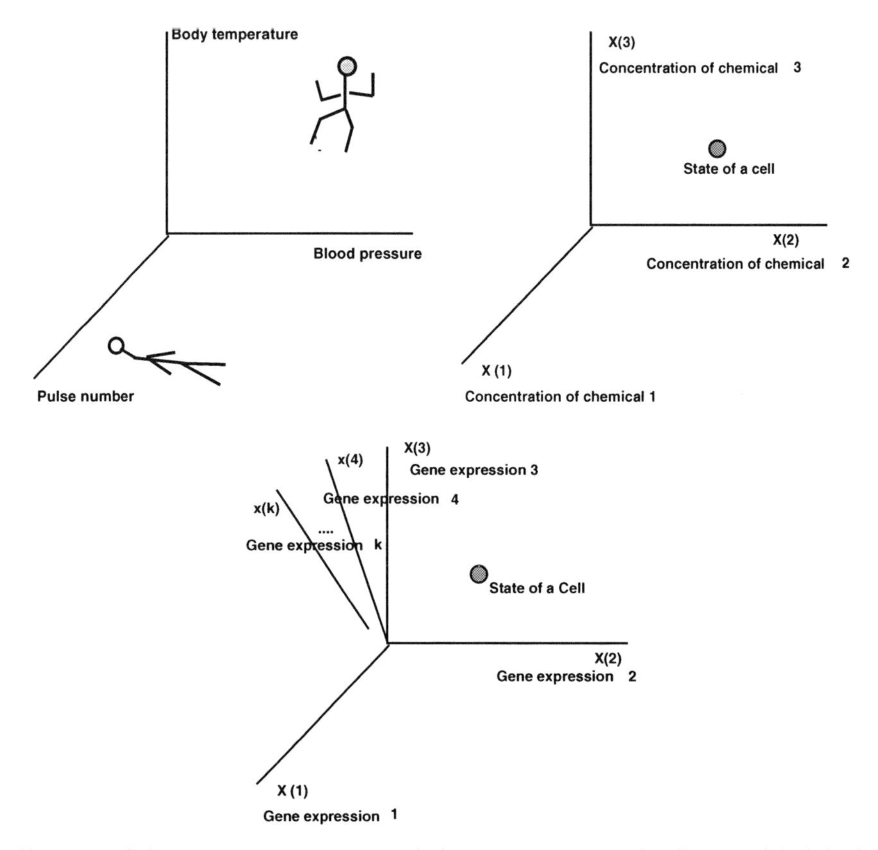


**Fig. 3.1.** Schematic representation of the state space: radically simplified body state; representation of cell's state by chemical abundances or gene expressions (i.e., the degree of production of mRNA from some gene to corresponding protein)

A technique to measure distribution of cell's state is provided by flow cytometry (or cell sorter). In flow cytometry each cell is separately measured with laser radiation. From the scattering of the data, information on a cell size, density, shape, etc., is measured.[2] With flow cytometry, one can represent a cell as a point in the state space. An example of such distribution, plotted by forward and side scattering intensity (which give a measure for the cell

[2] An apparatus that can separate cells according to these data is a cell sorter.

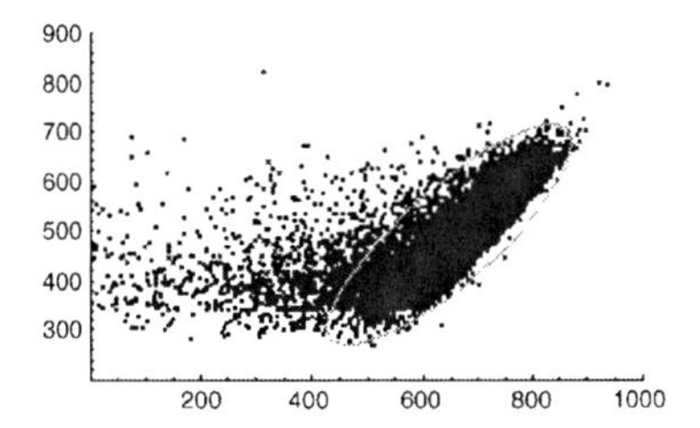


**Fig. 3.2.** Characterization of a cell's state by using flow cytometry. The figure is obtained from the forward scattering and side scattering from a bacterial cell radiated by laser beam in a flow cytometry. The two axes correspond to cell size and the magnitude of the fluorescence from the green-fluorescent protein

size and the density of cell's ingredients) is shown in Fig. 3.2. By using a cell sorter, one can take cells from some portion in this state space and cultivate them. In this sense, choice of "initial conditions" is now possible.

The flow cytometry can also give information on each protein's abundance by combining the technique of fluorescent protein. Recently, several proteins are designed to show the fluorescence with some color such as green, red, blue, etc. Then by inducing a gene to express such fluorescent protein (at the downstream of some gene), one can measure the degree of some gene expression by fluorescence. Since the fluorescence is easily measured by flow cytometry, one can obtain the distribution of abundances of such fluorescent proteins that correspond to specific gene expression. Hence the distribution of some gene expressions is available. With this technique, we have to assign the position to which this fluorescent protein gene is inserted, and it is not possible to measure many fluorescent proteins simultaneously. Hence the number of variables we can measure is usually limited to 3 or 7 or so (e.g., green, red, blue fluorescent proteins). Still, the direct measurement of distribution is a very powerful tool for the field of "statistical cell biology."

Of course, it is a critical issue how many degrees of freedom are necessary, and in principle, to give a detailed information on a cell, a huge number of degrees of freedom are necessary. How many degrees of freedom we need depends on what aspect of a state we are concerned with. If we need very much details of the state, the required number of degrees of freedom should be large, while for some coarse-grained representation, the number could be smaller. This is a rather difficult question in general, and for the moment we assume that there is some such number, and the state of a cell is represented by a set of variables $x_1, x_2, \dots x_k$. Still, it should be noted that biologists can distinguish cell types without knowing all gene expressions, and this classification is often valid. Also they can often distinguish if a cell is active or inactive, by just looking at it. As long as we do not assume "super" recognition ability of biologists, these empirical facts suggest that we may not need a such huge dimensional description of a cell state, and instead some coarse-grained description with a smaller number of variables will be effective.

### 3.1.1 Representation of a Cell Ensemble by the Distribution of Points in the State Space

By adopting this picture, the state of a cell is represented as a point in the $k$-dimensional state space. Then the state of an ensemble of cells is represented by the distribution of these cells.[3] As mentioned, the state of each unit (cell) can depend on the nature of its ensemble. Hence the state of each cell should be considered together with its ensemble. Indeed as the number of cells increases, the characteristics of each unit (cell) may change, as schematically shown in Fig. 3.3.

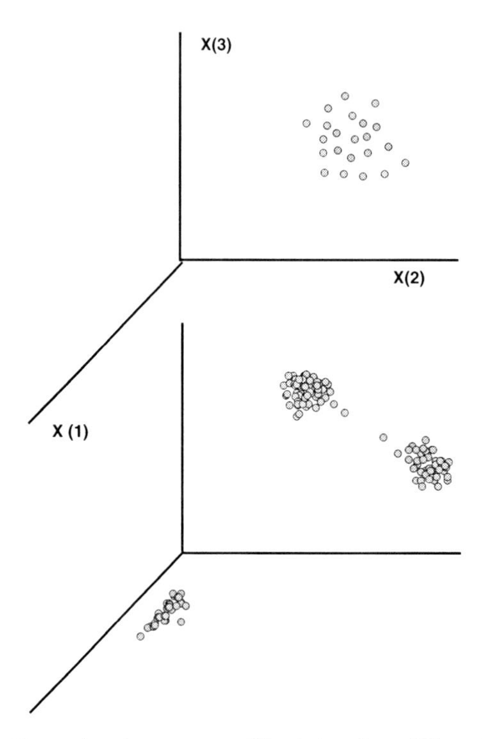


**Fig. 3.3.** As the cell number increases, cell's state often differentiates. In this case, the representation in the state space changes from the above to the below. See also Chap. 7

The technique of flow cytometry (together with fluorescent proteins) is ideal for such study. A state of a tissue can be characterized by the distribution of points in the state space obtained by flow cytometry, then. We will discuss several examples through this book.

[3] The spatial configuration of cells is discarded for the moment. This problem will be discussed later.

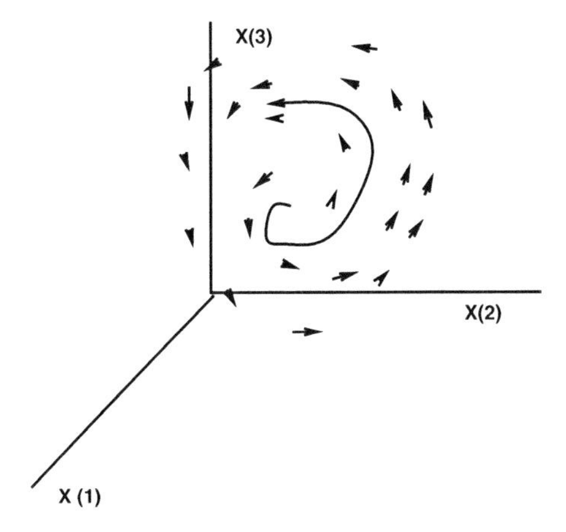


**Fig. 3.4.** In dynamical systems, the state change in time is represented by a flow in the state space (schematic diagram)

## Dynamical Systems

The temporal evolution of these variables is due to the reaction among these chemicals, which change their concentration in time. Since we adopt a macroscopic picture, the change of chemical concentrations is given by rate equations for chemical reactions. Then the change of chemical concentrations is given by a set of equations[4]

$$dx_i/dt = f_i(x_1, x_2, \ldots, x_k) \quad (i = 1, 2, \ldots, k) \tag{3.1}$$

As is seen in this example, the temporal change of state variables is given by a set of equations. In this "deterministic" picture, the state variable at later time is uniquely determined by the set of variables at a given time. The temporal change of the state is characterized by a flow in the $k$-dimensional phase space. When a point is given, the temporal evolution is shown by an orbit in the $k$-dimensional phase space (Fig. 3.4).

---

[4] The fluctuation is later included as a "noise term" in dynamical systems. Stability against such fluctuation is discussed later in Sect. 3.2. Another related problem against the present dynamical systems picture is the use of continuous variable representing, for example, the concentration rather than the number. If the number is large, this is fine, but sometimes the number of some molecules in a cell may be too low to be represented as a continuous quantity. The most typical case for this problem arises when the behavior of a system crucially depends on the existence (0 or 1) of some specific molecule. This problem is discussed in Chap. 4 as the origin of information.

### Dissipative System

When the state in concern is represented by a macroscopic variable (i.e., not at the level of atom or a molecule), the system is usually dissipative. In a dissipative system, information on initial conditions is lost through temporal evolution. For example, consider a frictional motion of a ball in a valley. Irrespective of initial conditions, it will stop at the bottom. This example may be too simple. Instead, for example, consider a motion of frictional pendulum with periodic forcing. Again, it will come to same pattern of periodic motion independent of initial conditions. Information on the initial condition is lost. This dissipation is represented as follows in the state space. Take a bunch of initial conditions of points covering a volume in the $k$-dimensional space, and see the temporal evolution of all orbits from the points. Then the volume covered by the points shrinks in time. This is a nature of dissipative system, in contrast with a Hamiltonian system (Newtonian dynamics without friction).

Note that this "dissipation" generally appears when we take a macroscopic (coarse grained) picture. At a macroscopic description that is robust against microscopic change, many microscopic states are projected on the same macroscopic state, and in this description, information on initial conditions is lost with some rate, giving rise to dissipation. When the change of macroscopic states is expressed by dynamical systems, it is generally dissipative. For example, the macroscopic dynamics of concentrations of biochemicals given by the rate equation of chemical reactions are generally dissipative.

### Attractor

In this dissipative case, an orbit does not necessarily return to the neighborhood of the starting point. With time, the orbit resides only within a restricted region in the state space. The region within which the orbit recurrently returns to its neighborhood is called an attractor. Roughly speaking, the attractor is the region to which the orbit is attracted as time passes. The regime before the orbit approaches an attractor is called transient (Fig. 3.5).

The abundance of gene expression or chemicals, of course, is not completely governed by the rate equation (3.1), which is deterministic. As the number of molecules is finite, there are generally fluctuations, as has been discussed in Sect. 1.9. (The concentration of some molecules is given by the number of molecules $N = nV$ divided by the volume $V$, while there can be generally fluctuations of the order of $\sqrt{N}$. Then, in the representation of concentration, fluctuations of the order of $1/\sqrt{N}$ generally appear, which is not necessarily very small.) When represented by "average equations" such as the rate equations of chemicals, there can be some deviations from this deterministic equation.

Even if a state is perturbed by such fluctuations, however, there can be a tendency to recover the original state, when the attraction to the state works.

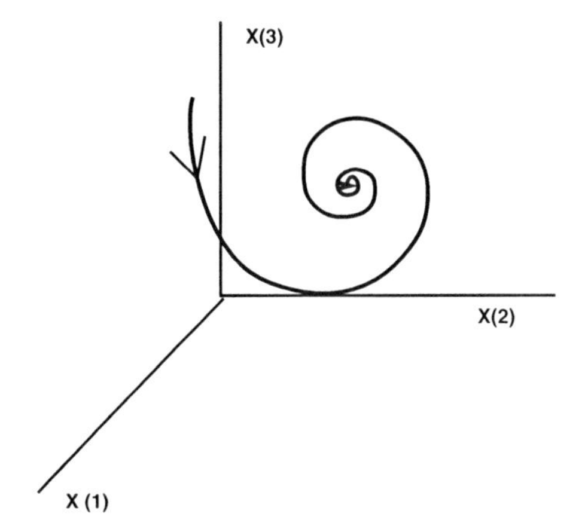


**Fig. 3.5.** Attraction to a fixed-point attractor

In this sense, the attractor can give an answer why a state is stable against (molecular) fluctuations.[5]

### Chaos; Strange Attractors

If a dissipative dynamical system has a potential and the orbit goes downhill to the valley of the potential, then the orbit finally is attracted to one of the minimal points of the potential, with dissipation. The attractor is a fixed point.

Generally speaking, there is no such potential, in which case, the attractor is not necessarily a fixed point. Sometimes, it is periodic cycle that leads to regular oscillation in time. In some other cases, the orbit is on a torus leading to a quasi-periodic oscillation, which consists of combination of perodic motions with several (incommensurate) frequencies.

Furthermore, there is a case called strange attractor. In this case, the attractor is not such a simple geometrical object in general, and the orbit is neither periodic nor represented by a combination of some cycles as in the quasi-periodic case.[6] The dynamic behavior for this type of orbits is called chaotic. When we try to represent the dynamics by Fourier modes, as a typical analysis for the time series, continuum spectra are required, in contrast with the discrete spectrum for a (quasi)periodic case.

[5] In some exceptional cases, there is an attractor that is not asymptotically stable in this sense (Milnor 1985, Kaneko 1997a, 1998a), though.

[6] Remark that there is a rather exceptional case, called strange nonchaotic attractor, where the torus is not smooth. In this case the attractor is geometrically fractal, but the motion is quasi periodic.

### Orbital Instability in Chaos

A chaotic attractor is characterized by orbital instability. A small difference in the initial condition of an orbit (on the attractor) is amplified. The amplification for tiny difference is exponential in time. Of course, the exponential deviation does not last forever. As the difference between the two orbits is larger, the difference will saturate. Indeed, the Lyapunov exponent, characterizing the exponential instability, is defined by taking the limit of the initial deviation to zero and of the time to infinity. Roughly speaking, the distance between two orbits with slight initial difference increases as $\exp(\lambda t)$, and this $\lambda$ gives the Lyapunov exponent.

Compatibility between instability and stability is often an important aspect of a biological system. To have diversity in a biological system, there should be some instability so that the original state of a unit is not maintained. On the other hand, a biological system is stable at least at a macroscopic level. Cells can differentiate but, the ensemble of cells formed through developmental process is rather robust against perturbations. To discuss such compatibility between ensemble stability and individual instability, chaos, or in general an attractor with orbital instability may give an insight for it, since each orbit in chaos is unstable in the above sense, but the orbits stay within an attractor, being robust against perturbations. This feature will be discussed later in the study of stability in the developmental process.

### Diversity of Orbits in Chaos

In contrast with the periodic or quasi-periodic orbit, chaotic orbit has diversity. For example, consider the famous chaotic orbit by Lorenz, as given in Fig. 3.6.[7] It rotates around the left "eye" for some times, and then right "eye" for some time. Now, to represent symbolically the dynamics, one can write down the symbol sequence like LLLRRRRLR.... If the orbit is periodic, after some time span, no new type of symbol sequence appears, (e.g., as in LRLRLR...), while in the chaos, possible diversity of symbol sequence increases as the number of symbol sequence increases. This diversity in chaos is also understood as a variety of unstable periodic orbits. Even in a system with chaotic dynamics, there is an infinite number of periodic orbits immersed within. These periodic orbits are all unstable, in the sense that any small perturbation to such periodic orbit kicks the orbit out of it. The chaotic orbit is also understood as a wandering over all such unstable periodic orbits. Hence the chaotic orbit has diversity (Fig. 3.7).[8]

Diversification is an important nature in a biological system. Chaos, or dynamics with orbital instability in general, gives one hint how such diversification occurs.

---

[7] Lorenz equation is written as $dx/dt = -s(x-y)$; $dy/dt = rx - y - xz$; $dz/dt = xy - bz$, where $s, r, b$ are positive parameters.

[8] The diversity of orbits is often measured as a Kolmogoirov–Sinai entropy.

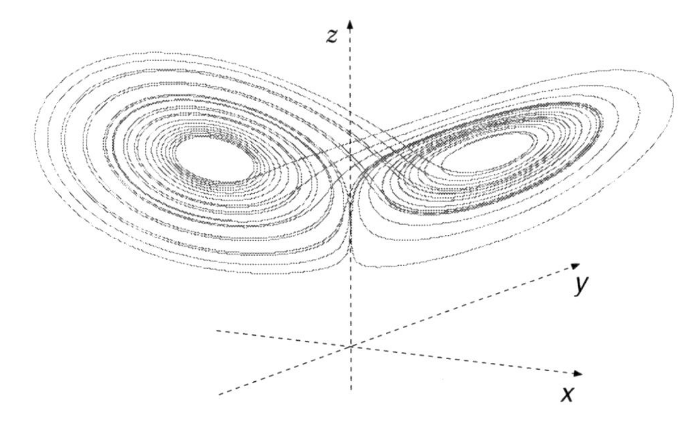


**Fig. 3.6.** Lorenz attractor: An example of an attractor with chaotic dynamics discovered in a 3-variable ordinary differential equation, as first discovered by Lorenz (1963)

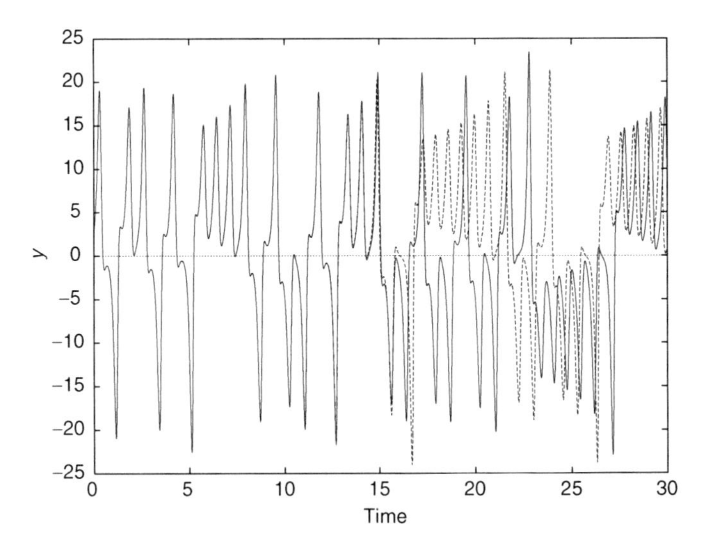


**Fig. 3.7.** The time series of $y(t)$ of the Lorenz equation, displayed with that when an initial condition is slightly perturbed. Against the time series represented by the *thick line*, the *dotted line* gives the time series started by perturbing $y$ with 0.000001 at time $t = 0$

### Stochastic Aspects

Although the future of an orbit is unpredictable as long as our precision is finite, the process may be described by introducing probability. Indeed, the long-time average of state variables is given by introducing some probability measure if the *ergodicity* is satisfied.

In a biological system, we often encounter with a situation intermediary between programmed behavior or probabilistic behavior, as in the differentiation from stem cell, some type of motion, or process of aging, etc. Chaos may provide a new look to this dualism.

### Statistical Description of Chaos

The orbital instability, of course, does not mean the instability of an attractor itself. Although the perturbed orbit departs from the original orbit, this orbit again stays within an attractor. Although an orbit is unstable, the statistical behavior is (in many cases) stable, in the sense that the average value of state, for example, is independent of initial conditions. At the level of long-term statistics, chaos (i.e., dynamics with orbital *instability*) is stable. A residence probability that an orbit is found in a given region in the phase space exists, and can be computed. Then the long-term average of states is replaced by the average of a given probability distribution assigned for the points in the phase space. This is an example of "ergodicity."

### Basin of Attraction

In some dynamical systems, there are several attractors. Depending on the initial condition, orbits are attracted into one of them. The set of initial conditions that are attracted to a given attractor is called a basin of attraction. Some attractors have a large volume of basin volume, while some others are not. The basin is not necessarily a simple geometrical object again, and sometimes it has a complicated structure. In this case, by small difference in initial conditions, the attractor reached through the evolution can be different ( see Figs. 3.8 and 3.9).

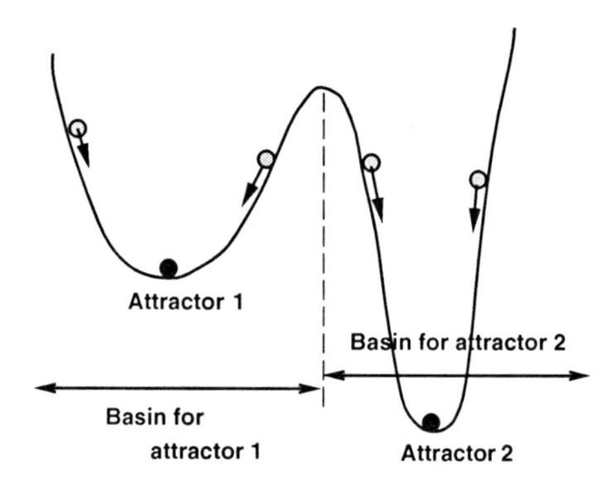


**Fig. 3.8.** An orbit in which the state is attracted toward the valley in the landscape. If there are multiple valleys, there are several attractors depending on initial conditions that give a basin for each attractor

## 3.2 The Role of Fluctuations

As briefly mentioned in Chap. 1, in the study of living systems, the occurrence of fluctuations cannot be avoided, because no matter what quantities we choose to express the states of a cells, under no circumstances will they exhibit perfectly constant values. Even if we consider cells of the same type, the values of the quantities characterizing them will vary from cell to cell, and for a single cell, these values will fluctuate in time. Now, suppose that we are considering the reactions taking place in a given cell. Then, even if it is possible to describe the changes undergone by the concentrations of the chemical constituents in these reactions by some rate equations, the number of molecules of each species contained in a cell is, at most, on the order of several thousand, and because cells exist in states that are by no means "low temperature," thermal fluctuations are necessarily significant. Furthermore, this description of the state of the cell does not include the positions and numbers of all molecules, and therefore there is necessarily a part missing from the description. Obviously, this missing part too will change in time, and thus with respect to this description of the cell's "state," this missing part is considered as *noise*, which leads to *fluctuations*.

### 3.2.1 Fluctuations and Stability

In the theory of equilibrium thermodynamics, fluctuations are represented as fundamental quantities, and within that framework, fluctuations and the stability of the system are intimately related. This can be understood by considering the potential depicted in Fig. 3.10 and the dynamics of a particle moving about the minimum of the potential. In particular, we are interested here in how these dynamics are affected by the gentle shaking of the system. Clearly, in the case that the slope around the minimum is steep, the state corresponding to this minimum is stable and the fluctuations about it are small, but when this slope is gentle, the state is relatively unstable and the fluctuations are large.

This kind of negative correlation between the degree of stability and the size of fluctuations is conjectured to hold in general. To understand this, let us consider the dynamics of a system as it evolves toward a stable state and the effect of fluctuations on these dynamics in the neighborhood of this state. If the deviation from the minimum of the potential is not large, the combined effects of the force acting to return the system to the minimum and the noise result in motion that can be described as follows (Fig. 3.10):

$$dx/dt = -\gamma x + \eta(t) \tag{3.2}$$

Here, $\eta(t)$ represents the "noise" produced by fluctuations. In other words, this term introduces a random influence on the state $x$. Recalling Brownian motion, let us closely study the situation in which the state of the system is

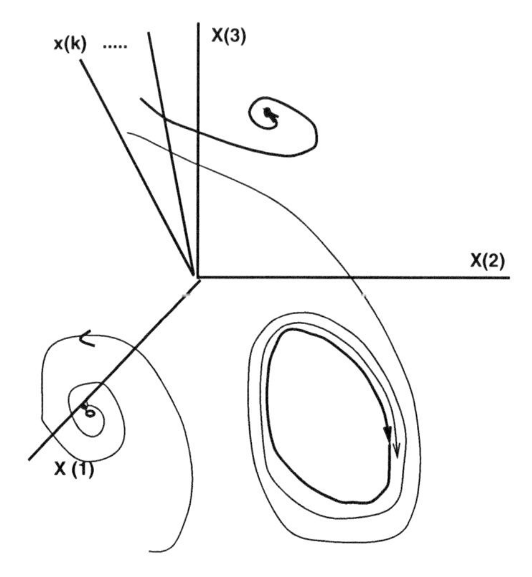


**Fig. 3.9.** A schematic representation of several attractors. Depending on initial conditions, orbits are attracted into one of them

**Fig. 3.10.** Schematic of a potential and fluctuations. When the slope about the potential minimum is large, the fluctuations are small, and when the slope is small, the fluctuations are large

caused to fluctuate randomly. We consider behavior in which there is a random effect that changes instant by instant and acts to increase or decrease the magnitude of $x$. If we add the contributions from this effect over a short time interval of length $\Delta t$, the result can be thought of as a sum of many random quantities. Therefore, the probability distribution corresponding to this sum can be assumed to take a Gaussian form. Also, we can assume that there is no correlation between such sums in different $\Delta t$ time intervals. Then, taking the limit that $\Delta t$ becomes infinitesimal, we obtain a stochastic differential equation like that above.[9] For our present purposes, however, it is only necessary to understand that $\eta(t)$ is a noise whose instantaneous values form a distribution of finite width. Here, $\gamma$ represents the strength with which the

[9] The type of random equations (stochastic differential equations) considered here, as well as the derivation of distributions from them, are treated rigorously in many mathematics and physics textbooks, and the reader should refer to these for a more complete exposition. (Readers familiar with the technical aspects of stochastic differential equations may wish to skip the treatment given below.)

system attempts to return to the stable state, $x = 0$. (Considering Fig. 3.10, the above equation corresponds to the case in which the minimum is located at $x = 0$, and the slope of the potential in the neighborhood of this point is given by $\gamma x$.) Finally, we assume that the width (variance) of the distribution corresponding to $\eta(t)$ is given by $\epsilon$.

Because the above equation describing the evolution of the state contains a probabilistic component, it is useful to consider the distribution of values of $x$. Referring to the distribution of such values obtained when we consider an ensemble of samples as $P(x)$, from the above, we obtain the following stationary form:

$$P(x) = \sqrt{\frac{\gamma}{2\pi\epsilon}} \exp\left(-\frac{\gamma x^2}{2\epsilon}\right) \tag{3.3}$$

(We do not give the derivation of this distribution, nor do we give a detailed discussion of its properties. We simply point out its basic features: It is peaked at $x = 0$, as $\gamma$ becomes large the distribution becomes sharp, and as $\eta(t)$ becomes large the distribution becomes flat.) With the notation $\langle \ldots \rangle$ expressing an average taken with respect to this distribution, its variance, defined as $\langle (x - \langle x \rangle)^2 \rangle$, is given by

$$\langle (\delta x)^2 \rangle = \langle (x - \langle x \rangle)^2 \rangle = \langle x^2 \rangle \propto \epsilon/\gamma \tag{3.4}$$

(Here we are considering the case in which the average, $\langle x \rangle$, is 0, and thus we see that the variance is given by $\langle x^2 \rangle$. However, here we have expressed it in the general form, $\langle (\delta x)^2 \rangle$.) Thus, as seen in Fig. 3.10, as the value of $\gamma$ increases (i.e., as the slope of the valley becomes steep), the fluctuations become smaller, and the state $x$ becomes more and more sharply determined. In this simple case, the relation between fluctuations and the stability of the $x = 0$ state can be represented by the ratio of the stability index, $\gamma$, to the variance.

Now, having considered a very simple case, let us turn to the question of how stability and fluctuations are related in the case of many interacting cells that each possesses more complicated internal states. This question is one of the main topics of this book.

### 3.2.2 The Relation Between Response and Fluctuations – the Fluctuation-Response Relation

Let us consider the relation between the change undergone by the state of a system when its boundary conditions are slightly altered and the fluctuations existing in the original system. It is reasonable to expect that, in general, when the conditions to which a system is subject are changed, the amount of the state change depends on the stability it originally possessed: For a very stable state with small fluctuations, the change should be small, while for a less stable state with large fluctuations, the change should be large.

As an example, let us consider a system described by (3.2) and the value of $x$ is altered slightly because of a change made to some externally applied condition. (Note that we need not assume that the change in $x$ is due to the application to the system of a "force" in the literal sense. Here, we are thinking in quite general terms, and this perturbation of $x$ could correspond, for example, to a change in the concentration of a chemical constituent or to the shift in a phenotype resulting from the addition of some selective pressure in a particular direction. In any case, whatever the nature of the system we consider, the effect of the perturbation is to shift the stable state.) We represent this perturbed system as follows:

$$dx/dt = -\gamma x + F + \eta(t) \quad (3.5)$$

(Here, if there is noise, we can include the term $\eta(t)$, but below we focus on average values, and therefore it need not be included.) With the perturbation represented by this force $F$, the stable state (its average if there is noise), corresponding to $dx/dt = 0$, is given by $x = F/\gamma$. In other words, the stable state has been shifted by the amount

$$\Delta x = F/\gamma$$

Then, comparing this with (3.3), we find

$$\Delta x/F = \langle(\delta x)^2\rangle/\epsilon \quad (3.6)$$

Thus, we have found that the amount by which the state of the system is changed because of the application of an external force is equal to the magnitude of this force multiplied by the ratio of the size of the fluctuations to the strength of the noise in the original system. For the time being, referring to the shift of the state per unit size of the applied force as the "response coefficient," we find that **the response coefficient under application of some operation to the system is proportional to the size of the fluctuations of the system without application of the operation**. This type of relations was first discussed by Einstein (1905) for the theory of Browinan motion, and in general corresponds to the Fluctuation Dissipation Theorem (Kubo et al., 1972) in thermodynamics and statistical mechanics. In that case, the formulation is such that the system is in the thermal equilibrium state before the application of the operation, $\epsilon$ represents the temperature, and the amount of work done by $F$ in changing $x$ by the value $\Delta x$ is $F\Delta x$.[10] However, in the present case, it is not important if the state of the original system can be considered "equilibrium" or if $F\Delta x$ can be considered "work." Rather, here, the necessary conditions are that the original state is in some general sense stable and that the deviations from this state are small (which implies that the distribution of the fluctuations is approximately Gaussian). In the case

[10] The explanation given here is greatly simplified. Rigorous treatments are given in nonequilibrium statistical mechanics textbooks (see, e.g., Kubo et al., 1972).

that deviations from $x = 0$ are small, all of the above equations are linear; that is, for example, in the equation for $dx/dt$, there are no terms beyond linear order in $x$ (i.e., no terms $\sim x^2$ or $\sim x^3$...). On the other hand, if the system is not linear, then, in general, the proportionality relation stated above will not hold, and the fluctuations will not follow a Gaussian distribution. However, even in that case, there will be a strong positive correlation between the size of the fluctuations and the response. Thus, if the system can be characterized by a "state", and if this state is stable, then we can expect quite generally that the following two conditions hold.

**Proposition (The fluctuation–response relation in a stable system).**

0. Fluctuations and the response coefficient are strongly correlated
1. If the externally applied operation is weak, considering the regime in which its "force" can be treated linearly, the size of fluctuations and the response coefficient are proportional.

For example, the above should hold in the case that the system we are considering is a cell, and the relation in question is between the fluctuations of its state and the response to an operation whose effect is to shift this state. Or, when one considers the phenotype of some organism, the above relation concerns that between its fluctuations and the response to some selective pressure which alters the genes in a particular manner, resulting in a shift of phenotype (Sato et al., 2003). (Of course, there are possible complicating factors here, for example, whether a description in terms of states is possible, whether the combined influence of all effects that are not included in the variables chosen to represent the state can be treated as noise, etc. A complete treatment must address such problems.) In any case, we believe that an important area of research in life science will be to investigate the validity of the weak relation 0 and the strong relation 1 in many living systems. The connection between these relations and the evolution is given in Chap. 10, while the relevance to development is discussed in Chap. 12.

In the field of cell biology, until recently, it had not been a mainstream approach to quantitatively measure fluctuations and derive results from these. However, recently, with the development of flow cytometry and other methods, it has become possible to measure the instantaneous distribution of cell states. It has also become much easier to alter cell states through the application of operations that change environmental conditions. Given these developments, we believe that **measuring fluctuations and seeking fluctuation–response relations** will become an influential experimental methodology in the future development of biology.

Of course, if we consider the response to changes in external conditions over long time scales, we encounter the problem of evolution. The extent to which the states of individuals fluctuate and the nature of the corresponding distributions will obviously have an effect on the changeability of descendants in adaptation to the environment. For this reason, the evolution speed, that is, the rate of a change of a state by genetic change per generations may be correlated with the extent to which their phenotypes fluctuate, as will be treated in Chap. 12.

### 3.2.3 Fluctuations Are Not due Entirely to Noise from the Environment But Result also from Internal Dynamics and Depend on the Internal State

In the above we considered a system fluctuating about a stable state. Now there are situations where the state of the system becomes unstable and, as a result, the system makes a transition to another state. Here, when the original state becomes unstable, the fluctuations in the direction of the instability grow. The strength of this instability and the degree of amplification of the fluctuations are proportional. It should be possible to express the directionality of this transition, discussed briefly in Sect. 3.5, in terms of those state variables whose fluctuations are amplified.

In general, consider a system that includes a positive feedback mechanism through which (as discussed in Sect. 3.4) a small discrepancy can grow and become significant. In the case that such a mechanism does exist, small fluctuations on the molecular level become amplified and, as a result, can exercise an influence on the macroscopic behavior. Furthermore, as in the case of chaos discussed in Sect. 3.3, the evolution of the cell's internal state can itself introduce irregular behavior that appears as spontaneous fluctuation. In this case, when some noise or a fluctuation in particle number is introduced from outside the cell, large fluctuations in the state of the cell can result. In general (even if there is no chaotic behavior), the degree of amplification of such influences depends on the state of the system, and thus the size of the fluctuations existing in a cell depends on its state.

Now, in the situation that the internal state selectively amplifies only certain types of fluctuations, the fluctuations appearing outside the cell too can be regarded to some degree as a product of the change of the internal state of the cell. Then, it can be the case that if the cell is in one state, fluctuations are amplified, but when the cell makes a transition to another state, fluctuations come to be suppressed. In such a situation, it is possible that within the fluctuations there exists some motion with a specific directionality, and they can therefore have some active influence on the system. This kind of selection of fluctuations can be found in the chaotic itinerancy as will be discussed in Sect. 3.6.

Of course, if the state of a system is not stabilized by some negative feedback, fluctuations of this type can introduce great change, as through their amplification, the state becomes unstable. As a result of such destabilization, in general, the state of the system will change. With regard to such state transitions, as in cell differentiation or speciation, it is then important to investigate how in the presence of positive feedback stable behavior can emerge.

## 3.3 Plasticity

Now we consider the change undergone by a system, originally in an equilibrium state, when subject to an externally applied operation. In this case, according to the Le Chatelier principle of thermodynamics, the system will exhibit a response in the direction in which the operation is negated. For example, if heat is added to one region and the temperature there increases, the response will be that heat flows from this region. Thus the manner in which the system responds cancels the increase in temperature resulting from the operation of heating. With a little thought, we realize that the reason that the response occurs in this direction is that both the specific heat and the temperature are positive quantities. At a more fundamental level, as discussed below, we can say that this behavior is due to the stability of an equilibrium state: As it is stable, when a system is perturbed from such a state, it will relax back toward this state.

The Le Chatelier principle is implied by the stability of the equilibrium state. To understand this, let us suppose, contrarily, that the response displayed by a system to an operation applied to it in a particular direction actually acts to strengthen the effect of this operation. First, note that there will be cases in which, even when no operation is applied, the fluctuations in the system cause it to move in the same direction as would the operation. For example, consider the case in which, because of fluctuations, the temperature in a particular region increases slightly. Here, we are assuming that the opposite of the Le Chatelier principle holds, and therefore, the change introduced by this fluctuation causes a response in the same direction. For example, heat would flow into this region of elevated temperature from the surrounding regions or an exothermic reaction would be promoted there. As a result, the temperature would increase more and more, becoming farther and farther removed from that of the original state. Thus, through this phenomenon, the originally stable equilibrium state would cease to exist. From these considerations, we see that if we deny the Le Chatelier principle, we are forced to conclude that the fluctuations in the system cause it to spontaneously move away from its original state. Hence, the original state cannot be stable. In other words, if we assume that the Le Chatelier principle does not hold, the basic premise of thermodynamics that a system isolated from external disturbances will evolve toward equilibrium itself cannot hold.

Reconsidering the above discussion, it is reasonable that the manner of thinking of the Le Chatelier principle (or Le Chatelier-Braun principle in a generalized form) would apply in a very broad context to systems existing in stable states, not just to those corresponding to thermal equilibrium. With this observation, we conjecture that, quite generally, the stability of a system introduces a limitation on the direction in which this system can evolve after some operation is applied to it. As an example of this idea, let us consider the case of a stem cell system that possesses the potential to differentiate into different types of cells. In general, as a result of differentiation from stem cells, some structure consisting of a distribution of cell types will form. For example, the hematogenetic stem cells are able to create a variety of cells in the blood, including red blood cells, white blood cells, and platelets. Now, within a given structure or tissue, there will be some appropriate distribution of these different types of cells. Then, if the proportions of certain types of cells within this structure come to deviate from their normal values, the structure will probably no longer function properly. Then, if the population of one species of cell decreases, we can expect that the stem cells will increase the rate of differentiation to this species, to return the system to its proper state. In Chap. 8, we present a theoretical model that displays just this type of behavior: In a system possessing stem cells, when cells of one type are removed, the stem cells adjust their rates of differentiation to make up for this loss. In other words, the conclusion derived from this model is that the response of the system is such that it cancels the effect of the externally applied perturbation. We can interpret this behavior as resulting from the stability of a system that has formed through the developmental process.

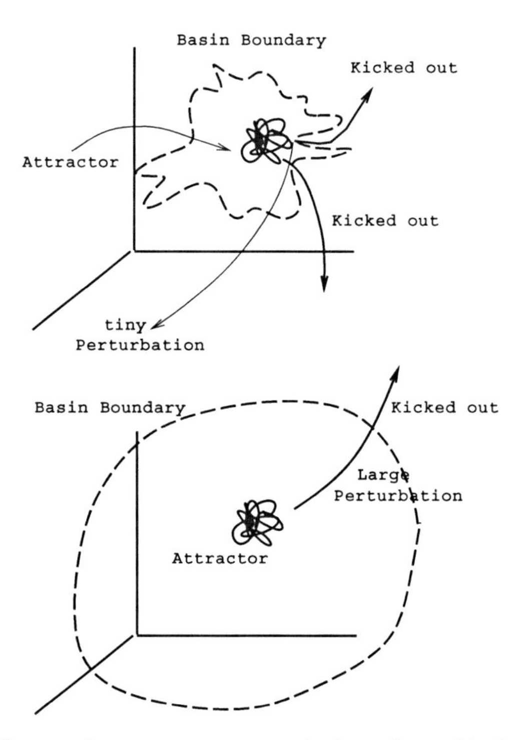


**Fig. 3.11.** The distance from an attractor to the boundary of its basin of attraction. (**a**) Here this distance is small, and thus a small disturbance is sufficient to cause the system to be drawn into another attractor. (**b**) Here the distance is large, and therefore unless a large perturbation is applied to the system, it will not escape from the original attractor

The stability of a state can be judged by testing whether the system returns to this state after being pushed slightly away. The amount by which it can be displaced and still returns is a measure of the degree of stability of the state. As a simple situation, let us consider the case in which the original state of the system corresponds to an attractor of the system dynamics. Then, the problem of whether or not the system will return to its original state upon perturbation reduces to that of whether or not it has been pushed over the boundary of this attractor's basin of attraction. If the system is pushed beyond this boundary, it will make a transition to another attractor. Thus in this case, the "distance" from the original attractor and the boundary of its basin of attraction is a measure of the stability of this state (see Fig. 3.11).

Now, if we apply a sufficiently large perturbation to a system, it will make a transition to a new state. It would seem that in such a situation, in general, the manner in which the system makes a transition is not random. Rather, there should correspond some structure to these transitions. Using large perturbations, it should be possible to obtain information not only about the stability of a given state but also about the relations between different states. In this way, we should be able to elucidate the paths taken in the processes of transition between states. Thus, studying the properties of (3.1)–(3.3) using both small and large perturbations should be useful in our pursuit of an understanding of biological systems.

In a typical developmental process, initially the system possesses the capacity to make any type of cell (the quality of *totipotency*). Then, through the differentiation process, this potentiality becomes limited, and eventually the system reaches the situation in which each cell is capable of creating only its own kind. That is to say, the normal process of development displays an irreversibility, because the potential for differentiation decreases as this process advances. However, this type of irreversible differentiation of cells has been reversed in certain cloning experiments from somatic cells. In fact, it has been found that the entire development process can be recreated beginning from committed somatic cells. This is realized by isolating cells, subjecting them to a state of starvation, and then using their nuclei. Thus the situation here is quite different from the natural process of development. Now, is it possible with this operation to reverse the process of differentiation and completely return to the original state? Or, when we carry out such an operation, is there a restriction introduced because of some kind of compensatory effect? In a broader context, if we appropriately combine a number of operations, is it possible to return to any kind of state?

Of course, at the present time, it is not possible to take an organism in a dead state and return it to a living state. (Or, perhaps we should define *dead* as the state from which an organism cannot be returned to a functioning state with any presently available operation [*point of no return*].) Thus, we see that there are states from which it is impossible to return to the original state, and this would appear to put a limitation on possible operations as well. From another point of view, clearly stipulating all possible processes step by step, we may be able to define the corresponding possible operations. Such a specification of impossibility should be the ultimate goal of research employing operational experiments.

As a first step, we must obtain a representation of the irreversibility of differentiation and then investigate the question of what kind of operation is necessary to reverse this. In the next section, we consider the representation of irreversibility as a loss of plasticity. There, we discuss experiments in which plasticity is restored to committed cells through the application of operations that make survival difficult, such as subjecting them to starvation. These experiments are related to the clone experiments mentioned above, and here too, it is worth contemplating the conditions on the operations applied to return the cells to the totipotency state.

## 3.4 Representation of "Softness"

In Sect. 1.9, we briefly discussed the "softness" of living organisms. Now we address the question of how this quality should be expressed. Intuitively, the idea that living systems are "flexible" and "plastic" is easily understood, but we wish to define this idea clearly and determine how to treat it in scientific terms. At the present time, it is not clear how this can be done, and, indeed, it may be said that the lack of a scientific understanding of "softness" is the reason that the study of life has largely been pursued by employing the analogy of a logical system or computer.[11] As a first step, we ask, What is the meaning of "softness" in everyday language? To most people, perhaps this term brings to mind something that is easily changed. Furthermore, this changeability may be something that at times reveals itself spontaneously and at times is revealed by some externally applied operation. Here, we refer to this quality as "flexibility."

The quality of flexibility in which we are interested could also be termed "plasticity." However, perhaps it is best to use the term "plasticity" with regard to the changeability of the internal state of a system and "flexibility" with regard to the variety of possible responses of a system to externally applied operations. But it should be pointed out here that, as stated in Sect. 3.2.2, for a system existing in a particular state, the size of the fluctuations that it exhibits and the magnitude of its response coefficient have a strong positive

[11] As pointed out in a previous footnote, Fumio Oosawa began a research program aiming at expressing the concept of the softness of living organisms as molecular machines (Oosawa & Hayashi, 1986; Oosawa, 1998), with the concept of "loose coupling." The softness on which we focus in this book exists on levels that are more macroscopic than that of molecules. However, the methodology that we employ to investigate this softness should be effective at the level of molecules as well. This is because the number of internal states of a polymer molecule can be so large that a single such molecule itself may well be regarded as a system.

correlation, and thus these qualities of flexibility and plasticity are closely related.

Now, the properties of flexibility and plasticity as described above exist on many different scales. For example, considering the scale at which cell differentiation takes place, cells that can differentiate into many types (e.g., embryonic stem cells) possess large flexibility, while committed cells, which can produce only their own type, have small flexibility. Thus, here, there is a natural progression from more flexible to less flexible states. However, it is also interesting to note that in some cases such flexibility can be regained. For example, when some tissue is injured because of some external influence, it is possible for committed cells to dedifferentiate and then differentiate into different types of cells. In this way, the damaged tissue can be repaired.

Now, let us consider an example illustrating flexibility on a slightly larger scale. In the development process, it sometimes happens that the properties of one group of cells will be changed through the influence of another group. This phenomenon is referred to as “induction.” Through this phenomenon, there is a successive formation of systems within an organism. Of course, induction takes place when two groups of cells encounter one another, and therefore it is just one side of the interaction between the two groups. However, in such a situation, of the two interacting groups, the state of one is more easily changed than that of the other. In other words, one of the groups is more flexible with respect to differentiation. This is the group that becomes induced.

How should we understand these qualities of flexibility and plasticity?

- *Fluctuations*: In Sect. 3.2.2, we described why for a state with large fluctuations the response coefficient with respect to an externally applied operation is also large. In other words, a system in such a state is highly susceptible to change through alteration of the environmental conditions. Such a state is characterized by a large flexibility. Thus the size of the fluctuations exhibited by a system (in the absence of an external influence) can be regarded as indicating both the degree of plasticity and the degree of flexibility.
- *The distance between an attractor and the boundary of its basin of attraction*: The property of flexibility should regard not only “quantitative” fluctuations, as described above, but also responses in which the qualitative nature of the system changes. In fact there are many situations in which such “qualitative flexibility” is quite important. For example, systems sometimes suddenly jump from one state to another, and as a result, the qualitative nature of the system behavior changes. As one such type of transition, a very typical situation is that in which a system consists of a number of attractors and because of some operation applied to the system it is caused to jump from one to another.

  As a measure of the susceptibility of a system to transition, the distance from the attractor to the boundary of its basin of attraction is a natural choice (Kaneko, 1997a, 1998a). The basin of attraction of an attractor

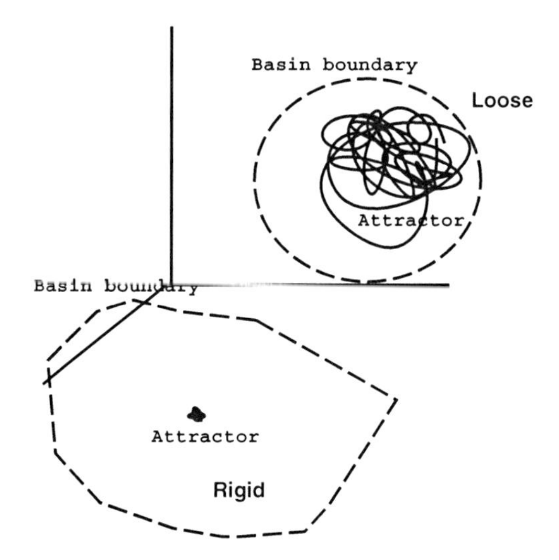


**Fig. 3.12.** A soft state and a hard state. For the soft state on the upper right, the fluctuations are large, and the state is readily changeable over a wide region up to the boundary of its basin of attraction

consists of the set of all initial points whose trajectories are attracted to it. Thus if a system that is initially in the state corresponding to some attractor is perturbed, it will return to this attractor as long as the perturbation does not carry it out of its basin. However, if the perturbation causes the system to leave this basin, it will be pulled into the state corresponding to another attractor. It is thus easy to understand that in the case that the distance from an attractor to the boundary of its basin of attraction is small, a weak perturbation can cause the system to make a transition to a different attractor, while if this distance is large, even a strong perturbation may not cause such a transition (see Fig. 3.12). Hence, the distance from an attractor to the boundary of its basin represents a measure of flexibility.

- *Variety of responses and directionality*: A characteristic aspect of living systems is that they can exhibit a wide variety of responses when the external conditions to which they are subjected are changed only slightly. In other words, living systems display flexibility in their responses. In some sense, this flexibility leads to one-to-many correspondence between input and output, as, for example, only any slight changes in the operation applied to a cell could make it move to the left or to the right or remain still. One possible explanation for this behavior is that the transition paths connecting the various states of the system form a highly tangled structure, and thus perturbations that differ only very slightly can cause the system to take different paths and result in transitions to completely different states. (Here dimensionality of such paths can be one measure). In any case, it is meaningful to investigate both experimentally and theoretically how different operations can result in the selection of different transitions.

From the present point of view, we would have to regard logical systems like those represented by computers as having the lowest degree of flexibility. We refer to such systems as "rigid." Considering processes within a cell, for example, we can say that a system whose response consists of the on–off expression of genes is more rigid than that whose response is the more continuous change in the quantity of protein molecules present.

The flexibility and plasticity of a state are most naturally expressed in terms of the dynamic nature of that state. Actually, one of the most important goals of observing states and making measurements of their time-dependent behavior as well as their response to external operations is to obtain a way of representing this flexibility and plasticity.

To summarize the above discussion, in the study of the concept of "softness," the relations among the following phenomena must be investigated:

- Flexibility: The potentiality to change due to external influence
- Plasticity: The potentiality to change of a cell's internal state
- Fluctuations in the internal state
- Dynamics of the internal state
- Change in flexibility or plasticity of a group of cells due to interaction

We must also study the nature of the change undergone by the state of a cell when it is subjected to an external operation, including the extent of this change and whether it appears in some sense deterministic or with a degree of unpredictability. Then, we must elucidate how the nature of the changes undergone by the state of a cell is related to the state of the larger system to which it belongs. Investigating such behavior of the cell in relation to interactions among cells and the state of the larger system, we hope to discover rules governing the manner in which flexibility and plasticity are determined and can change. We consider these problems in the next section.

Next, let us consider the situation in which there exists a readily changeable state $A$ and an unchangeable state $B$ and a system makes a transition from $A$ to $B$. If we suppose that this system exists in isolation, then, beginning in state $A$, the possible transitions are $A \rightarrow A$ and $A \rightarrow B$, and beginning in $B$, the only possible transition is $B \rightarrow B$. Thus, because a system can jump from $A$ to $B$ but not from $B$ to $A$, it is natural that eventually all systems will be in state $B$. We can think of the same type of conditions existing more generally, and it is not difficult to imagine that in such a situation, as long as the external influence does not change, the number of elements that exist in the unchangeable state (or states) will gradually increase. Thus, we see how flexibility can be lost over time (see Fig. 3.13).

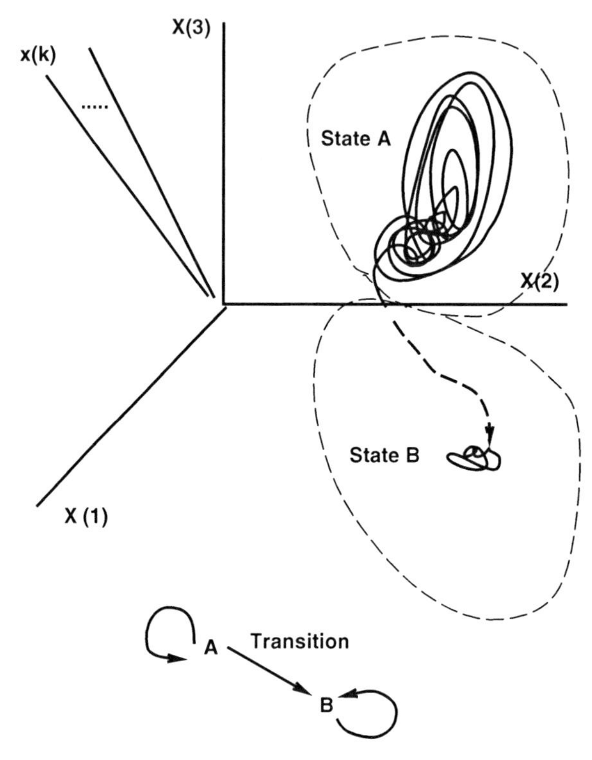


**Fig. 3.13.** In general, transitions in which the degree of plasticity decreases are more likely

In the actual process of development, the normal, irreversible progression is from embryonic stem cells to stem cells to committed cells. In this process, stem cells exhibit various types of behavior, sometimes producing other stem cells and sometimes differentiating into committed cells. By contrast, committed cells can produce only their own type. Thus, in the process of development there is a gradual loss of plasticity.

The type of decrease in plasticity seen in the process of development will be further clarified below by the use of theoretical models. We will see in a variety of models how, as the number of cells increases, and as a result, as the role of interactions becomes more important, cells begin to differentiate, creating committed cells. There, using the measure of flexibility defined in the previous section, we will show directly how its magnitude decreases in stages. These results support the following hypothesis.

**Hypothesis:** Consider a set of elements (cells) that is evolving in time. Then, suppose that the environment in which this set exists is (almost) time independent. Furthermore, assume that with regard to the behavior of this set of elements, interactions with other external systems are insignificant. In other words, we assume that this is an almost closed reproducing system. Then, the plasticity of this system gradually decreases in time, with that of each element at each time either decreasing or remaining constant.

- *Change in plasticity due to an external operation*: Clearly, it is possible for the plasticity of a system to change as a result of an external operation. In such a situation, it is reasonable that the amount by which a state changes is positively correlated with its plasticity. Then, the transition from one rigid state to another can be facilitated by first causing the original state to make a transition to a state with greater flexibility and then allowing it to gradually change in the direction of increasing rigidity.

Let us apply this idea to the case of cells. In a natural state, the metabolic dynamics, directly coupled with the external condition, is easier to be changed through the interaction with the environment, while it is difficult to directly change the gene expression. Therefore, to alter the genes, it is necessary to first cause the metabolic state of the cell to change through some externally applied conditions or interaction with surrounding cells. Once such a change has been realized, it can become fixed through a corresponding change of the gene expressions. Thus, here, the plasticity of the cell's state is exploited to change the rigidity in the change in the genetic expression.

The above considerations suggest the following as a possible scenario. Even though it would appear as if the change in the state of a cell are controlled by the genes, it may be the case that, first, the flexibility of the cell state is utilized and, then, through the transition that this allows, the expression state of the genes is changed. This change, in turn, influences the state of the cell, and in this way, the states of the cell and the gene expression patterns come to be fixed. Although it often appears that the cause of a transition undergone by a cell lies in the genes, the information that we can obtain experimentally is only that regarding correlations. Hence we cannot rule out the possibility that change in the states of the genes and the cell are mutually strengthened through their interaction, and the combined system settles into a stable state made possible by this interaction. While an organism may appear in some cases to change in accordance with some blueprint, any system that exhibits fluctuations as strong as those seen in living systems could not behave like a computer. Thus, we must keep in mind the possibility that even when the state of the system seems to be changing in a "rigid" on–off manner, behind this apparent rigidity is a flexible state and the transition it makes. This new state is then caused to be consolidated through the influence of the genetic system that is relatively more rigid, pushing it toward a new stable attractor.

## 3.5 Coupled Dynamical Systems for the Study of Cell System

As a theoretical side, our plan is to understand a biological system in which there is a dynamic circulation between the behavior of each element (unit) and the whole behavior. In physics, such relationship between element and the whole is regarded as a relationship between microscopic level and macroscopic level. In thermodynamics, we focus our attention on the macroscopic phenomena. By restricting our interest to equilibrium systems and the transition among equilibrium states, thermodynamics has succeeded in extracting out universal behavior in macroscopic systems, independent of each microscopic details. The laws and formulation of thermodynamics do not depend on what molecules the system is composed of. Theoretically, the thermodynamics is possible, since the phenomena at microscopic levels are separated out from macroscopic levels, under some restrictions imposed on equilibrium systems. Microscopic properties influence on the parameters in the system, but not on the theory itself.

On the other hand, the aim of statistical mechanics is to bridge between microscopic level and macroscopic level. From microscopic (molecular) characteristics, thermodynamic parameters at a macroscopic level are obtained. In setting a boundary condition of the theory, however macroscopic information is necessary, but once it is given, there is a formulation to extract macroscopic behavior from microscopic quantities. (Note that both thermodynamics and statistical mechanics are formulated on equilibrium states and close to them, while in thermodynamics, restriction on the transitions between equilibrium states are also formulated as the second law).

If each element at a microscopic level has internal degrees of freedom, the behavior at a macroscopic level may influence the characteristic feature of a microscopic element itself. The dynamics of internal degrees of freedom of the unit may be influenced by the interaction with the environment, which is influenced by all the units. For example, consider a cell constituted by several chemicals whose concentrations change through catalytic chemical reactions. The behavior of each cell is given by the composition of the chemicals. Here these cells can interact with each other by transport of chemicals through cell membranes. Then the state of each cell can be influenced by environment, which is influenced by the states of all cells. The internal state can be changed by the cell–cell interaction. Through the interaction, the behavior of a unit (cell) may be different. Hence there is mutual dependence of each element and the average behavior of the total cell systems. This type of mutual dependence between unit and the whole can generally appear if each unit has internal dynamics and the interaction between these units are not negligible.

If each state is fixed in time, this mutual dependence may not be so serious. Since the states of the total system (say the chemical concentration of the environment) and of each unit are constant in time, we can solve the behavior of each unit, under a fixed boundary condition (say the chemical concentration of the environment) and then obtain the total behavior (including cell–cell interactions). By checking the self-consistency of the states, we can solve the behavior of each state.

In a biological system, however, the state often changes in time. In the case of cell, the number of cells changes through cell division and cell death, and the interaction with other cells cannot be fixed in time. Internally, chemical concentrations or gene expressions change in time. Hence we need to seriously consider dynamics of each unit. Furthermore, since each unit does not exist in isolation, but many units interact with each other to form a total system.

Other examples are seen in an ecological system consisting of a population of organisms, neural network system, or more microscopically, a population of interacting polymers within a cell. In such biological system, constituting units have their own internal dynamics, and they are interacting. Here, both the internal dynamics and interaction are important. Often, neither of the two is regarded as a weak perturbation to the other. Hence it is important to study a coupled dynamical system, where the interference between internal dynamics and interaction lead to phenomena that cannot be understood by the dynamics of each element only. Hence a study of coupled dynamical system is necessary.

Indeed, dynamical systems for interacting elements with internal dynamics have been investigated extensively and intensively. There, it is clarified how elements synchronize with each other, and how identical elements form a few distinct groups with different behaviors. Long-term dynamical change in relationship between unit dynamics and ensemble behavior are also investigated. Some concepts developed in the coupled dynamical systems will be relevant as a basis for our study on complex systems biology.

Here, as an illustrating example, we consider a system of interacting units which have internal nonlinear dynamics. As the simplest case, we consider a system in which the units interact only thorough the "mean field," that is simply the average of state variables over all elements. (See Fig. 3.14 for schematic representation).

Even in this simple setting, we have found that elements can come to take different states, when there is some instability in the dynamics to make the homogeneous state unstable. Then, the phenomenon called "clustering"

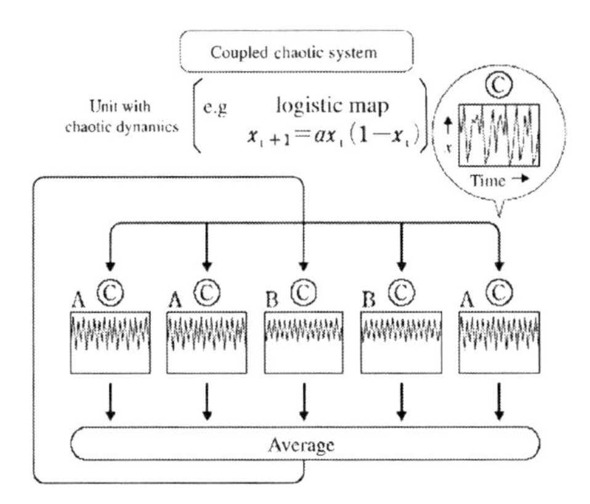


**Fig. 3.14.** Schematic representation of a globally coupled dynamical systems (in particular globally coupled map)

appears in general, where elements differentiate into several clusters. Elements belonging to a same cluster take the same state, while elements from different clusters take a different state, for example, different phase of oscillation. It is remarkable that elements with identical dynamics rule start to show different states.

Besides the interaction and internal dynamics, the unit in a biological system often divides after growth or dies. Sometimes the attractor as dynamical system is not necessarily a state with the highest growth speed for replication of the unit. Hence the logic for the fittest and the logic for dynamical systems do not necessarily match. In some chapters, we will see some examples for such interplay between dynamics and growth. Also, in a dynamical system with many variables, there often are many attracting states, depending on initial condition of the unit. Here, how appropriate initial condition is selected will also be an important issue, since in a biological system, choice of initial condition is not given externally, but has to be done autonomously. This problem will be discussed in later chapters, while we also study how a state with recursive growth is generated, together with the potentiality to differentiate into alternate states.

In later chapters we often study a system consisting of (i) internal dynamics of a unit (by reaction network), (ii) interaction between units, and (iii) the growth of unit depending on its state. Note that the clustering mentioned above implies that the cell with identical genes (i.e., rule for dynamics) can be spontaneously differentiated, through cell–cell interaction. With the set–up of this minimal condition, the logic for the irreversibility and robustness in development will be discussed in Chaps. 7 and 8.

Also following this general process in coupled dynamical systems, one can imagine that even bacteria may differentiate into distinct groups if put in a state with strong interaction (see Fig. 3.15). Experiments on differentiation or diversification of bacteria will be discussed in Chaps. 7, 8, and 11.

## 3.6 Chaotic Itinerancy

In biological systems, the change often occurs through successively visiting some quasi-stationary states. Cell differentiation process occurs through successively visiting several types that themselves are quasi stationary. A cellular process with the change of gene expression often occurs through visiting several discrete states. There, some rule for switching over several states which are generated from dynamics involving gene expression and metabolic process exists. To discuss such switching dynamics and rule generation, some concept on dynamics with successive transitions will be necessary.

About a decade ago, chaotic itinerancy (CI) was proposed as a universal dynamical concept in high-dimensional dynamical systems. In the study of high-dimensional chaos, there is often a state that switches back and forth

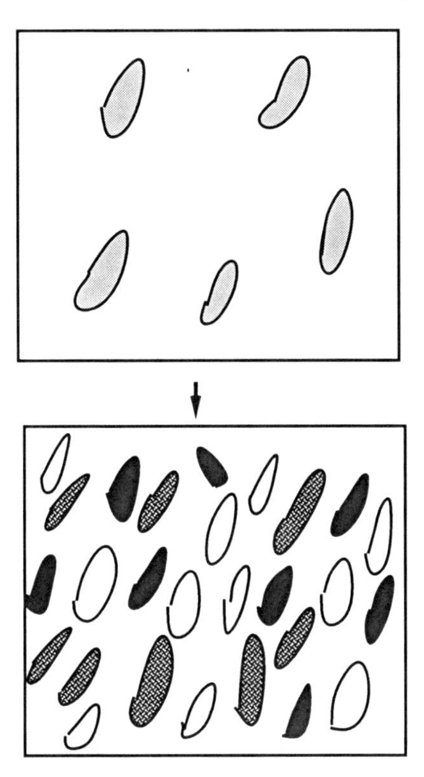

**Fig. 3.15.** A schematic representation of bacteria cell differentiation. As the population density of bacteria is higher, they start to take some distinct states. See Chaps. 7 and 9 for details

between fully disordered state and several ordered states. Here fully disordered state can be approximated partly by "random motion" which may be described as the motion consisting of many degrees of freedom. The ordered states are given by effectively low degrees of freedom and located in a low-dimensional region in the phase space. There are several such ordered states. The itinerant motion among varieties of ordered states through high-dimensional chaotic motion is commonly observed. The term for this is *chaotic itinerancy*.

The dynamics are represented as an itinerancy over several ordered states through disorganized, desynchronized motion. It consists of a quasi-stationary high-dimensional state and exits to ordered states with low effective degrees of freedom, residence therein, and chaotic exits from them.

In CI, an orbit successively itinerates over ordered motion expressed by a few effective degrees of freedom. Considering attraction to, and the residence at the ordered motion state, each of such states is called "attractor-ruin." The motion at "attractor-ruins" is quasi stationary in the sense that it is close to that in low-dimensional attractor.

After staying at one attractor-ruin, the orbit eventually exits from it. This escape from an attractor-ruin stems from instability of the ruin. (See Fig. 3.16) for schematic representation). For example, if the effective degrees of freedom is 2, the dynamics are in the vicinity of two-dimensional subspace in the original high-dimensional phase space. Such low-dimensional motion is not described by a stable attractor, even though orbits are attracted to its vicinity. After staying at an attractor-ruin, an orbit exits from it. This exit arises from a certain kind of instability

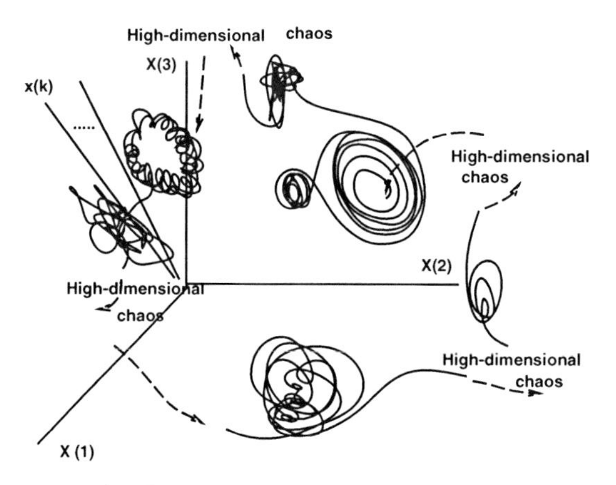


**Fig. 3.16.** A schematic representation of chaotic itinerancy

With this instability the orbits enter into a high-dimensional chaotic motion, losing coherence or correlation among variables. This high-dimensional dynamic state is also quasi stationary, although after this chaotic wandering the orbit is again attracted to one of the attractor ruins which again possesses low dimensionality. In other words, there are some "holes" connecting to attractor-ruins from the high-dimensional chaotic state. Once the orbit is trapped at a hole, it is suddenly attracted to one of attractor ruins, that is, low-dimensional ordered states.

The exit process from ordered state, on the other hand, uses a limited region in the state space (phase space), and the path lies in a low-dimensional region in the phase space. The low-dimensionality of this path (unstable manifold) gives a constraint for the itinerancy. For example, if the number of effective degrees of freedom is 3, the system wanders around a three-dimensional space, and the exit is typically lower dimensional. In a class of coupled dynamical systems, each ordered state is represented by the motion in a low-dimensional subspace. The state is weakly unstable, and the orbit exits from it (Fig. 3.17).

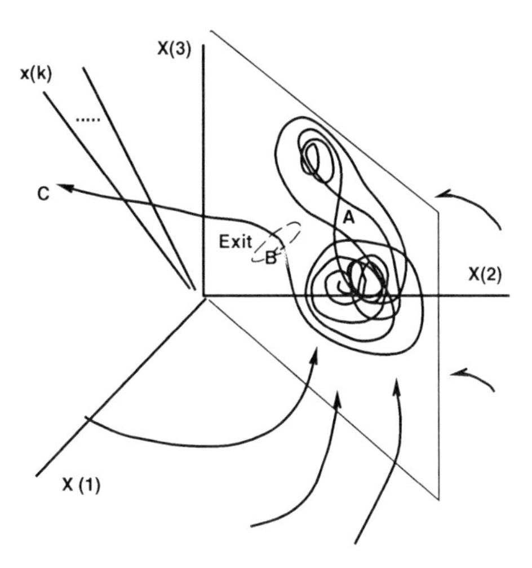


**Fig. 3.17.** A schematic representation of chaotic itinerancy; once an orbit stays at a state characterized by a few degrees of freedom, it goes out to a disordered state, and then is trapped to another ordered state

### 3.6.1 Relevance to Biological System

To date, chaotic itinerancy is considered to be a rather universal phenomenon in a system with many degrees of freedom (Kaneko & Tsuda, 2003). In particular, relevance to brain dynamics has been discussed extensively. Mathematical definition of CI is still under discussion, but here we do not go into this problem, but only briefly mention possible relevance of CI to biological system. Note also that we use the term "CI" in a rather broad sense to include the successive transitions over several quasi-stable states, depending on the state of the system itself. In this sense, even the term "chaotic" does not necessarily involve the mathematical sense of deterministic chaos.

"CI" provides a novel viewpoint in such "dynamic change of relationships," which may shed a new light on how module-type structures are formed spontaneously, and also on how rules on dynamic change among these structures are generated.

- *Formation of rules of switching*: The ordering of successive visits of states at CI is not random, but there is some restriction. In the representation of phase space, an orbit has to pass through some region to some other region, which leads to a restriction of itinerancy.

  In a biological system, there generally exists some rule for successive changes of states. For example, after some group of genes are expressed, some other groups are expressed. In this case, one standard case is that such rule is preprogrammed in genes. Still, CI suggests another alternative possibility that without such pre-programmed rule, some rule can appear from dynamics. The rule thus generated can change in accordance with external environmental condition without further programmed rules for it. With some change in environment, the rules are changed flexibly. In this sense, CI may be relevant to understand flexible, context-dependent response of a biological system.

- *Formation and reorganization of groups of correlated action*: In a biological system, often several elements show correlated motion or work together. Activities of neurons are often correlated, while a group of genes is sometimes expressed together. For example, search of a group of genes that express together in time has been a hot topic in cell biology. Alon et al. have studied how a group of genes express together, by using fluorescent proteins (Alon et al., 1999, Ronen et al., 2002).

  When groups of correlated motion are generated, it is often assumed that couplings are prepared to be strong within these elements forming correlated motion. Accordingly, existence of predefined "module" is often assumed in neuroscience and in cellular biology. Still, there are increasing reports to suggest that these groups may not be predefined, but can change in time. Groups of neurons or genes that work correlatively often change in time or change depending on external condition.

  In CI, groups of correlated action are formed at each attractor-ruin, as a result of dynamics, even without predefined strong couplings among the elements. The connection strength among elements in each group is not necessarily strong. The connection among elements may give restriction for the transition among attractor-ruins, but groupings are not necessarily determined by them. For example, some elements $x_1, x_2, \ldots x_5$ osculate synchronously while other elements $x_6, x_7, \ldots x_9$ oscillate synchronously, even though the coupling among elements is identical, in a system of globally coupled map [Kaneko 1990, 1991, Kaneko & Tsuda 2000].

  In this sense, CI gives a new insight into how module-type structures are formed spontaneously, and also into how rules on dynamic change among these structures are generated.

- *Evolvability and stability*: Biological system consists of a huge number of elements, say chemical species for a cell, species for ecosystem. How such high-dimensional dynamical systems keep stability is one of the key questions in a biological system. Furthermore, a biological system, although it is stationary over some timespan, can also change in a longer timespan, as are common in developmental dynamics and in evolution. Biological system satisfies both recursiveness to maintain its macroscopic state and changeability to a novel state as evolution. Here, the time regimes for recursiveness and evolution are sometimes separated, as seen in metamorphosis in development, and as discussed as punctuated equilibrium for evolution. Chaotic itinerancy gives a theoretical mechanism to stability for a recursive state and also to evolution. Stability of ecosystem is discussed with population dynamics of a variety of species (Kaneko & Ikegami, 1992) in relationship with CI, while recursive production of biochemical states and their evolution are studied with itinerant dynamics (see Chap. 5).

### 3.6.2 Experimental Method in Studying Itinerancy

To see the temporal ordering in genetic expression, one can use the technique of a fluorescent protein. Then by observation of a single cell, one can detect the temporal change of protein number in a cell. Of course, it is rather difficult to embed and measure several fluorescent proteins simultaneously, still one can detect the dynamics of several degrees of freedom. On the other hand, the detection of cell states at an ensemble level is carried out by the use of cell sorter, where selection of initial cell state will be possible by taking cells within a specified range of values measured by the cell sorter. At an ensemble level, the use of microarray is useful where a large number of variables (concentrations of mRNAs) are measured even though the time resolution is limited. At any rate, there are several possibilities to measure the successive changes of cell states.

# 4

# Origin of Bioinformation

## 4.1 Question to Be Addressed

**Question: A cell consists of several replicating molecules that mutually help the synthesis and keep some synchronization for replication. At least a membrane that partly separates a cell from the outside has to be synthesized, keeping some degree of synchronization with the replication of other internal chemicals. How is such recursive production maintained, while keeping diversity of chemicals? Furthermore, this recursive production is not complete, and there appears a "mutational" change over generations, which leads to evolution. How is evolvability possible in spite of recursive production? In a cell, among many chemicals, only some chemicals (e.g., DNA) are regarded to carry genetic information. Why do only some specific molecules play the role to carry the genetic information? How has such separation of roles in molecules between genetic information and metabolism progressed? Is it a necessary course of a system with internal degrees and reproduction?**

Of course, this problem was addressed in the study on the origin of life or origin of replicating system. Here we are not necessarily interested in revealing "what happened in the past," but rather, we intend to unveil the logic of the heredity. Still, it is relevant to review the earlier studies.

To consider the origin of replication system, one needs to discuss how genetic information is faithfully transferred to the next generation. A typical standpoint is seen in the recent study of "RNA world," while such approach has much longer history. For example, Spiegelman et al. (Millis et al., 1967; Safhill et al., 1970) set up an experiment RNA replication, by using a solution of RNA and enzyme. In this experiment, one has to supply some enzyme, from outside, and in this sense it is not an autonomous replication system. Still, his group found that RNA molecules with proper sequences are reproduced with some error (see Fig. 4.1). The rate of replication (i.e., selectivity) depends on

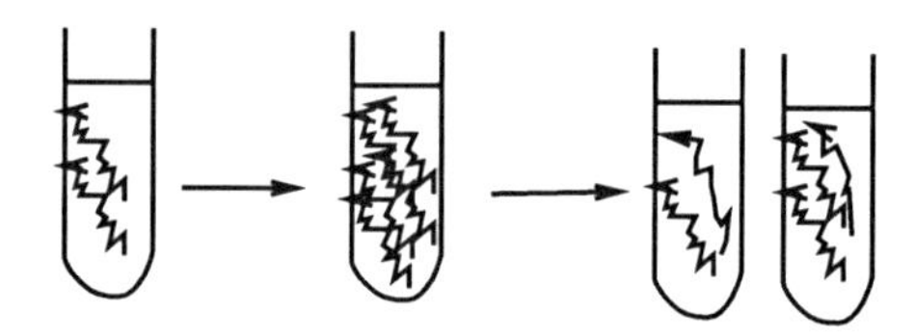

**Fig. 4.1.** Schematic illustration of Spiegelman's experiment on in vitro replication system of RNA molecules

the sequence of RNA, and with error to the original sequence, this rate will be decreased.

Following this experimental study of Spiegelman on replication of RNA, Eigen's group started theoretical study on the replication of molecules (Eigen & Schuster, 1979) (see Fig. 4.2). The replication process of polymer in biochemical reaction is generally carried out with the aid of enzymes. The enzyme is given by a polymer, while its catalytic activity strongly depends on its sequence. For most sequences of the polymers, the catalytic activity is very small, but few of them may have high catalytic activity. Depending on the sequence, some have a much higher catalytic activity, and the replication rate of polymers depends on the sequence.

Now, as a theoretical argument, consider replication of polymers whose replication rate depends on its sequence. Here, assume that a "good" sequence has replication rate $\alpha$ times larger than its mutant with a substitution of a monomer from the original sequence. Here, the replication progresses under some error. Without fine machinery for error correction, the error rate is not negligible. Assume that in each replication process, a monomer is substituted by another monomer with the rate $\mu$. Then the probability that a polymer

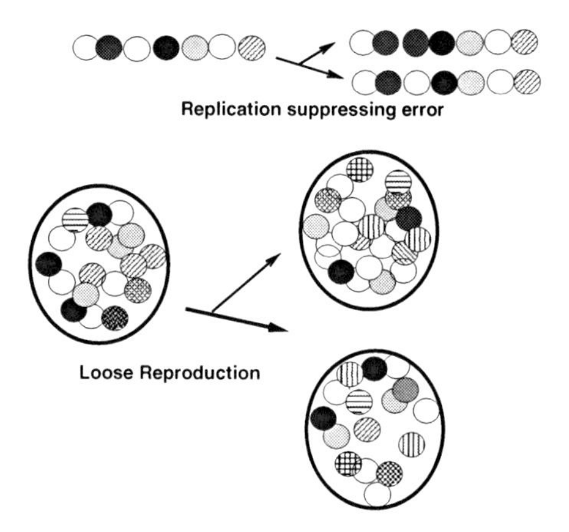


**Fig. 4.2.** Two standpoints on the origin of life: (*upper*) faithful replication system initially (*lower*) loose reproduction system. Schematic illustration

consisting of $N$ monomers can produce itself is given by $(1-\mu)^N \approx \exp(-N\mu)$, by assuming that $\mu$ is small.

Now, let us examine whether the good polymer can continue to replicate, maintaining the good sequence, so that the information of this sequence is transferred. Once it is changed to its mutant sequences, the probability that the original good sequence comes back by mutation is negligible, as long as the number of monomers is not small. Then, the condition that the good sequence dominates in populations in the soup of polymers is given by

$$(1-\mu)^N \alpha > 1 .$$

Using the expression of small $\mu$, we get $\exp(-N\mu)\alpha > 1$. Accordingly, the condition that a good sequence is preserved is given by

$$N < ln(\alpha)/\mu \qquad (4.1)$$

Here, ln $(\alpha)$ is typically $O(1)$, while the error rate in the replication of monomer is estimated to be around $0.01 \sim 0.1$, in usual polymer replication process. Then the above condition gives $N < 100$ or so. In other words, information using a polymer with a sequence longer than this threshold $N$ is hardly be sustained. This problem was first posed by Eigen and is called "error catastrophe." On the other hand, information for the replication for a minimal life system must require much larger information. Of course, the error rate could be reduced once some machinery for faithful replication as in the present life emerges. However, such machinery requires much more information to be transmitted by the polymer. Thus it cannot solve the problem on the origin of bioinformation raised by Eigen.

Summing up, for replication to progress, catalysts are necessary, and information in a polymer to produce it must be preserved. However, the error rate in replication must have been high at a primitive stage of life, and accordingly, it is recognized that the information carried on a selected sequence will be lost within few generations. In other words, faithful replication system requires larger information, while a larger information requires faithful replication system. Thus there appears a catch-22-type paradox.

To resolve this problem of inevitable loss of catalytic activities through replication errors, Eigen and Schuster proposed hypercycle (Eigen & Schuster, 1979), where replicating chemicals catalyze each other forming a cycle, as A catalyzes the synthesis of B, B catalyzes the synthesis of C, C catalyzes the synthesis of A (see Fig. 4.3). In this case, each chemical mutually amplifies the synthesis of the corresponding chemical species in this cycle. There occurs a variety of mutations to each species, but this mutant is not generally catalyzed in some other species in the cycle. For example, consider A′, a mutant of A. Then, such mutant is not catalyzed by C. Then the growth rate of A′ is much smaller than A.

This is also understood by writing out the rate equation for the increase of the population. In this hypercycle the population increase is given by the

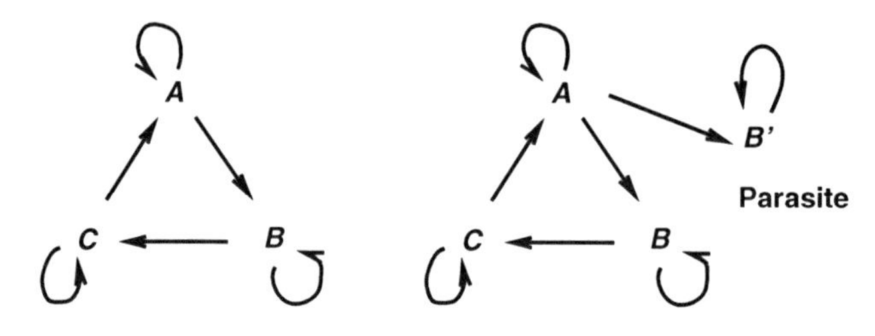


**Fig. 4.3.** Eigen's hypercycle. Replicating molecules catalyze each other, forming a cycle. The arrows indicate what molecule catalyze which molecule. By forming a cycle as in the left column, this replication is sustained, while in the right column, a parasite molecule $B'$ appears, which is replicated with the aid of $A$, but does not catalyze others. With the emergence of such parasite molecule species, the replication of hypercycle is destroyed

product of the populations of molecules such as $N_A \times N_B$, $N_B \times N_C$, $N_C \times N_A$, while the growth of the population of the mutants is linear to each population $N_A$, $N_B$, $N_C$. In the previous estimate for error catastrophe, both the good and mutant sequences increase in a manner linear to the number. Then the number of possible varieties of mutants dominates to surpass the difference in the growth rate. In the present case, once the populations of the good sequence in the hypercycle dominates, they can sustain the population against possible emergence of mutants, because of the difference in the order in the growth term. With this hypercycle, the original problem of error accumulation is avoided.

Since the proposal of hypercycle, the field of population dynamics of molecules in catalytic networks has been developed (see e.g., Eigen et al. 1989, Stadler et al. 1993). On the other hand, discovery of RNA with catalytic activities (Zaug & Cech, 1986) has led to the RNA world hypothesis, where a replication system closed only by RNA molecules is discussed as a possible candidate for the origin of life.

However, the hypercycle itself turned out to be weak against parasitic molecules, that is, those that are catalyzed by a molecule in the cycle and replicated rather fast (in contrast to the previous mutant with low replication rate), but do not catalyze those in the cycle. The growth rate of the population of these molecules is again the product of the populations of two species, and such parasitic molecules can invade. For example, the population of the molecule B′ in Fig. 4.3 grows, catalyzed by molecule $A$. Once such parasite molecule dominates in population, this hypercycle collapses, since it does not catalyze other molecules. Indeed, such parasite problem has been pointed out by several authors (see e.g., Maynard-Smith 1979, Eigen 1992,).

Although the hypercycle itself may be weak against parasitic molecules, that is, those that are catalyzed but do not catalyze others, it is then discussed that compartmentalization by a cell structure may suppress the invasion of parasitic molecules (Maynard-Smith, 1979; Szathmary & Demeter, 1987; Eigen, 1992; Szathmary & Maynard-Smith, 1997), or that the reaction–diffusion system at spatially extended system resolves this parasite problem (Boerlijst & Hogeweg, 1991; Hogeweg, 1994; Altmeywer & McCaskill, 2001). Chemically, it is not so surprising that a compartment structure is formed. Still, as an origin of life, this means that more complexity and diversity in chemicals are required other than a set of information-carrying molecules (e.g., RNA).

If initially there is a variety of chemicals that form a complex network of mutual catalyzation, this system may be robust against the invasion of parasitic molecules. Such discussion is similar with that in the problem of the stability of ecosystem, where coexistence of several species organizing complex ecological network (e.g., food web) may resist to invasion of external species. Still, the increase of stability by the complexity in the ecological network is not necessarily true (May, 1973). Hence if replication of complex reaction network or not is an issue to be elucidated. For such approach of study, we assume, from the beginning, there are many molecule species that mutually catalyze, allowing for the existence of many parasitic molecules. Here, complete replication of the system is probably difficult (see Fig. 4.2). Then the question we have to address is whether such complex network can maintain molecules that catalyze the synthesis of the network species. This question was addressed by Dyson (1985), as a possibility of loose reproduction system for an origin of life.

Dyson, noting the experiment of Oparin (1967) on the formation of cell-like structure, considered a collection of molecules with proteins and others. These molecules cannot replicate themselves like DNA or RNA. They, on the other hand, can have enzymatic activities and catalyze the synthesis of other molecules albeit not faithful reproduction. Still, they may keep similar compositions. Although accurate replication of such variety of chemicals is not possible, chemicals, as a set, may continue reproducing themselves loosely, while keeping catalytic activity. Indeed, the accurate replication must be difficult at the early stage of life, but loose reproduction could be easier. However, whether this collection of molecules can keep catalytic activity through reproduction or not is not evident.

Dyson obtained a condition for the sustainment of catalytic activities in these collection of molecules, by taking an abstract model. For simplicity he classified molecules into two states, depending on whether they have catalytic activity or not. Furthermore, he assumed that the ratio of the synthesis of catalytic molecules is amplified as the fraction of catalytic molecules increases, that is, a positive feedback process is assumed. This model is mapped to a kind of Ising model, studied extensively in statistical physics for phase transition. With the aid of mean-field analysis in statistical physics, he showed that the catalytic activities can be sustained depending on the numbers of molecules and their species. Although his model is abstract, the results he obtained probably can be applied to any system with a set of catalytic molecules, be it protein, lipids, or other polymers. It could be applied even to a system with carbon hydrates in deep hot sphere as Gold (1998) claimed, or peptide replication system as Ghadiri constructed (Lee et al., 1997).

From the standpoint of constructive biology (Chap. 2), it is important to construct a loose reproduction system consisting of catalytic molecules in vitro, as will be discussed in Sect. 4.5. On the other hand, construction of cell-like system with membrane for compartmentalization will be discussed in the next chapter. Here we should again note that the main purpose of these experiments is not to reproduce or imitate what might have occurred in the origin of life in the Earth, but to unveil the logic for the sustainment of loose reproduction without fine machinery for it.

It is important to study if such loose reproduction as a set is possible in a mutually catalytic reaction network. If this is possible, and if these chemicals also include molecules forming a membrane for compartmentalization, reproduction of a primitive cell will become possible (Ganti, 1975). In fact, from the chemical nature of lipid molecules, it is not so surprising that a compartment structure is formed.

Still, in this reproduction system, any particular molecules carrying information for reproduction do not exist, in contrast to the presently existing cells that have specific molecules (DNA) for it. As for a transition from early loose reproduction to later accurate replication with genetic information, Dyson did not give an explicit answer theoretically. He referred only to "genetic takeover" that was originally proposed by Cairns-Smith (1986), who proposed the loose reproduction system by using the clay, and that later a precise replication system by nucleic acids took over it. Indeed, Dyson wrote that his idea is based on "Cairns-Smith theory minus clay," and adopted the idea of "take-over."[1] However, the logic for this "take-over" is not unveiled.

Considering these theoretical studies so far, it is important to study how recursive production of a cell is possible, with the appearance of some molecules to play a specific role for heredity.[2]

To consider this problem, we start from a simple prototype cell that consists of mutually catalyzing molecule species whose growth in number leads to division of the protocell. In this protocell, the molecules that carry the genetic information are not initially specified. The first question we discuss here is how heredity to maintain production of the protocell emerges.

The second question related with the first one is whether there appears some specific molecules to carry information for heredity, to realize continual reproduction of such protocell. We note that in the present cells, it is generally

[1] It seems that clay is not liked by many researchers, but the idea itself is taken over, say by the use of pyrite (iron–sulfur) by Wačhtershauser (1990), which is now discussed seriously as a candidate of the origin of life.

[2] Condition for recursive production in a system with a variety of chemicals is discussed in the next chapter

believed that information is encoded in DNA, which controls the behavior of a cell.

In the present book, we do not necessarily take a "geno-centric" standpoint, in the sense that gene is the only one to determine the course of a cell. In fact, even in these cells, proteins and DNA mutually influence their replication process. Still, it cannot be denied that there exists a difference between DNA and protein molecules with regards to the role as information carrier. In spite of this mutual dependence, why is DNA molecule usually regarded as the carrier of heredity? Is there any general rule that some specific molecules play the role of carrier of genetic information so that the recursive production of cells continues?

Now, the origin of genetic information in a replicating system is an important theoretical topic that should be studied, not necessarily as a property of certain molecules, but as a general property of replicating systems. To investigate this problem, we need to clarify what the term "information" really means. In considering information, one often tends to be interested in how several messages are encoded on a molecule. (In fact, a polymer such as DNA would be suited to encode many bits of information.) One might point out that DNA molecules would be suited to encode many bits of information, and hence would be selected as an information carrier. Although this "combinatorial" capacity of an information carrier is important, what we are interested here is a basic property that has to be satisfied prior to that, that is, origin of just "1 bit" information.

As Shannon beautifully demonstrated, information means selection of one branch from several possibilities (Shannon & Weaver, 1949; Brillouin, 1969). Assume that there are two possibilities in an event, each of which can occur with the probability 1/2. In this case, when one of these possibilities turns out to be true, then this choice of a branch is regarded to have 1 bit information (see Fig. 4.4). In this sense, if a specific cell state is selected from several possible states, this selection process has information, and a molecule to control such process carries information.

Now, a molecule that carries the information is postulated to play the role to control for the choice of cellular state. Furthermore, to play the role to carry the information for heredity, the molecules must be transmitted to next generations relatively faithfully. These two features, that is, control and preservation, are nothing but the problem of heredity.

Let us reconsider what "heredity" really means. The heredity causes a high correlation in phenotype between ancestor and offspring. Then, for a molecule to carry heredity, we identify the following two features as necessary.

(1) If this molecule is removed or replaced by a mutant, there is a strong influence on the behavior of the cell. We refer to this as the **"control property."**
(2) Such molecules are preserved well over generations. The number of such molecules exhibits smaller fluctuations than that of other molecules, and

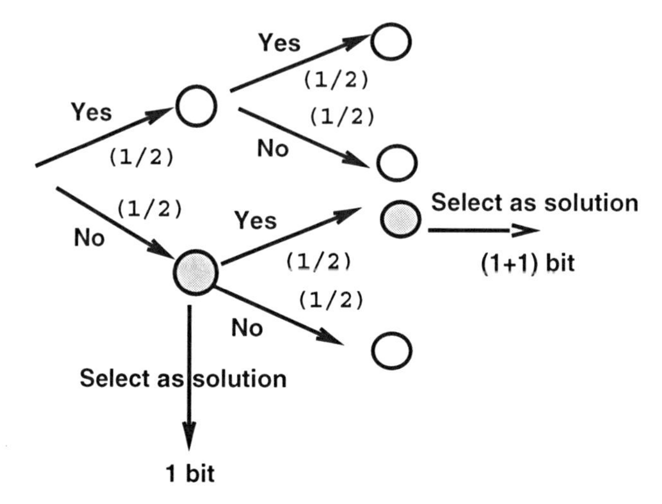


**Fig. 4.4.** Information according to the definition by Shannon. When one event is selected from several possibilities, some information is gained. In the figure, selection from two equally possible events gives 1-bit information. By successive independent choices of events of equally probable events, another bit of information is gained

their chemical structure (such as polymer sequence) is preserved over a long timespan, even under potential changes by fluctuations through the synthesis of these molecules. We refer to this as the **"preservation property."**

These two conditions are regarded as a fundamental condition for a molecule to establish the heredity. Now, the problem of "information" at a minimal level, that is, 1-bit information is nothing but the problem of the origin of heredity. As the origin of heredity, we study how a molecule starts to have the above two properties in a protocell. In other words, we study how 1-bit information starts to be encoded on a single molecule in a replicating cell system. After we answer this basic question, we will then discuss how a protocell with the heredity in the above sense has incentive to evolve genetic information in today's sense.

To sum up, the question we address in the present chapter is restated as follows. Consider a protocell with mutually catalyzing molecules. Then, under what conditions, recursive production continues maintaining catalytic activities? How are recursiveness and diversity in chemicals compatible? How is evolvability of such protocells possible? To answer these questions, are molecules carrying heredity necessary? Under what conditions, does one molecule species begin to satisfy conditions (1) and (2) so that the molecule carries heredity? We show, under rather general conditions in our model of mutually catalyzing system, that a symmetry breaking between the two kinds of molecules takes place, and through replication and selection, one kind of molecule comes to satisfy conditions (1) and (2).

## 4.2 Logic: Minority Control Hypothesis

Let us start from consideration of a prototype of cell, consisting of molecules that catalyze each other. As the reaction progresses, the number of molecules in this protocell will increase. Then, considering the physical nature of membrane, this cell will be divided, when its volume (the total number of molecules) is beyond some threshold. Then the molecules split into two "daughter cells." Then our question in the last section is restated as follows: How are the compositions transferred to the offspring cells? Do some specific molecules start to carry heredity in the sense of control and preservation so that the reproduction continues?

Before considering the logic for the origin of heredity, it may be relevant to recall the difference of roles between DNA (or RNA) and protein. According to the present understanding of molecular biology (Alberts et al., 1983), changes undergone by DNA molecules are believed to exercise stronger influences on the behavior of cells than other chemicals. Also, a DNA molecule is transferred to offspring cells relatively accurately, compared with other constituents of the cell. Hence a DNA molecule satisfies (at least) properties (1) and (2).

In addition, a DNA molecule is stable, and the time scale for the change of DNA, for example, its replication process as well as its decomposition process, is much slower. Because of this relatively slow replication, the number of DNA molecules is smaller than the number of protein molecules. At each generation of cells, single replication of each DNA molecule typically occurs, while other molecules undergo more replications (and decompositions).

With these natures of DNA in mind, while without assuming the detailed biochemical properties of DNA, we seek a general condition for the differentiation of the roles of molecules in a cell and study the origin of the control and preservation of some specific molecules.

To clarify the logic, we consider the simplest case (see Fig. 4.5), assuming that only two kinds of molecules $X$ and $Y$ exist in this protocell, and they catalyze each other for the synthesis of the molecules.

$$X + Y \to 2X + Y; Y + X \to 2Y + X$$

Here this "catalytic reaction" is not necessarily a single reaction. In general, there can be several intermediate processes for each "reaction." The model simply states that there are two molecules that help the synthesis of the other, directly or indirectly. In general, the catalytic activities as well as the synthesis speeds differ by types of molecules. Without losing generality, one can assume that $X$ is synthesized faster than $Y$.

With this synthesis of molecules, the total number of molecules in the protocell will increase, until it divides into two. As long as the molecules catalyze each other, this synthesis continues, as well as the division (reproduction) of protocell. However, some structural changes in molecules can occur through replication ("replication error"). These structural changes in each

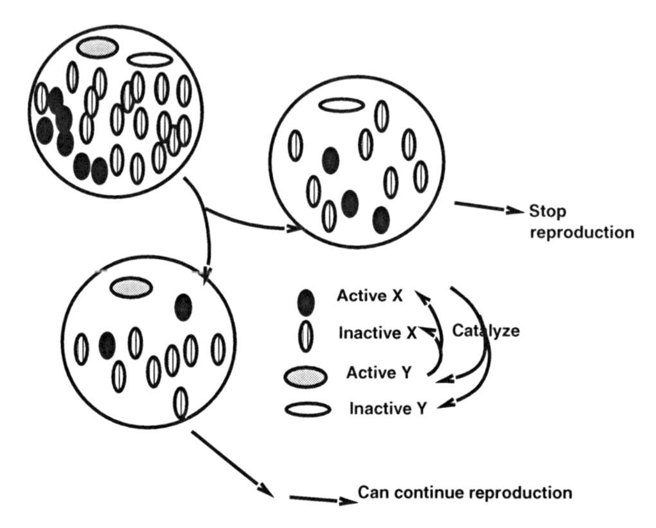


**Fig. 4.5.** Schematic representation of our logic. Once an active molecule of each molecule species is lost, the reproduction does not continue

kind of molecules may result in the loss of catalytic activity. Indeed, the molecules with catalytic activity are not so common. On the other hand, molecules without catalytic activity can increase their number, if they are catalyzed by other catalytic molecules. Then, as discussed in the parasite problem for hypercycle, the maintenance of reproduction is not so easy.

For example, each of $X$ and $Y$ has several types $0, 1, \dots, F$, and only the type 0 has catalytic activity. Now the question is how the active form 0 is maintained through reproduction even if $F$ is large.

Recall, because of the difference in the speed of synthesis, in each protocell, the number of $X$ molecules will be much larger than that of $Y$. Then, the $X$ molecules will dominate in the protocell. As long as the number of molecules in the cell is not large, finally, there comes to a stage that the catalytically active $Y$ molecule (that is, $Y^0$) goes extinct. Then, $X$ molecules are no longer synthesized. Inactive $Y$ molecules ($Y^j$, $j > 0$) can still be synthesized as long as active $X^0$ molecules remain. However, after each division, the number of $X^0$ molecules becomes half, and sooner or later the cell stops dividing. Hence, once the number of $Y^0$ becomes 0, the reproductivity of the protocell will be lost.

Now consider a situation that these protocells reproduce, competing for chemical resources for reproduction. Then, some cells, because of fluctuations in molecule numbers, may keep active $Y$ molecules. Since there is little room for $Y$ molecules in protocells, because of its slow synthesis, the total number of $Y^j$ molecules (for all $j = 0, 1, \dots, F$) is small. Hence, typically, to keep the active $Y$ molecules, the number of all the other $Y$ molecules should be suppressed, probably to zero. As long as such situation is realized, the protocell can keep growing.

Furthermore, if the inactive $Y$ molecules go extinct, they do not reappear so often, since the number of $Y$ molecules is small. The expectation value for the reappearance is the product of the number of $Y$ molecules and the error rate in replications. This expectation value given by the product of these is much smaller than 1.

Hence, a cell with very few $Y^0$ molecules and almost zero $Y^j$ ($j > 0$) can keep reproduction. In the next section, we will discuss the validity of the logic presented in this scenario, by carrying out numerical simulations of the stochastic model based on the above argument.

Now, this $Y$ molecule satisfies a condition for heredity-carrying molecules, that is, **"preservation property,"** and **"control property."**

First, these active $Y$ molecules should be preserved well over generations, since otherwise they cannot replicate. Next, assume that there occurs a structural change in $Y$ molecule, to change its catalytic activity. Since the number of active $Y$ molecules is few and all the $X$ molecules are catalyzed by them, this influence is enormous. The synthesis speed should drastically change. On the other hand, a change to $X$ molecules has a weaker influence on the average, since there are many active $X$ molecules, and influence of one molecule change is averaged out. Unless almost all molecules change in the same manner, the average catalytic activity of $X$ molecules would not change, but such coherent change is not possible, as deduced from the law of large numbers. Hence the change of $Y$ molecule has a crucial influence on the cell behavior, compared with $X$ molecules.

*Remark.* Note that the compartmentalization is essential to the above argument. Indeed, the rare cell state with null inactive $Y$ molecules is selected, since a "cell" with inactive $Y$ molecules cannot reproduce itself repeatedly. The necessity of compartmentalization to eliminate inactive states is already discussed in the "stochastic corrector model" by Szathmary (Szathmary & Demeter, 1987). The importance of minority molecules, on the other hand, is pointed out by Koch (1984), with regards to the evolvability to be discussed below (but not with regards to the controllability here).

## 4.3 Toy Model

On the basis of the argument of the last section, we consider a very simple protocell system (Kaneko & Yomo, 2002a), consisting of two species of replicating molecules that catalyze each other (see Fig. 4.6). Each species has active and inactive molecule types, with only the active types of one species catalyzing the replication of the other species for the replication. The rate of replication is different for the two species. We consider the behavior of a system with such mutually catalyzing molecules with different replication rate, as a first step in answering the question posed above.

To study the general features of a system with mutually catalyzing molecules, we consider the following minimal model. First, we envision a (proto)cell

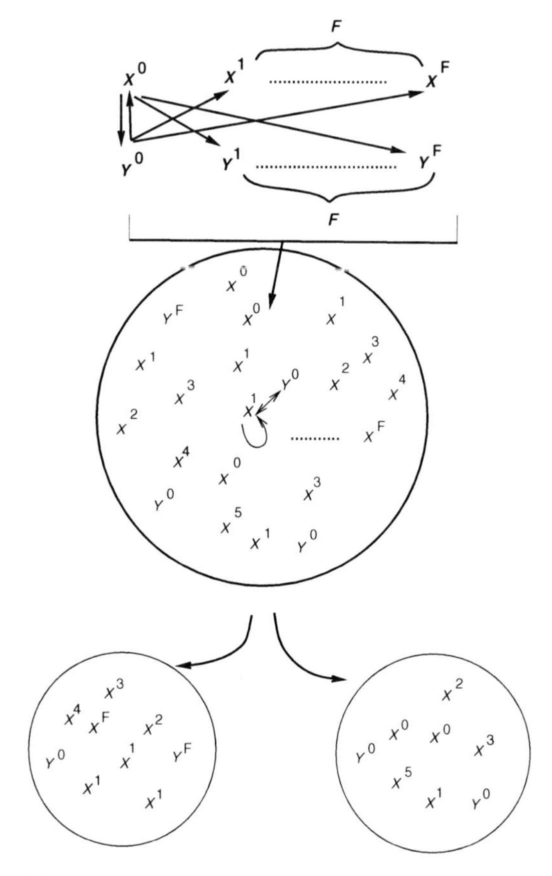


**Fig. 4.6.** Schematic representation of our model

containing molecules. With a supply of chemicals available to the cell, these molecules replicate through catalytic reactions, so that their numbers within a cell increase. When the total number of molecules exceeds a given threshold, the cell divides into two, with each daughter cell inheriting half of the molecules of the mother, chosen randomly. Regarding the chemical species and the reaction, we make the following simplifying assumptions:

(i) There are two species of molecules, X and Y, which are mutually catalyzing.
(ii) For each species, there are active and inactive (I) types. It is natural to assume that the active molecule type is rather rare. With this in mind, we assume that there are $F$ types of inactive molecules per active type. For most simulations, we consider the case in which there is only one type of active molecules for each species.

    Active types are denoted as $X^0$ and $Y^0$, while there are inactive types $X^I$ and $Y^I$ with $I = 1, 2, \ldots, F$. The active type has the ability to catalyze the replication of both types of the other species of molecules. The catalytic reactions for replication are assumed to take the form[3]

$$X^J + Y^0 \rightarrow 2X^J + Y^0 \quad (\text{for } J = 0, 1, \ldots, F)$$

and

$$Y^J + X^0 \rightarrow 2Y^J + X^0 \quad (\text{for } J = 0, 1, \ldots, F)\,.$$

(iii) The rates of synthesis (or catalytic activity) of the molecules $X$ and $Y$ differ. We stipulate that the rate of the above replication process for $Y$, $\gamma_y$, is much smaller than that for $X$, $\gamma_x$. This difference in the rates may also be caused by a difference in catalytic activities between the two molecule species.

(iv) In the replication process, there may occur structural changes that alter the activity of molecules. Therefore the type (active or inactive) of a daughter molecule can differ from that of the mother. The rate of such structural change is given by $\mu$, which is not necessarily small, because of thermodynamic fluctuations. This change can consist of the alternation of a sequence in a polymer or other conformational change, and may be regarded as replication "error." Note that the probability for the loss of activity is $F$ times greater than for its gain, since there are $F$ times more types of inactive molecules than active molecules. Hence, there are processes described by

$$\begin{array}{llll} X^I \rightarrow X^0; & \text{and} & Y^I \rightarrow Y^0 & (\text{with rate } \mu) \\ X^0 \rightarrow X^I; & \text{and} & Y^0 \rightarrow Y^I & (\text{with rate } \mu \text{ for each})\,, \end{array}$$

resulting from structural change.

(v) When the total number of molecules in a protocell exceeds a given value $2N$, it divides into two, and the chemicals therein are distributed into the two daughter cells randomly, with $N$ molecules going to each. Subsequently, the total number of molecules in each daughter cell increases from $N$ to $2N$, at which point these divide.

(vi) To include competition, we assume that there is a constant total number $M_{\text{tot}}$ of protocells, so that one protocell, randomly chosen, is removed whenever a (different) protocell divides into two.

With the above-described process, we have basically four sets of parameters: the ratio of synthesis rates $\gamma_y/\gamma_x$, the error rate $\mu$, the fraction of active molecules $1/F$, and the number of molecules $N$. (The number $M_{\text{tot}}$ is not important, as long as it is not too small.)

We carried out simulation of this model, according to the following procedure. First, a pair of molecules is chosen randomly. If these molecules are of different species, then if the $X$ molecule is active, a new $Y$ molecule is

[3] More precisely, there is a supply of precursor molecules for the synthesis of $X$ and $Y$, and the replication occurs with catalytic influence of either $X^0$ or $Y^0$.

produced with the probability $\gamma_y$, and if the $Y$ molecule is active, a new $X$ molecule is produced with the probability $\gamma_x$. Such replications occur with the error rates given above. All the simulations were thus carried out stochastically, in this manner.

We consider a stochastic model rather than the corresponding rate equation, which is valid for large $N$, since we are interested in the case with relatively small $N$. This follows from the fact that in a cell, often the number of molecules of a given species is not large, and thus the continuum limit implied in the rate equation approach is not necessarily justified (Hess & Mikhailov, 1994 Stange et al., 1998). Furthermore, it has recently been found that the discrete nature of a molecule population leads to qualitatively different behavior than in the continuum case in a simple autocatalytic reaction network (Togashi & Kaneko, 2001, 2003, 2004, 2005).[4]

## 4.4 Result

If $N$ is very large, the above-described stochastic model can be replaced by a continuous model given by the rate equation. Let us represent the total number of inactive molecules for each of $X$ and $Y$ as

$$N_x^I = \sum_{j=1}^{F} N_x^j, N_y^I = \sum_{j=1}^{F} N_y^j$$

Then the growth dynamics of the number of molecules $N_x^j$ and $N_y^j$ is described by the rate equations, using the total number of molecules $N^t$,

$$dN_x^j/dt = \gamma_x N_x^j N_y^0/N^t; dN_y^j/dt = \gamma_y N_x^0 N_y^j/N^t \; . \tag{4.2}$$

From these equations, under repeated divisions, it is expected that the relations $\frac{N_x^0}{N_y^0} = \frac{\gamma_x}{\gamma_y}$, $\frac{N_x^0}{N_x^I} = \frac{1}{F}$, and $\frac{N_y^0}{N_y^I} = \frac{1}{F}$ are eventually satisfied. Indeed, even with our stochastic simulation, this number distribution is approached as $N$ is increased.

However, when $N$ is small, and with the selection process, there appears a significant deviation from the above distribution. In Fig. 4.7, we have plotted the average numbers $\langle N_x^0 \rangle$, $\langle N_x^I \rangle$, $\langle N_y^0 \rangle$, and $\langle N_y^I \rangle$. Here, each molecule number is computed for a cell just prior to the division, when the total number of

[4] They studied autocatalytic reaction system with a small number of molecules, numerically by stochastic particle simulations, and found a novel state because of fluctuation and discreteness in molecular numbers, that is characterized as extinction of molecule species alternately in the autocatalytic reaction loop. Phase transition to this state with the change of the system size and flow is clarified, where a single-molecule switch of the molecule distributions is important. This gives a clear example in which a discreteness in molecule number leads to a novel type of phase that is not observed from a continuous rate equation of chemical reaction.

molecules is $2N$, while the average $\langle \ldots \rangle$ is taken over all cells that are divided throughout the simulation. (Accordingly, a cell removed without division does not contribute to the average.) As shown in the figure, there appears a state satisfying $\langle N_y^0 \rangle \approx 2 - 10$, $\langle N_y^I \rangle \approx 0$. Since $F \gg 1$, such a state with $\frac{\langle N_y^0 \rangle}{\langle N_y^I \rangle} > 1$ is not expected from the rate equation (4.1). Indeed, for the $X$ species, the number of inactive molecules is much larger than the number of active ones. Hence, we have found a novel state that can be realized because of the smallness of the number of molecules and the selection process.

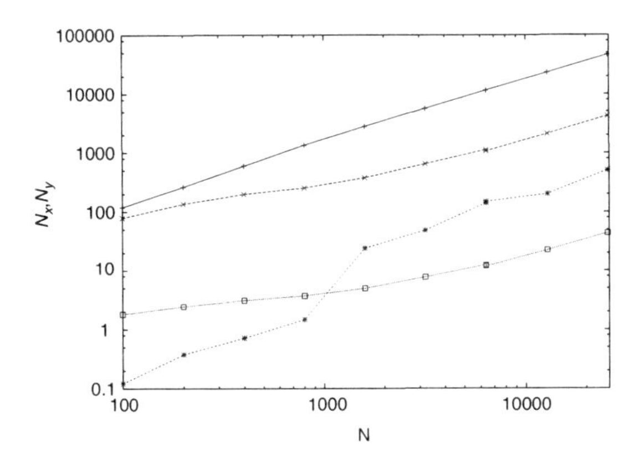


**Fig. 4.7.** Dependence of $\langle N_x^0 \rangle(\times)$, $\langle N_x^I \rangle(+)$, $\langle N_y^0 \rangle(\square)$, and $\langle N_y^I \rangle(*)$ on $N$. The parameters were fixed as $\gamma_x = 1$, $\gamma_y = 0.01$, and $\mu = 0.05$. Plotted are the averages of $N_x^0$, $N_x^I$, $N_y^0$, and $N_y^I$ at the division event, and thus their sum is $2N$. In all the simulations conducted for this chapter, we used $M_{\text{tot}} = 100$, and the sampling for the averages were taken over $10^5 - 3 \times 10^5$ steps, where the number of divisions ranges from $10^4$ to $10^5$, depending on the parameters. (Reproduced from [Kaneko and Yomo 2002a] with permission)

In Fig. 4.7, $\gamma_y/\gamma_x$ and $F$ are fixed to 0.01 and 64, respectively, while the dependence of $\{\langle N_x^0 \rangle, \langle N_x^I \rangle, \langle N_y^0 \rangle, \langle N_y^I \rangle\}$ on these parameters is plotted in Fig. 4.8. As shown in these figures, the above-mentioned state with $\langle N_y^0 \rangle \approx 2 - 10$, $\langle N_y^I \rangle < 1$ is reached and sustained when $\gamma_y/\gamma_x$ is small and $F$ is sufficiently large. In fact, for most dividing cells, $N_y^I$ is exactly 0, while there appear a few cells with $N_y^I > 1$ from time to time. It should be noted that the state with almost no inactive Y molecules appears in the case of larger $F$, that is, in the case of a larger possible variety of inactive molecules. This suppression of $Y^I$ for large $F$ contrasts with the behavior found in the continuum limit (the rate equation). When $\frac{\langle N_y^0 \rangle}{\langle N_y^I \rangle}$ is plotted as a function of $F$ (Kaneko & Yomo, 2002a) from the data of Fig. 4.8, we can observe this suppression clearly. Indeed, up to some value of $F$, the proportion of active $Y$ molecules decreases, in agreement with the naive expectation provided by (4.1), but this proportion increases with further increase of $F$, in the case that $\gamma_y/\gamma_x$ is small ($\lesssim .02$) and $N$ is small.

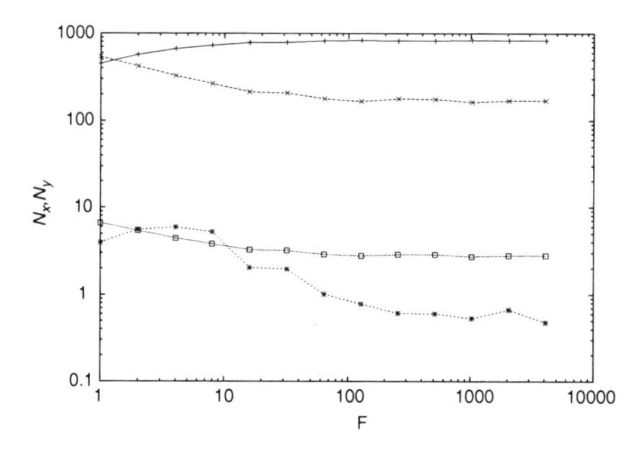


**Fig. 4.8.** Dependence of $\langle N_x^0 \rangle(\times)$, $\langle N_x^I \rangle(+)$, $\langle N_y^0 \rangle(\square)$, and $\langle N_y^I \rangle(*)$ on $F$. The parameters were fixed as $\gamma_x = 1$, $\gamma_y = .01$, $\mu = .05$, and $N = 1000$. Plotted are the averages of $N_x^0$, $N_x^I$, $N_y^0$, and $N_y^I$ at the division event, and thus their sum is $2N = 2000$. (Reproduced from [Kaneko and Yomo 2002a] with permission)

This behavior of the molecular populations can be understood from the viewpoint of selection: In a system with mutual catalysis, both $X^0$ and $Y^0$ are necessary for the replication of protocells to continue. The number of $Y$ molecules is expected to be rather small, since their synthesis speed is much slower than that of $X$ molecules. Indeed, the fixed point distribution given by the continuum limit equations possesses a rather small $N_y^0$. In fact, when the total number of molecules is sufficiently small, the value of $\langle N_y^0 \rangle$ given by these equations is less than 1. However, in a system with mutual catalysis, both $X^0$ and $Y^0$ must be present for replication of protocells to continue. In particular, for the replication of $X$ molecules to continue, at least a single active $Y$ molecule is necessary. Hence, if $N_y^0$ vanishes, only the replication of inactive $Y$ molecules occurs. For this reason, divisions producing descendants of this cell cannot proceed indefinitely, because the number of $X^0$ molecules is cut in half at each division. Thus, a cell with $N_y^0 < 1$ cannot leave a continuing line of descendant cells. Also, for a cell with $N_y^0 = 1$, only one of its daughter cells can have an active $Y$ molecule. Hence a cell with $N_y^0 = 1$ has no potentiality to multiple through division, and for this reason, given the presence of selection, protocells with $N_y^0 > 1$ are selected.

The total number of $Y$ molecules is limited to small values, because of their slow synthesis speed. This implies that a cell that suppresses the number of $Y^I$ molecules to be as small as possible is preferable under selection, so that there is a room for $Y^0$ molecules. Hence, a state with almost no $Y^I$ molecules and a few $Y^0$ molecules, once realized through fluctuations, is expected to be selected through competition for survival.

Of course, the fluctuations necessary to produce such a state decrease quite rapidly as the total molecule number increases, and for sufficiently large numbers, the continuum description of the rate equation is valid. Clearly then, a state of the type described above is selected only when the total number of molecules within a protocell is not too large. In fact, a state with very small $N_Y^I$ appears only if the total number $N$ is smaller than some threshold value depending on $F$ and $\gamma_y$[5].

## 4.5 Minority-Controlled State

We showed that in a mutually catalyzing replication system, the selected state is one in which the number of inactive molecules of the slower replicating species, $Y$, is drastically suppressed. In this section, we first show that the fluctuations of the number of active $Y$ molecules is smaller than those of active $X$ molecules in this state. Next, we show that the molecule species $Y$ (the minority species) becomes dominant in determining the growth speed of the protocell system. Then, considering a model with several active molecule types, the control of chemical composition through specificity symmetry breaking is demonstrated.

### 4.5.1 Preservation of Minority Molecule

First, we computed the time evolution of the number of active $X$ and $Y$ molecules, to see whether the selection process acts more strongly to control the number of one or the other. We computed $N_x^0$ and $N_y^0$ at every division to obtain the histograms of cells with given numbers of active molecules.

The fluctuations in the value of $N_y^0$ are found to be much smaller than those of $N_x^0$. The numbers $N_y^0$ and $N_y^I$ are more nearly conserved than $N_x^0$ and $N_x^I$, and the selection process discriminates more strongly between different concentrations of active $Y$ molecules than between those of active $X$ molecules. Hence the active $Y$ molecules are well preserved with relatively smaller fluctuations in the number.

### 4.5.2 Control of the Growth Speed

Now, it is expected that the growth speed of our protocell has a stronger dependence on the number of active $Y$ molecules than the number of active $X$ molecules. We have found that the division time is a much more rapidly decreasing function of $N_y^0$ than of $N_x^0$. Even a slight change in the number of active $Y$ molecules has a strong influence on the division time of the cell. Of course, the growth rate also depends on $N_x^0$, but this dependence is much weaker (see Fig. 4.9). Hence, the growth speed is controlled mainly by the number of active $Y$ molecules.

[5] In the model considered here, we have included a mechanism for the synthesis of molecules, but not for their decomposition. To investigate the effect of the decomposition of molecules, we have also studied a model including a process to remove molecules randomly at some rate. We found that the above-stated conclusion is not altered by the inclusion of this mechanism.

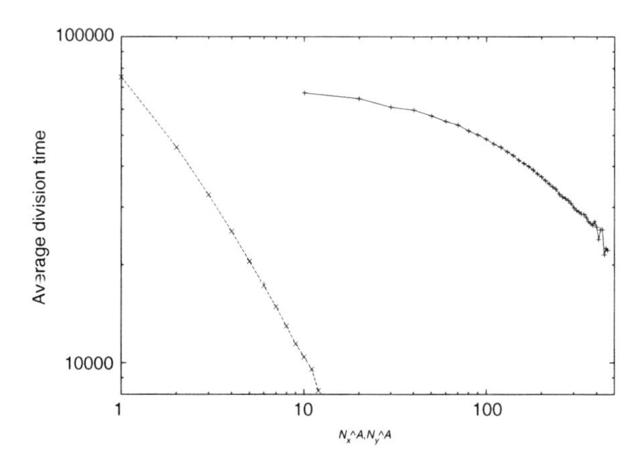


**Fig. 4.9.** The histogram of division time. The left graph displays that with regards to $N_y$, with integrating out the values of $N_x$, while the right graph displays that with regards to $N_x$, by integrating out the value of $N_x$. The division time sensitively depends on $N_y$, but weakly on $N_x$. (Reproduced from [Kaneko and Yomo 2002a] with permission)

### 4.5.3 Control of Chemical Composition by the Minority Molecule

As another demonstration of control, we study a model in which there is more specific catalysis of molecule synthesis. Here, instead of single active molecule types for $X$ and $Y$, we consider a system with $k$ types of active $X$ and $Y$ molecules, $X^{0_i}$ and $Y^{0_i}$ ($i = 1, 2, \dots k$). In this model, each active molecule type catalyzes the synthesis of only a few types ($m < k$) of the other species of molecules. Graphically representing the ability for such catalysis using arrows as $i_x \to j_y$ for $X \to Y$ and $i_y \to j_x$ for $Y \to X$, the network of arrows defining the catalyzing relations for the entire system is chosen randomly, and is fixed throughout each simulation. Here we assume that both $X$ and $Y$ molecules have the same "specificity" (that is, the same value of $m$) and study how this symmetry is broken (see Fig. 4.10).

As already shown, when $N$, $\gamma_y$, and $F$ satisfy the conditions necessary for realization of a state in which $N_y^I$ is sufficiently small, the surviving cell type contains only a few active $Y$ molecules, while the number of inactive ones vanishes or is very small. Our simulations show that in the present model with several active molecule types, only a single type of active $Y$ molecule remains after a sufficiently long time. We call this "surviving type," $i_r$ ($1 \le i_r \le k$). Contrastingly, at least $m$ types of $X^0$ species that can be catalyzed by the remaining active $Y$ molecule species remain. Accordingly, for a cell that survived after a sufficiently long time, a single type of $Y^0$ molecule catalyzes the synthesis of (at least) $m$ kinds of $X$ molecule species, while the multiple types of $X$ molecules catalyze this single type of $Y^{0_i}$ molecules. Thus, the original symmetry regarding the catalytic specificity is broken as a result of the difference between the synthesis speeds.

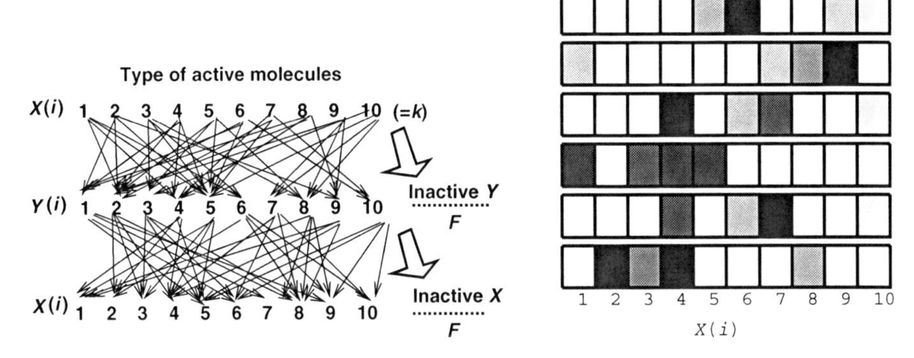


**Fig. 4.10.** (**a**) Catalytic network between $X(i)$ and $Y(i)$. The arrows from the top column points to types of the $Y$ species that each $X(i)$ catalyzes, while those from the middle points to types of the $X$ species that each $Y(i)$ catalyzes. Each of the $k = 10$ active species catalyzes the synthesis of four chemicals. (**b**) The logarithm of the average population of $X(i)$ displayed wish a gray scale. From top to bottom, six samples resulting from different initial conditions are plotted. The type $i_r$ of $Y(i)$ with nonvanishing population $Y(i)$ corresponding to each column is as follows (*top to bottom*): $i_r = 10, 8, 4, 5, 4$, and 2. The parameters were chosen as $F = 16 \times k = 160$, $\mu = 0.03$, $\gamma_x = 1$, $\gamma_y = 0.01$, $N = 1000$. (Reproduced from Kaneko & Yomo, 2002a)

Because of autocatalytic reactions, there is a tendency for further increase of the molecules that are in the majority. This leads to competition for replication between molecule types of the same species. Since the total number of $Y$ molecules is small, this competition leads to all-or-none behavior for the survival of molecules. As a result, only a single type of species $Y$ remains, while for species $X$, the numbers of molecules of different types are statistically distributed as guaranteed by the uniform replication error rate.

Although $X$ and $Y$ molecules catalyze each other, a change in the type of the remaining active $Y$ molecule has a much stronger influence on $X$ than a change in the types of the active $X$ molecules on $Y$, since the number of $Y$ molecules is much smaller.

With the results so far, we can conclude that the $Y$ molecules, that is, the minority species, control the behavior of the system, and are preserved well over many generations. We therefore call this state the minority-controlled (MC) state.

### 4.5.4 Evolvability

An important characteristic of the MC state is evolvability, that is, the capacity for evolution. To discuss this problem, we consider a variety of active molecules instead of one type for $X$ and $Y$, and assume that they have different catalytic activities. Then the synthesis rates $\gamma_x$ and $\gamma_y$ for each molecule depend on the catalytic activities that catalyze it. Hence, $\gamma_x$ can be written in terms of the molecule's inherent growth rate, $g_x$, and the activity, $e_y(i)$, of the corresponding catalyzing molecule $Y(i)$, and $\gamma_y$ similarly, as

$$\gamma_x = g_x \times e_y(i); \gamma_y = g_y \times e_x(i) .$$

Since such a biochemical reaction is entirely facilitated by catalytic activity, a change in $e_y$ $e_x$, for example by the structural change of polymers, will be important. Given the occurrence of such a change to molecules, those with greater catalytic activities will be selected through competition evolution, leading to the selection of larger $e_y$ and $e_x$. As an example to demonstrate this point, we have extended the model in Sect. 4.2 to include $k$ kinds of active molecules with different catalytic activities. Then, molecules with greater catalytic activities are selected through competition.

Since only a few molecules of the $Y$ species exist in the MC state, a structural change to them strongly influences the catalytic activity of the protocell. On the other hand, a change to $X$ molecules has a weaker influence, on the average, since the deviation of the *average* catalytic activity caused by such a change is smaller, as can be deduced from the law of large numbers.[6] Hence, the MC state is important for a protocell to realize evolvability.

Note that the difference between the synthesis speeds between $X$ and $Y$ molecules is maintained through the evolution. Hence the MCS is stable evolutionarily.[7] This is because that the evolution of the catalytic activity in $Y$ molecules is accelerated because of its minority, so that the replication rate of $X$ is increased, and the difference between the synthesis speeds of $X$ and $Y$ molecules is maintained.

## 4.6 Experiment

Recently, there have been some experiments to construct minimal replicating systems in vitro. As an experiment corresponding to this problem, we describe an in vitro replication system, constructed by Yomo's group (Matsuura et al., 2002).

In general, proteins are synthesized from the information on DNA through RNA, while DNA are synthesized through the action of proteins. DNA and protein are both necessary for their replication. They, in addition to other molecules, autonomously replicate themselves. Now simplifying this replication process, Matsuura et al. constructed a replication system consisting of DNA and DNA polymerase, that is, an enzyme for the synthesis of DNA. This DNA polymerase is synthesized by the corresponding gene in the DNA, while

[6] See Koch (1984) for this kind of statistical argument, while the kinetic suppression of fluctuation of $Y$ molecules is stronger than expected by the standard statistical analysis.

[7] This is in contrast to the result from the stochastic corrector model (Grey et al., 1995).

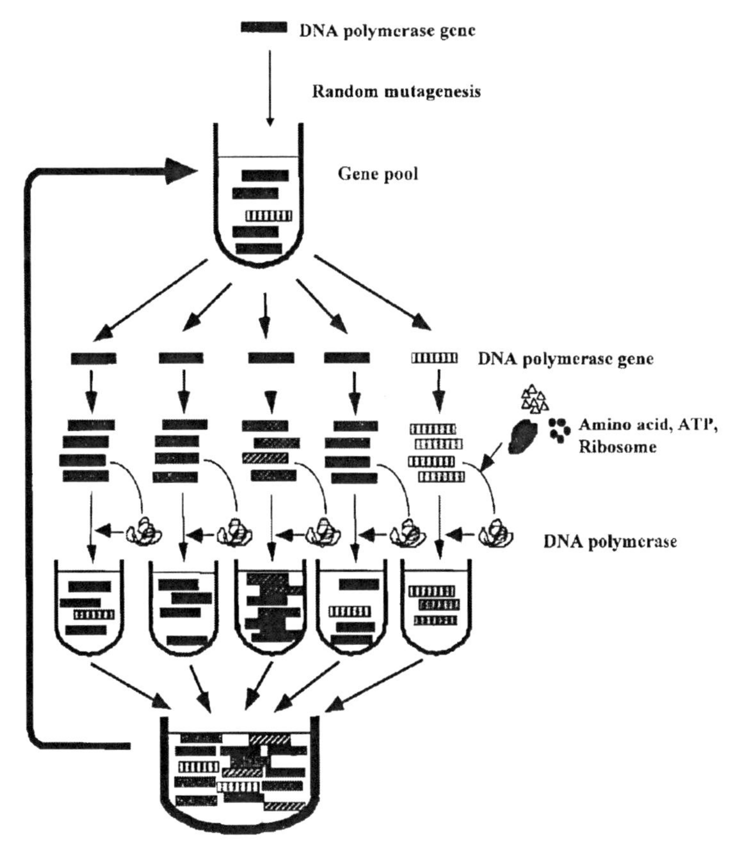


**Fig. 4.11.** Illustration of in vitro autonomous replication system consisting of DNA and DNA polymerase. See text and Matsuura et al. (2002) for details. Supplied with the courtesy of Yomo, Matsuura et al.

it works as the catalyst for the corresponding DNA. Through this mutual catalytic process the chemicals replicate themselves (see Fig. 4.11).

To amplify the number of DNA, the method PCR (polymerase chain reaction) is widely used, and is a standard tool for molecular biology. In this case, however, enzymes that are necessary for the replication of DNA must be supplied externally. In this sense, it is not a self-contained autonomous replication system. In the experiment by Yomo's group, while they use PCR as one step of experimental procedures, the enzyme (DNA polymerase) for DNA synthesis is also replicated in vitro within the system. Of course, some (raw) material, such as amino acid or ATP, and even Ribosome have to be supplied, but otherwise the DNA and protein replicate by themselves. (see Fig. 4.12 for experimental step).

In this experiment, there is a mutual synthetic process between gene and enzymes. Roughly speaking, the polymerase (protein) in the experiment

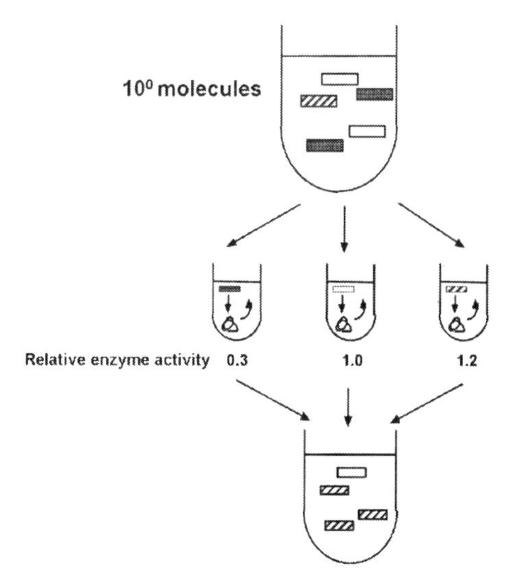


**Fig. 4.12.** Procedure of experiment: In each of 10 test tubes containing a single DNA molecule, autonomous replication progresses. The components of the tubes are mixed in a pool, from which a single DNA is chosen to a tube, to repeat the procedure. See text and Matsuura et al. (2002) for details. Supplied with the courtesy of Yomo, Matsuura et al.

corresponds to $X$ in our model, while the polymerase gene (DNA) corresponds to $Y$.

Now, at each step of replication, about $2^{30} \sim 2^{40}$ DNA molecules are replicated.[8] Here, of course there are some errors that can occur in the synthesis of enzyme, and also in the synthesis of DNA. With these errors, there appear DNA molecules with different sequences. Now a pool of DNA molecules with a variety of sequences is obtained as a first generation.

From this pool, the DNA and enzymes are split into several tubes. Then, materials with ATP and amino acids are supplied, and the replication process is repeated (see Fig. 4.12). In other words, the "test tube" here plays the role of "cell compartmentalization." Instead of autonomous cell division, splitting into several tubes is operated externally. (Spontaneous split of cells will be discussed in the next chapter.)

In this experiment, instead of changing the synthesis speed $\gamma_y$ or $N$ in the model, one can control the number of genes, by changing how the pool is split into several test tubes. Indeed, they studied two distinct cases, that is, splitting to tubes containing a single DNA molecule in each and to tubes containing 100 DNA molecules. Recall that in the theory, the evolvability by minority control is predicted. Hence, the behavior between the cases with a single DNA, that is, minority, and 100 DNA may be drastically different.

[8] This high amplification rate is different from the theoretical model, but for the later discussion this difference is not essential.

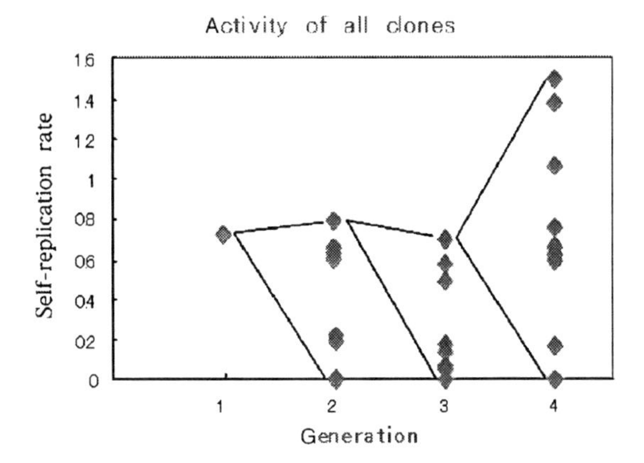


**Fig. 4.13.** Change of self-replication activity from a system with single DNA. The activities for 10 tubes are shown, The next generation is produced mostly from the top DNA. Although activities vary by each tube, higher ones are selected, so that the activities are maintained. See text and Matsuura et al. (2002) for details. Supplied with the courtesy of Yomo, Matsuura et al.

First, we describe the case with a single DNA in each tube. Here, as given in the middle of Fig. 4.13, the pool of chemicals is split into 10 tubes each of which has a single DNA molecule, and replication process described already progresses in each tube. The sequence of DNA molecules could be different for each tube, since there is replication error. Then the activity of DNA polymerase in each tube is also different, and the number of DNA molecules synthesized in each tube is different. In other words, some DNA molecules can produce more offspring, but others cannot. The variation of self-replication activity by tubes is shown in Fig. 4.13, as the values in the generation 2. Then the contents of each tube are mixed as in the lower column of Fig. 4.13. This soup of chemicals is used for the next generation. Then in this soup, the DNA molecules that have higher replication rates, as well as the mutants generated from them, are included with a larger fraction. Now a single DNA is selected from the soup in each of 10 tubes, and the same procedures are repeated. Hence, there is a larger probability that a DNA molecule with a higher reproduction activity is selected for the next generation. In other words, Darwinian selection acts at this stage. The self-replication activity from this soup is plotted in the third generation. Successive plots of the self-replication activity are given in Fig. 4.14. As shown, the self-replication activity is not lost (or can evolve in some case), although it varies in each tube in each generation.

One might say that the maintenance of replication is not surprising at all, since a gene for the DNA polymerase is included in the beginning. However, enzyme with such catalytic activity is rare. Indeed, with mutations some proteins that lost such catalytic activity but are synthesized in the present system could appear, which might take over the system. Then the self-replication activity would be lost. In fact, this is nothing but the error catastrophe by Eigen,

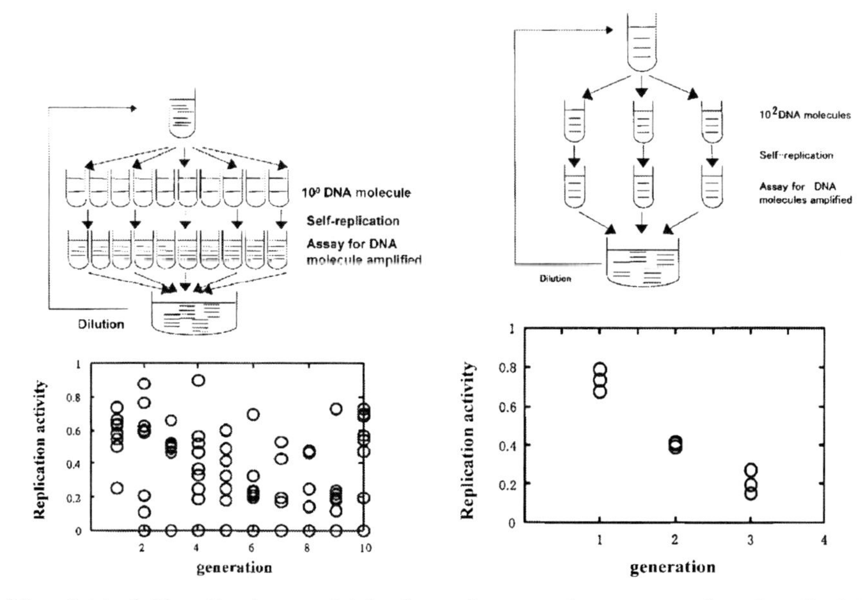


**Fig. 4.14.** Self-replication activities for each generation, measured as described in the text. Result from a single DNA; lower: result from 100 DNA molecules. Adapted from Matsuura et al. (2002) with the authors' courtesy

discussed in the beginning of this chapter. Then, why is the self-replication activity maintained in the present experiment?

For those who read the theory part, the answer must be clear. In the model of this chapter, mutants that lost the catalytic activity are much more common (that is, $F$ times larger in the model). Still, the number of such molecules is suppressed. This was possible first because the molecules are in a cell. In the experiment also they are in a test tube, that is, in a compartment. Now the selection works for this compartment, not for each molecule. Hence the tube (cell) that includes a gene giving rise to lower enzyme activity produces less offspring. In this sense, compartmentalization is one essential factor for the maintenance of catalytic activity (see also (Maynard-Smith 1979, Szathmary & Demeter 1987, Eigen 1992, Szathmary and Maynard-Smith 1997, Boerilst & Hogeweg, 1991, Altmeyer & McCaskill 2001)). Here, another important factor is that in each compartment (cell) there are very few DNA molecules (in the model $Y$ molecule). In the theory, if the number of $Y$ molecules is larger, inactive $Y$ molecules surpass the active one in population.

To confirm the validity of our theory, Matsuura et al. carried out a comparison experiment. Now, they split the chemicals in the soup so that each tube has 100 DNA molecules instead of a single one. Otherwise, they adopt the same procedure. In other words, this corresponds to a cell with 100 copies of genome. Change of self-replication activity in the experiment is plotted in the lower column of Fig. 4.14. As shown, the self-replication activity is lost by each generation, and after the fourth generation, capability of autonomous replication is totally lost. This result shows that the number of molecules to carry genetic information should be small, which is consistent with the theory.

When there are many DNA molecules, there can be mutation to each DNA molecule. In each tube, the self-replication activity is given by the average of the enzyme activities from these 100 DNA molecules. Although catalytic activity of molecules varies by each, the variance of the average by tubes should be reduced drastically. Recall that the variance of the average of $N$ variables with the variance $\mu$ is reduced to $\mu/N$, according to the central limit theorem of probability theory (Koch, 1984). Hence the average catalytic activity does not differ much by tube. Here, the mutant with a higher catalytic activity is rare. Most changes in the gene lead to smaller or null catalytic activity. Hence, on the average, the catalytic activity after mutations to original gene gets smaller, and the variance by tubes around this mean is rather small (see Fig. 4.14).

By the selection, DNA from a tube with a higher catalytic activity could be selected, but the variation by tubes is so small that the selection does not work. Hence deleterious mutations remain in the soup, and the self-replication activity will be lost in a few generations. In other words, the selection works because the number of information carrier in a replication unit (cell) is very small and is free from the statistical law of large numbers.

Summing up, in the experiment, it was found that replication is maintained even under deleterious mutations (that correspond to structural changes from active to inactive molecules in our model), only when the population of DNA polymerase genes is small and competition of replicating systems is applied. When the number of genes (corresponding to $Y$) is small, the information containing in the DNA polymerase genes is preserved. This is made possible by the maintenance of rare fluctuations, as found in our study. The system has evolvability only if the number of DNA in the system is small. Otherwise, the system gradually loses its activity to replicate itself. These experimental results are consistent with the minority control theory described.

## 4.7 Relevance to Biology

### 4.7.1 Heredity from a Kinetic Viewpoint

In this chapter, we have shown that in a mutually catalyzing system, molecules $Y$ with the slower synthesis speed tend to act as the carrier of heredity. Through the selection under reproduction, a state, in which there is a very small number of inactive $Y$ molecules, is selected. This state is termed the "minority controlled state." Between the two molecule species, there appears separation of roles, that with a larger number, and that with a greater catalytic activity. The former has a variety of chemicals and reaction paths, while the latter works as a basis for the heredity, in the sense of the two properties mentioned in Sect. 4.1, "preservation" and "control." We now discuss these properties in more detail.

*Preservation property*: A state that can be reached only through very rare fluctuations is selected, and it is preserved over many generations, even though the realization of such a state is very rare when we consider the rate equation obtained in the continuum limit.

*Control property*: A change in the number of $Y$ molecules has a stronger influence on the growth rate of a cell than a change in the number of $X$ molecules. Also, a change in the catalytic activity of the $Y$ molecules has a strong influence on the growth of the cell. The catalytic activity of the $Y$ molecules acts as a control parameter of the system.

Once this MCS is established, it is rather straightforward to assume the following scenario for the evolution of genetic information. First, a new evolutionary incentive, that is, a new selection pressure is now able to emerge, to evolve a machinery to ensure that the minority molecule makes it into the offspring cells, since otherwise the reproduction of the cell is highly damaged. Hence a machinery to guarantee the faithful transmission of the minority molecule evolves. In other words, the origin of heredity is established. Note that this heredity can evolve before the appearance of any specific metabolic or genetic contents transmitted faithfully. This heredity evolves just as a result of kinetic phenomenon and is a rather general phenomenon in a reproducing protocell consisting of mutually catalytic molecules.

Once this faithful transmission of minority molecule is evolved, it provides a basis for critical information for reproduction of the protocells. Since the molecule is well taken care of and is guaranteed to be transmitted, other chemicals that are synthesized in connection with it are likely to be transmitted, albeit not always faithfully. Hence there appears a further evolutionary incentive to package life-critical information into the minority molecule. Now more information ("many bits" of information) are encoded on the minority molecule. Now, the molecules work as a carrier of genetic information in today's sense. With this evolution, having more molecules catalyzed by the minority molecule, it is then easier to further develop the machinery to better take care of minority molecules, since this minority molecule is essential to many reactions for the synthesis of many other molecules.

Hence the evolution of faithful transmission of minority molecules and of coding of more information reinforce each other. At this point one can expect a separation of metabolism and genetic information.

To sum up, how a single molecule starts to control the heredity is understood from a kinetic viewpoint.[9] We first show that the MCS is a rather

[9] Of course, even in the present organisms, some genes exist in a multiple copy – plasmids in bacteria, genes in macronucleate in ciliates, and so forth. These genes are transferred to their descendants. In this sense, not all genes are minority. However, even in these cases, genes in a multiple copy are often less important than the genes in minority, with regards to the controllability.

general consequence of kinetic process of mutually catalytic molecules. This provides a basis for heredity. Taking advantage of the evolvability of MCS, then, preservation mechanism of the minority molecule evolves, which allows for more information encoded on it, leading to separation of genetic information and metabolism. In this sense, the minority molecule species with slower synthesis speed, leading to the preservation of rare states and control of the behavior of the system, acts as an information carrier. The important point of our theory is that heredity arises prior to any metabolic information that needs to be inherited.

### 4.7.2 Accessibility to MCS

One important consequence of the existence of the MC state is evolvability. Mutations introduced to the majority species tend to be canceled out on the average, in accordance with the law of large numbers. Hence, the catalytic activity of the minority species ($Y$ in our model) not only is sustained but has a greater potentiality to increase through evolution as well.

The evolution and stability of the MC state with respect to mutation was discussed in Sect. 4.4.3. If the initial difference between the catalytic abilities $e_x$ and $e_y$ (and other parameters) satisfies the conditions stated in Sect. 4.4, it is shown that the MC state once realized is stable over generations against mutations.

A question may still remain. How does the difference in the catalytic activity necessary to realize the MC state generally come to exist? Of course, it is quite natural in a complex chemical system that there will be differences in synthesis speeds or catalytic activities, and, in fact, this is the case in the biochemistry of present-day organisms. Still, it would be preferable to have a theory describing the spontaneous divergence of synthesis speeds without assuming a difference in advance, to provide a general model of the possible "origin" of bioinformation from any possible replication system.

There are several possibilities for spontaneous evolution to this differentiation of roles in the molecules.

1. *Higher order catalysis*: In the first toy model considered in this chapter, in order to realize the MC state, the difference between the time scales of the two kinds of molecules often must be rather large. For example, the ratio $\gamma_y/\gamma_x$ should typically be less than 0.05 when the number of molecules is in the range 500–2000.

   The condition for this MC state is drastically reduced by considering Higher order catalytic reaction processes in the replication of molecules. Consider, for example, a replication of molecules described by the following:

$$X + X' + Y \rightarrow 2X + X' + Y ; Y + Y' + X \rightarrow 2Y + Y' + X \, . \qquad (4.3)$$

   In complex biochemical reaction networks, such higher order catalytic reactions often exist. Indeed, proteins in a cell are catalyzed not solely by nucleotides but with collaboration of proteins and nucleotides. Nucleotides, similarly, are catalyzed not solely by proteins but with collaboration of nucleotides and proteins.

In this higher order catalytic reaction, difference between the numbers of $X$ and $Y$ molecules is amplified, and it is shown (Kaneko & Yomo, 2002a) that the MC state is reached for almost all values of $\gamma_y/\gamma_x$.[10]

2. *Network structure*: The catalytic network in a cell is generally quite complex, with many molecules participating in mutual catalysis for replication. The evolution of replication systems with such catalytic networks have been studied. We will study this problem in the next chapter.
3. *Spatial structure of a cell*: Considering chemical condition to support a cell, the roles of molecules are often different from the beginning in a different context.

   For example, in our model, collisions of molecules occur randomly. However, if one of the molecule species forms a membrane, and the other are in a contained medium, then the collision rate between the different molecule species is enhanced.

   Also, in the present model we have assumed that the division of the protocell occurs when the total number of molecules doubles. However, it may be more natural to have a threshold that depends on the number of molecules of one species (or, more generally, of some subset of all species), rather than the total number. For example, consider the case that division occurs when the size of a membrane synthesized by biochemical reactions is larger than some threshold.

It is discussed (Kaneko & Yomo, 2002a) that by such changes of collision or division conditions MC state is easier to evolve.

### 4.7.3 Closing Remark

In the beginning of this chapter we described two standpoints on the origin of life, that is, genetic information first or complex metabolism first.[11] We showed that there is a difficulty in each standpoint. In the former picture, there was a problem on the stability against parasites, while the latter has

---

[10] In the continuum limit, the rate equation corresponding to the reaction (4.2) is given by $dN_x^j/dt = \gamma_x N_x^j N_x^A N_y^A$ and $dN_y^j/dt = \gamma_y N_y^j N_y^A N_x^A$. Consider the equation $dx/dt = \gamma_x x^2 y$ and $dy/dt = \gamma_y y^2 x$. Then the relation

$$x^{1/\gamma_x}/y^{1/\gamma_y} = const. = x_0^{1/\gamma_x}/y_0^{1/\gamma_y}$$

holds with the initial values $x_0$ and $y_0$ (where $x_0 + y_0 = C$). Then, consider the following division process: if $x(t) + y(t) = 2C$, then $x \to x/2$ and $y \to y/2$. With the continued temporal evolution and the division process, $y$ approaches 0 if $\gamma_y < \gamma_x$. (Recall that the curve $y(x)$ given by the above equation is concave.)

[11] Roughly speaking, one may rephrase these standpoints as egg first or chicken first.

not solved how genetic information took over. The minority control gives a new look to these problems.

The first problem was the appearance of parasitic molecules to destroy the hypercycle, that is, mutually catalytic reaction cycle. If only the replication process of molecules is concerned, it is not so easy to resolve the problem. Here we consider the dual level of replication, that is, molecular and cellular replication.

In the theory for the origin of information we have presented here, existence of a cell, unit that reproduces itself is required. Two levels of reproduction, both molecules and cells, are assumed. Hence a cell with parasitic molecules cannot grow, and is selected out. Relevance of this type of two-level reproduction to avoid molecular parasites has been discussed (Maynard-Smith, 1979; Szathmary & Demeter, 1987; Eigen, 1992; Szathmary & Maynard-Smith, 1997; Boerilst & Hogeweg, 1991). Here, importance of cellular compartment to the *origin of genetic information* is more important.

This two-level selection works effectively, with the aid of minority control of specific molecules for a cell. Indeed, surviving cells satisfy the minority control. With the selection pressure for reproduction of cells, there appears a state that is not expected by the rate equation for reaction of molecules, where the number of inactive $Y$ molecules that are parasitic to the catalytic reaction is suppressed. Furthermore, resistance against parasitic (inactive) $Y$ molecules is established by this MCS.

This minority control also resolves the question of the genetic takeover, the problem in the standpoint of metabolism first. Among several molecules, a specific minority molecule species controls the behavior of a cell and is well preserved. The possible scenario mentioned in the beginning of this section gives one plausible scenario how genetic take-over progresses from this MCS.

The differentiation of role between the two molecules looks like "symmetry breaking." In physics, how initially symmetric states become unstable to break the symmetry has been studied in depth. When initially two (or more) states are equally possible, and later only one of them is selected, it is said that the symmetry is broken. Indeed, in Chap. 7, we discuss cell differentiation process as a kind of symmetry breaking.

In the differentiation of roles of molecules, however, the molecules initially have different characters as to the replication speed. Here a difference in one character (that is, the replication speed) is "transformed" into the difference in the control behavior, and difference as the carrier of heredity. In other words, a characteristics with already broken symmetry is transformed into a different type of symmetry breaking. This kind of transformation of one character's difference to another is often seen in biology. Indeed we will see such examples in Chaps. 9 and 11. The result of this chapter is a simple example for it.

# 5

# Origin of a Cell with Recursive Growth

## 5.1 Question to Be Addressed

**Question: A cell consists of several replicating molecules that mutually help the synthesis of each other and keep some synchronization for replication. How is such recursive production maintained, while preserving the diversity of chemicals? Furthermore, this recursive production is not completely faithful, and there appears a slow "mutational" change over generations, which leads to evolution. How is evolvability possible in spite of recursive production?**

In the discussion of Chap. 4, we considered a system consisting of two kinds of molecules. In a cell, however, a variety of chemicals form a complex reaction network to synthesize themselves each other. It is natural to wonder how such a cell with a huge number of components and complex reaction network can sustain reproduction, keeping similar chemical compositions. This "protocell" must produce catalytic molecules that sustain the metabolism, while for the growth of the cell, it has to synthesize membrane molecules (Ganti, 1975). Several reaction processes have to proceed with some balance, to sustain reproduction.

On the other hand, the total number of molecules in a cell is limited. If there are a huge number of chemical species that catalyze each other, the number of some molecules species may go to zero. Then molecules that are catalyzed by them no longer are synthesized. Then, other molecules that are catalyzed by them cannot be synthesized, either. In this manner, the chemical compositions may vary drastically, and the cell may lose its reproduction activity.

Of course, the cell state is not constant, and a cell may not divide for ever. Still, a cell state is preserved to some degree to keep producing similar offspring cells. We call "recursive production" or "recursiveness" the set of such conditions necessary for the reproduction of a cell. The question we address here is whether the distribution of chemicals or the structure of reaction network must fulfill specific conditions.

In a cell, the number of each molecule changes in time through reaction and must increase on average to allow cell replication. As mentioned in Chap. 3, a positive feedback process underlying the replication process may lead to large fluctuations in the molecule numbers. With such large fluctuations and complexity in the reaction network, how is recursive production of cells sustained?

## 5.2 Logic

Let us start by discussing what conditions must be satisfied by the biochemical processes underlying cell growth. In a cell, there is a complex catalytic network for the synthesis of molecules. These molecules are spatially distributed, and in some cases such spatial arrangement is crucial to determine cellular function, while for some others, the discussion on just the composition of chemicals in a cell is sufficient to determine a state of a cell, since they are not rigidly fixed in space but can move around randomly to some degree.[1] Hence, as a first approximation we ignore the spatial structure within a cell, and consider just the composition of chemicals.

To study the general features of a system with mutually catalyzing molecules, it is important to consider a system with a variety of chemicals ($k$ molecule species), forming a mutually catalyzing network (Eigen & Schuster, 1979; Eigen et al., 1989; Stadler et al., 1993; Kauffman, 1986, 1993). The molecules replicate through catalytic reactions so that their numbers within a cell increase. However, as already discussed in Sect. 4.1 parasitic molecules may disrupt the growth process: Those that are catalyzed but do not catalyze others may increase their number, and the cell may ultimately lose the ability of growth.

A minority-control mechanism discussed in Chap. 4 is one solution, while in a network of chemical reactions, there is a possibility that an ensemble of molecule species that catalyze each other may resist to the invasion of parasitic molecules. Even if one molecule species is attacked by some parasite molecule, when there are several paths that catalyze the attacked molecule, the decrease of this molecule may be compensated by the synthesis from other paths. Hence some complex network can be more robust against the parasite attack. As will be shown numerically, intermingled hypercycle network is thus robust against the appearance of parasitic molecules by replication errors. The presence of a complex network can promote recursive growth.

At this stage, however, we need to seriously consider the fluctuation of the number of molecules. First consider a possible range of the number of total molecules in a cell. If the number is too small, the fluctuations in each

[1] When modeling spatial inhomogeneity, a reaction–diffusion equation is used. When the concentration of some molecule is low, the molecules exist in a discrete manner in space, in which case amplification of some molecule concentration occurs (Shnerb et al., 2000; Togashi & Kaneko, 2001, 2003, 2004, 2005).

molecule number are so large that there will be no possibility to keep the recursive growth. If the number is too large, the dynamics will be basically represented by a continuum equation. Then it is often difficult to have a higher rate of replication, or even if it is found, the chemical compositions are completely fixed in the beginning, and no evolution may be possible. The appropriate range for the number of molecules and the number of types of molecules must be determined.

As long as the total number of molecules is not very large, fluctuation in each molecule concentration is inevitable. Of course, negative feedback process to stabilize the concentration is one possible solution, to reduce the fluctuations. For reproduction of molecules, however, some positive feedback process is needed to amplify the number of each molecule species. Such positive feedback process leads to autocatalytic process to synthesize each molecule species.

To increase the growth speed, it is better to strengthen this amplification rate, which, however, may also amplify the fluctuations. We need to study some general features of the fluctuations inherent in such positive feedback system. As a very simple illustration, let us consider a process where a molecule $x_m$ is replicated with the aid of other catalytic molecules.

Then, the growth of the number $N(m)$ of the molecule species $x_m$ is given by $dN(m)/dt = AN(m)$. Here $A$ involves the rate of several reaction processes to synthesize the molecule $x_m$. Such synthetic reaction process depends on the number of the molecules involved in the catalytic process. Recall, however, that all chemical reaction processes are inevitably accompanied with fluctuations arising from stochastic collision of chemicals. Thus, although the reactions to synthesize a specific chemical and convert it to other chemicals is balanced in a steady state, the fluctuation terms remain. Accordingly, the above rate $A$ has fluctuations $\eta(t)$ around its temporal average $\overline{a}$. Then the above rate equation is written as $dN(m)/dt = N(m)(\overline{a} + \eta(t))$, one gets

$$dlogN(m)/dt = \overline{a} + \eta(t) \ . \tag{5.1}$$

In other words, the logarithm of chemical abundances shows Brownian motion around its mean. Then, the logarithm of chemical abundances is expected to obey normal (Gaussian) distribution if $\eta(t)$ is approximated by a Gaussian noise, Accordingly, the logarithm of the number molecules is suggested to obey normal (Gaussian) distribution. Such a distribution is known as log-normal distribution.

In general, each term of reaction process may include multiplicative stochastic contribution written as $\eta(t)x_m$. Hence, the log-normal distribution, rather than Gaussian distribution, may be common in a cell that reproduces itself recursively.

In contrast to the Gaussian distribution, the log-normal distribution is characterized by a longer tail for large abundances, if plotted in the original, normal scale without taking logarithm. This large fluctuation in number may destroy the recursive growth. Is there some control mechanism to suppress some fluctuations there? Note that in a synthesis of molecules with mutual catalysis, there can appear some key-stone chemicals, as expected from the minority-control mechanism. These key-stone molecules work as a controlling part. As discussed in Sect. 4.4, this key-stone molecule has a higher connection with others in reaction network. The increase of the number of such molecules by fluctuations results in a drastic change in the number of other molecules, especially the decrease of molecules that catalyze this key molecule, since the total number of molecules in a cell is limited. Then the key molecule is not synthesized so much. In other words, increase of the key-stone molecule later brings about its decrease. Thus, the dynamics of this key-stone molecule species has a negative feedback leading to the stabilization of its concentration. Accordingly, some key-stone molecules have relatively smaller fluctuations, in contrast to the large fluctuations of other molecules whose abundances follow a log-normal distribution. With the existence of such key-stone molecule species, the fluctuations will be suppressed, and recursive production of a cell is stabilized. Furthermore, with a combination of such key-stone chemicals, information on recursive growth is generated, which works as a controller for synchronized growth. A specific example of this mechanism will be discussed in Sect. 5.4.

## 5.3 Model

### 5.3.1 Modeling Strategy for the Chemical Reaction Networks

Now we introduce our standpoint when modeling a cell on the basis of the discussion of the last section. What type of model of a cell is best suited to provide an answer to the question in Sect. 5.1? With all the current biochemical knowledge, we can say that one could write down several types of models. Because of the complexity of a cell, there is a tendency of building a complicated model when trying to capture the essence of a cell. However, doing so only makes it more difficult to extract new concepts, although simulation of the model may produce phenomena similar to those in living cells. Therefore, to avoid such pitfalls, it may be more appropriate to start with a simple model that encompasses only the essential properties of living cells. Simple models may not produce all the observed natural phenomena, but are comprehensive enough to bring us new thoughts on the course of events taken in nature.

In setting up a theoretical model here, we do not impose many conditions to imitate the life process. Rather we wish to introduce as few postulates as possible and study the universal properties of the resulting system. For example, as a minimal condition for a cell, we consider a system consisting of chemicals separated by a membrane. The chemicals are synthesized through catalytic reactions, and accordingly the amount of chemicals increases, including the membrane component. As the volume of this system becomes larger, the surface tension for the membrane can no longer sustain the system, and it

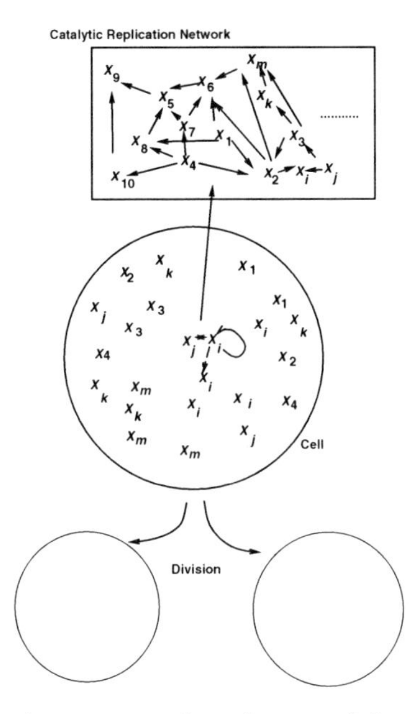


**Fig. 5.1.** Schematic representation of our modeling strategy of a cell

will divide.[2] Under such minimum setup as will be discussed later, we study the condition for the recursive growth of a cell, as well as for the differentiation of the cell.

As discussed, we assume a chemical reaction network in a cell, while spatial arrangement of molecules is discarded. As discussed above, a cell is modeled as a chemical reaction network, where the spatial arrangement of molecules is neglected. Hence, if there are $k$ chemical species in a cell, the cell state is characterized by the number of molecules of each species as $N_1, N_2, \dots N_k$. These molecules change their number through reaction among these molecules each other. Since most reactions are catalyzed by some other molecules, the reaction dynamics consist of a catalytic reaction network.

Through the membrane, some chemicals may flow in, which are successively transformed to other chemicals through this catalytic reaction network. For a cell to grow recursively, a set of chemicals has to be synthesized for the next generation. As the number of molecules is large enough, the membrane is no longer sustained, even just due to the constraint of surface tension. Then, when the number of molecules is larger than some value, the cell is expected to divide. The basic picture for a simple toy model of a cell is shown in Fig. 5.1.

[2] After the division of this protocell systems, the daughter cells may interact with each other, since they share resource chemicals. Interaction among these dividing cells will be discussed in Chap. 7.

Of course, it is impossible to include all possible chemicals in a cell. As our constructive biology is aimed at neither making complicated realistic model for a cell nor imitating specific cellular function, we set up a minimal model with reaction network, to answer the questions raised in Sect. 5.1. Now, there are several levels for the modeling depending on what question we try to answer.

(0) By taking reversible two-body reactions, including all levels of reactions, ranging from metabolites, proteins, nucleic acids, and so forth. Such level of modeling is, for instance, desirable when investigating how nonequilibrium conditions may be sustained in a cell (Awazu & Kaneko, 2004).

(1) Assuming that some reaction process are fast, fast reaction processes can be adiabatically eliminated.[3] Most fast reversible reactions can also be eliminated by assuming that they are already balanced. Then we need to include only molecule species whose concentration (number) changes relatively slowly. For example by assuming that an enzyme is synthesized and decomposed fast, its concentration can be eliminated, to give an effective catalytic reaction network dynamics consisting of reactions such as

$$X_i + X_j \to X_\ell + X_j \tag{5.2}$$

where $X_j$ catalyzes the reaction. The study of catalytic reaction networks in relationship with the origin of life was pioneered by Kauffman (Farmer et al., 1986; Kauffman, 1986, 1993), and then adopted for the study of intracellular reaction dynamics for a replicating cell in (Kaneko & Yomo, 1994; Furusawa & Kaneko, 2003). If the catalysis progresses through several steps, this process is replaced by

$$X_i + mX_j \to X_\ell + mX_j \tag{5.3}$$

leading to higher order catalysis (Furusawa and Kaneko, 1998).

For a cell to grow, some resource chemicals must be supplied through the membrane. Through the above catalytic reaction network, the resource chemicals are transformed to others, and as a result, the cell grows (see also Ganti, 1975). Indeed, this class of model will be adopted to study the condition for cell growth, to unveil universal statistics for such cells, and also as a model for cell differentiation, in Chaps. 7–9.

(2) Model focusing on the dynamics of replicating units (e.g., Hypercycle): For a cell to grow effectively, there should be some positive feedback process to amplify the number of each molecule species. Such positive feedback process leads to autocatalytic process to synthesize each molecule species. For reproduction of a cell, (almost) all molecule species are somehow synthesized.

---

[3] When there are variables of fast change and slow change, the fast variables relax rapidly to an equilibrium state given by slow variables. By solving a stationary solution for fast variables, they are eliminated so that the equations only for slow variables are derived. See, e.g., Haken, 1979.

Then, it would be possible to take a replication reaction from the beginning as a model. For example, consider a reaction

$$S + X + Y \to X' + Y : S' + X' \to 2X \ .$$

Then as a total, the reaction is represented as $S + S' + X + Y \to 2X + Y$. Assuming the resources $S$ and $S'$ are constantly supplied, we can consider the replication reaction

$$X + Y \to 2X + Y, \tag{5.4}$$

catalyzed by $Y$. At this level, we can take a unit of replicator and consider a replication reaction network. This type of model was first discussed in the hypercycle by Eigen and Schuster, and was adopted in Chap. 4.

(3) coarse-grained (phenomenological) level: Some other reduced model is adopted for the study of gene expression or signal transduction network. The modeling at this level is relevant to understand specific function of a cell.

In this present chapter we use the modeling of the level (2). This class of model can be obtained by reducing from the level (1) model, by restricting our interest only to take into account of replicating units. In this sense, the model is a bit simpler than the level (1) model. On the other hand, it may not be suitable to discuss the condition for cell growth, since at the level (2) model, the supply of resource chemicals is automatically assumed, and one cannot discuss how transported chemicals are transformed into others. Study of the level (1) model will be given in the next chapter.

### 5.3.2 Specific Model

We envision a (proto)cell containing $k$ molecular species with some of the species possibly having a zero population. Following the discussion in the last subsection, a chemical species is assumed to catalyze the synthesis of some other chemical species as

$$[i] + [j] \to [i] + 2[j] \ , \tag{5.5}$$

with $i, j = 1, \ldots, k$, according to a randomly chosen reaction network. (The reaction is set at far-from-equilibrium) In (5.5), the molecule $i$ works as a catalyst for the synthesis of the molecule $j$, while the reverse reaction is neglected as discussed in the hypercycle model. For each chemical the rate for the path of catalytic reaction (5.1) is given by $\rho$, that is, each species has about $k\rho$ possible reactions.

Furthermore we assume that catalytic activity for the above reaction (5.5) depends on each molecule $i$, and this activity $c_i$ is given by a random number over [0,1] (Once it is chosen, it is fixed). Assuming an environment with an ample supply of chemicals available to the cell, the molecules then replicate leading to an increase in their numbers within a cell.

During the replication process, structural changes, for example, the alternation of a sequence in a polymer, may occur, which alter the catalytic activities of the molecules. The rate of such structural changes is given by the replication "error rate" $\mu$. As a simplest case, we assume that this "error" leads to all other molecule species with equal probability (i.e., with the rate $\mu/(k-1)$), and could thus regard it as a background fluctuation. In reality, of course, even after a structural change, the replicated molecule will keep some similarity with the original molecule, and this equal rate of transition to other molecule species is a drastic simplification. Some simulations where the errors in replication lead to only a limited range of molecule species, however, show that the simplification does not affect the basic conclusions presented here. Hence we use the simplest case for most simulations.

The model is simulated as follows: At each step, a pair of molecules, say, $i$ and $j$, is chosen randomly. If there is a reaction path between species $i$ and $j$, and $i$ ($j$) catalyzes $j$ ($i$), one molecule of the species $j$ ($i$) is added with probability $c_i$ ($c_j$), respectively. The molecule is then changed to another randomly chosen species with the probability of the replication error rate $\mu$. When the total number of molecules exceeds a given threshold (denoted as $N$), the cell divides into two such that each daughter cell inherits half ($N/2$) of the molecules of the mother cell, chosen randomly. To take the importance of the discreteness in the molecule numbers into account, we adopted a stochastic rather than the usual differential equations approach (see also Segreet et al., 2000).

## 5.4 Result

In our model there are four basic parameters: the total number of molecules $N$, the total number of molecule species $k$, the mutation rate $\mu$, and the reaction path rate $\rho$. By carrying out simulations of this model, choosing a variety of parameter values $N, k, \mu, \rho$, also by taking various random networks, we have found that the behaviors are classified into the following three types:

(1) Fast switching states without recursiveness
(2) Achievement of recursive production with similar chemical compositions
(3) Switch over several quasi-recursive states

In the first phase, there is no clear recursive production and the dominant molecule species changes frequently. At one time step, some chemical species are dominant but only a few generations later, this information is lost, and the number of the molecules in these species go to zero (see Fig. 5.2a). No stable set of catalytic networks is formed. Here, the time required for reproduction of a cell is much larger than in the second case.

In the second phase, on the other hand, a recursive production state (or recursive state in short) is established, and the chemical composition is stabilized such that it is not altered much by the division process. Generally, all the observed recursive states consist of 5–10 species, except for those species with one or two molecule numbers, which exist only as a result of replication

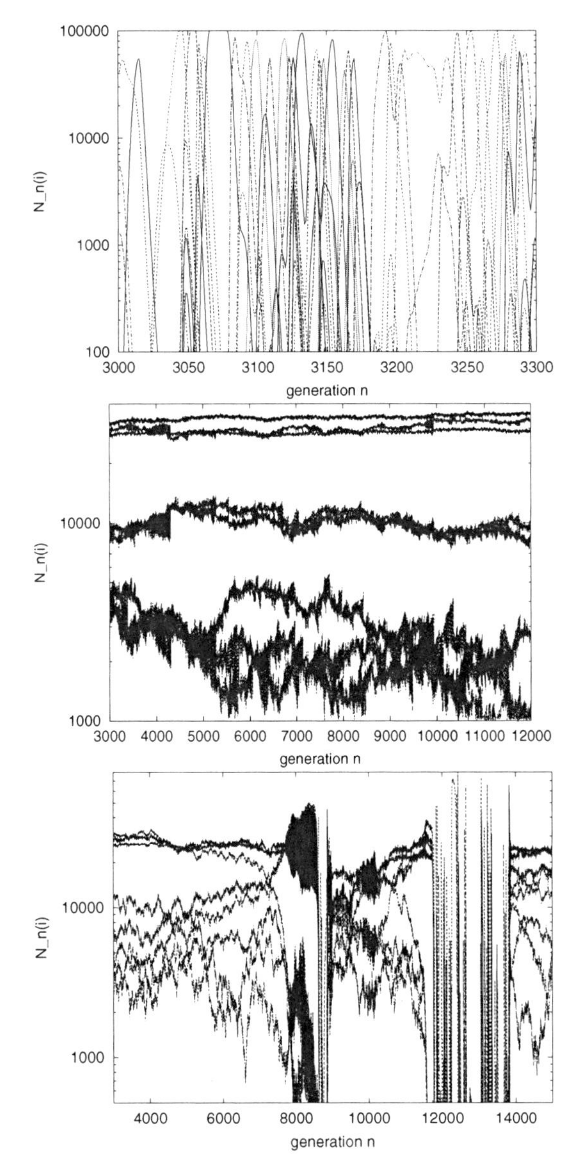


**Fig. 5.2.** The number of molecules $N_n(i)$ for the species $i$ is plotted as a function of generation $n$ of cells, that is, at each successive division event $n$. In (**a**), a random network with $k = 500$ and $p = 0.2$, and in (**b**) that with $k = 200$ and $p = 0.2$, and in (**c**) that with $k = 200$ and $p = 0.2$ was adopted, with $N = 64,000$ and $\mu = 0.01$. Only some species (whose population becomes large enough at some generation) are plotted. In (a), dominant species change successively in generation, while in (b) recursive production with similar chemical compositions is achieved. In (c) three quasi-recursive states are observed with switching

errors (see Fig. 5.2b). These 5–10 chemicals mutually catalyze, by forming a catalytic network as will be discussed later. The set of these 5–10 species does not change between generations, and the chemical compositions are transferred to the offspring cells. Once reached, this state is preserved throughout whole simulations, lasting more than 10,000 generations. Even in this case, the number of each molecule shows relatively large fluctuations, since the total number of molecules $N$ is not large (typically we choose $N \sim (10^2 \sim 10^5)$ in our simulations) [4]

In the third phase, after one recursive state lasts over many generations (typically a thousand generations), a fast switching state appears until a new (quasi-)recursive state appears. As shown in Fig. 5.2c, for example, each (quasi-)recursive state is similar to that in phase (2), but in this case, its lifetime is finite, and it is replaced by the fast switching state as in the phase (1). Then the same or different (quasi-)recursive state is reached again, which lasts until the next switching occurs. In the example of Fig. 5.2(c), around the 12,000th generation, the core network is taken over by parasites to enter the phase (1) like fast switching state, which in turn gives way for a new quasi-recursive state around the 14,000th generation.

### 5.4.1 Dependence of Phases on the Basic Parameters

Although the behavior of the system depends on the choice of the network, there is a general trend with regards to the phase change from (1), to (3), and then to (2) with the increase of $N$, or with the decrease of $k$, as schematically shown in Fig. 5.3. By choosing a variety of networks, however, we find a clear dependence of the fraction of the networks on the parameters, leading to a rough sketch of the phase diagram. Generally, the fraction of (2) increases and the fraction of (1) decreases also with the decrease of $\rho$ or $\mu$. For example, the fraction of (1) (or (3)) gets larger as $k$ is decreased from $k \lesssim 300$ for $N = 50,000$ (with $\rho = 0.1$ and $\mu = 0.01$), while the dependence on $\rho$ will be discussed below.

For a quantitative investigation, it is useful to classify the phases by the similarity of the chemical compositions between two cell division events. To check the similarity, we first define a $k$-dimensional vector $\vec{V_n} = (p_n(1), \ldots, p_n(k))$ with $p_n(i) = N_n(i)/N$. Then, we measure the similarity between $\ell$ successive generations with the help of the inner product as

$$H_\ell = \vec{V_n} \cdot \vec{V_{n+\ell}} \,/(|V_n||V_{n+\ell}|) \tag{5.6}$$

(see Fig. 5.4).

[4] The recursive state observed here is not necessarily a fixed point with regards to the population dynamics of the chemical concentrations. In some case, the chemical concentrations oscillate in time, but the nature of the oscillation is not altered by the process of cell division.

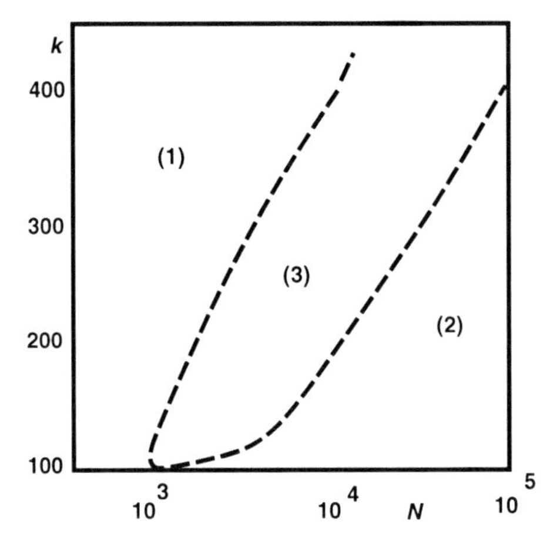


**Fig. 5.3.** Schematic representation of the phase diagram of the three phases, plotted as a function of the total number of molecules $N$, and the total possible number of molecule species $k$

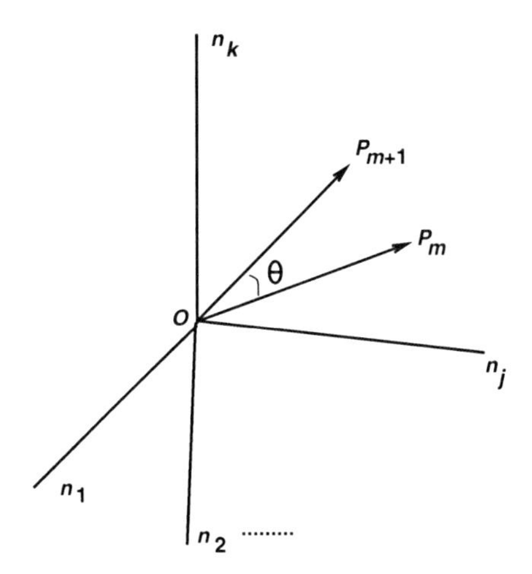


**Fig. 5.4.** Similarity as an inner product of composition vectors

In Fig. 5.5, the average similarity $\overline{H_{20}}$ and the average division time are plotted for 50 randomly chosen reaction networks as a function of the path probability $\rho$. Roughly speaking, the networks with $\overline{H_{20}} > 0.9$ belong to (2), and those with $\overline{H_{20}} < 0.4$ to (1), empirically. Hence, for $\rho > 0.2$, the phase (1) is observed for nearly all the networks (e.g., 48/50), while for lower path rates, the fraction of (2) or (3) increases. The value $\rho \sim 0.2$ gives the phase boundary in this case. Generally speaking, a positive correlation between the growth speed of a cell and the similarity $H$ exists. In Fig. 5.5, the division time is also plotted, where to each point with a high similarity $H$ corresponds to a lower division time. The network with higher similarity (i.e., in the phase

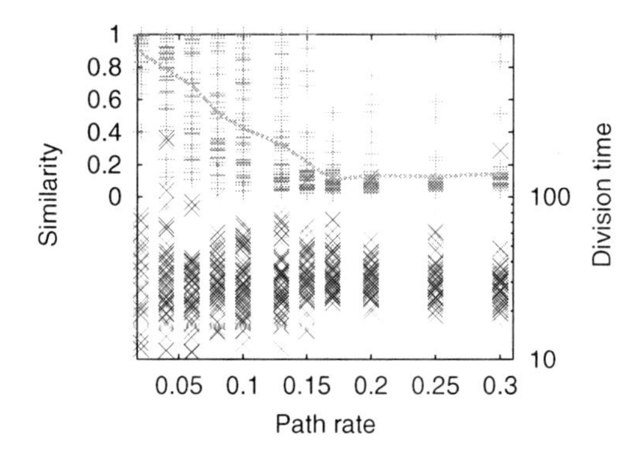


**Fig. 5.5.** The average similarity $\overline{H_{20}}$ (+) and the average division time (×) are plotted as a function of the path rate $\rho$. For each $\rho$, data from 50 randomly chosen networks are plotted. The average is taken over 600 division events. The *dotted* line indicates the average of $\overline{H_{20}}$ over the 50 networks for each $\rho$. For $\rho > 0.2$, networks over 98% have $H < 0.4$, and they show fast switching, while for $\rho = 0.08$, about 95% belong to the phase (2) or (3) At $\rho = 0.02$, 25 out of 50 networks cannot support cell growth, 4 cannot at $\rho = 0.04$ (Adapted from Kaneko, 2005a)

(2)) gives a higher growth speed. Indeed, the recursive states maintain higher growth speeds since they effectively suppress parasitic molecules.

In Fig. 5.5, by decreasing path rates, the variations in the division speeds of the networks become larger, and some networks that reach recursive states have higher division speeds than networks with larger $\rho$. On the other hand, when the path rate is too low, the protocells generally cannot grow since the probability to have mutually catalytic connections in the network is nearly zero. Indeed there seems to exist an optimal path rate (e.g., around 0.05 for $k = 200$, $N = 12,800$ as in Fig. 5.5 for having a network with high growth speeds)(Kaneko, 2005a).

Besides the correlation between the growth speed and similarity, the correlation with the diversity of the molecules also exists. Protocells with higher growth speed and similarity in the phase (2) have higher chemical diversity also. In the phase (1), one (or a very few) molecule species is dominant in the population, while about 10 species have higher population in the phase (2), where the chemical diversity is maintained.

### 5.4.2 Maintenance of Recursive Production

How is the recursive production sustained in the phase (2)? In Sect. 4.1, we have discussed already the danger of parasitic molecules that have lower catalytic activities and are catalyzed by molecules with higher catalytic activities. As discussed there, such parasitic molecules can invade the hypercycle. Indeed, under the structural changes and fluctuations, the recursive production state could be destabilized. To answer the question, we have examined several

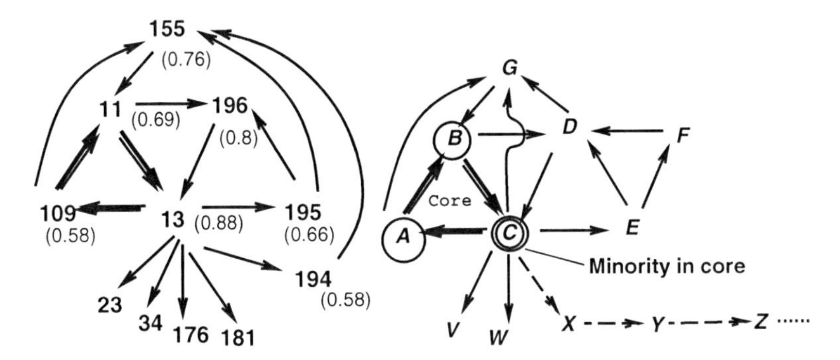


**Fig. 5.6.** The catalytic network of the dominant species that constitutes the recursive state. The catalytic reaction is plotted by an arrow $i \to j$, representing the replication of the species $j$ with the catalytic species $i$. The numbers in () denote $c_i$ of the species. Only species that continue to exist with a population larger than 10 are plotted. (Note many other species can appear thanks to replication errors at each generation.) (**a**): Corresponding to the recursive state of Fig. 5.2a, where the three species connected by thick arrows are the top 3 species in Fig. 5.2b, obtained universally. (**b**) A schematic example of mutually catalytic network in our model. The core network for the recursive state is shown by circles, while parasitic molecules ($X$,$Y$,..) connected by broken arrows, are suppressed in the (quasi-)recursive state

reaction networks. The unveiled logic for the maintenance of a recursive state is summarized as follows.

(a) **Stabilization by intermingled hypercycle network**: The 5–12 species in the recursive state form a mutually catalytic network, for example, as in Fig. 5.6. This network has a core hypercycle network, as shown in thick arrows in Fig. 5.6). As shown in the figure, such core hypercycle has a mutually catalytic relationship, as "$A$ catalyzes $B$, $B$ catalyzes $C$, and $C$ catalyzes $A$." However, they are connected with other hypercycle networks such as $G \to D \to B \to G$, and $D \to C \to E \to D$, and so forth. The hypercycles are intermingled to form a network. Coexistence of a core hypercycle and other attached hypercycles are common to the recursive states we have found in our model. This intermingled hypercycle network (IHN) leads to stability against parasites and fluctuations.

Assume that there appears a parasitic molecule to one species in the member of IHN (say $X$ as a parasite to $C$ in Fig. 5.6b). The species $X$ may decrease the number of the species $C$. If there were only a single hypercycle $A \to B \to C \to A$, the population of all the members $A, B, C$ would be easily decreased by this invasion of parasitic molecules, resulting in the collapse of the hypercycle. In the present case, however, other parts of the network (say, that consisting of $A, B$,$G, D$ in Fig. 5.6b), compensate the decrease of the population of $C$ by the parasite, so that the population of $A$ and $B$ do not decrease so much. Then, through the catalysis of the species

B, the replication of the molecule $C$ progresses so that the population of $C$ is recovered. Hence the complexity in the hypercycle network leads to stability against the attack of parasite molecules.

Next, IHN is also relevant to the stability against fluctuations. It is known that in the population dynamics of a simple hypercycle the population of one (or a few) molecule species may approach 0, and then grow again. For a continuum model, the population density can come arbitrarily close to zero, and then grow again. This type of recurrent dynamical behavior is known as a heteroclinic cycle (Hofbauer & Sigmund, 1988). In a stochastic model, however, the number of the some molecule species may sometimes become exactly zero, because of fluctuations. Once this species becomes extinct, its recovery by replication error would be hopeless. Hence, to achieve stability against fluctuations, oscillations in which some of the molecule numbers become very low should be avoided. Indeed, by forming an IHN, such oscillatory instability is avoided or reduced. Thanks to the coexistence of several hypercycle processes, instability in each hypercycle cancels out, leading to fixed-point dynamics or to oscillation with a smaller amplitude. Thus the danger that the population of some molecules in the hypercycle goes to zero because of fluctuations is reduced. Stabilization obtained from the coexistence of many species is discussed as "homeochaos" (Kaneko & Ikegami, 1992), while the possibility of stable reproduction in reaction network is also discussed in (Ikegami & Hashimoto, 1996).

(b) **Minority in the core hypercycle**: Now we study more closely the population dynamics of a core hypercycle. Here, the number of molecules $N_j$ of molecule species $j$ is in the inverse order of their catalytic activity $c_j$, that is, $N_A > N_B > N_C$ for $c_A < c_B < c_C$. Because a molecule with higher catalytic activity helps the synthesis of others more, this inverse relationship is expected. Here, the $C$ molecule is catalyzed by a molecule species with higher activities but larger populations ($A$). Hence, the parasitic molecule species cannot easily invade to destroy this mutually catalytic network. Since the minority molecule ($C$) is catalyzed by the majority molecule ($A$) (with the aid of another molecule ($B$)), a large fluctuation in molecule numbers is required to destroy this network.

The stability in the minority molecule is also accelerated by the complexity in IHN. If the catalytic activity of $C$ is highest, the recursive state here is mainly achieved by catalysis of the molecule $C$. On the other hand, this also implies that $C$ is the minority in the core network. (The population of the molecule $C$ is usually larger than $D$, $E$, etc. in Fig. 5.6b, though.) Hence the attack to $C$ molecule is most relevant to destroy this recursive state. In the IHN, this minority molecule species is involved in several hypercycles as in $C$ in Fig. 5.6b. On the one hand, this confirms the prediction made in Sect. 4.5, according to which more species are catalyzed by the minority molecules. On the other hand, this also leads to the suppression of fluctuations in the number of minority molecules, as will be discussed in Sect. 5.4.4. With the decrease of the fluctuation, the probability that the minority molecules becomes extinct is reduced so that the recursive state is robust.

### 5.4.3 Switching

Next, we discuss the mechanism of switching. In the phase (3), the recursive production state is destabilized, when the number of parasitic molecules increases. For example, the number of the molecule $C$ may decrease because of fluctuations, while the number of some parasitic molecules ($X$) that are not originally in the catalytic network but are catalyzed by $C$, may increase. The frequency of such a fluctuation is larger when the total population of molecules in a cell is smaller. If such a fluctuation appears, other molecule species in the original network successively lose the main source of molecules that catalyze their synthesis. Then the new parasitic molecule $X$ occupies a large portion of populations. However, the molecule's main catalyst ($C$) soon disappears, the synthesis of $X$ is stopped, and this species $X$ is taken over by some molecules $Y$ that are catalyzed by $X$. Then, within a few generations, dominant species changes, and recursive production does not continue. Indeed, this is what occurred in the phase (1). Then the parasitic molecule $X$ is taken over by some other $Y$. This take-over by parasites continues successively, until a new (or same) recursive state with hypercycle network is formed. Hence the fluctuation in the minority molecule in the core network is relevant to the switching process.

As discussed in Sect. 3.6, switching over several quasi-stationary states has been also studied as chaotic itinerancy (Kaneko, 1990, 1994a; Tsuda, 1991ab; Kaneko & Tsuda, 2003), while Jain and Krishna (2002) analyzed transitions over autocatalytic sets in a linear reaction network, by introducing the mutation of the network only after the temporal evolution of growth dynamics reaches a stationary state.

### 5.4.4 Fluctuations

Finally, we investigate the fluctuations of the molecule numbers of each of the species. Since the number of molecules is not very large, the fluctuations over the generations can possibly have a significant impact on the dynamics of the system. To quantify the sizes of these fluctuations, we have measured the distribution $P(N_i)$ for each molecule species $i$, by sampling over division events. Our numerical results are summarized as follows:

(I) For the fast switching states, the distribution $P(N_i)$ satisfies the power law $P(N_i) \approx N_i^{-\alpha}$, with $\alpha \approx (1 \sim 2)$.[5]

[5] The exponent $\alpha$ depends on the parameter and species [Kaneko, 2005a].

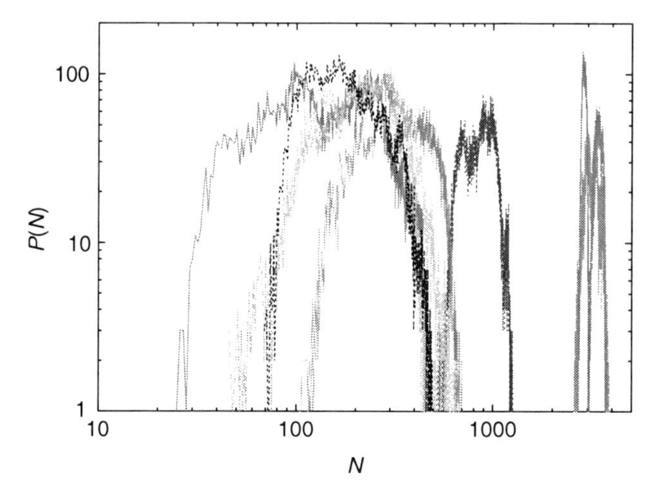


**Fig. 5.7.** The number distribution of the molecules corresponding to the network in Fig. 5.2. The distribution is sampled from 1000 division events. From right to left, the plotted species are 11,109,13,155,176,181,195,196,23 of Fig. 5.6a. Log–Log plot

(II) For recursive states, the fluctuations in the core network (i.e., 13,11,109 in Fig. 2) are typically small. On the other hand, species that are peripheral to but catalyzed by the core hypercycle have log-normal distributions $P(N_i) \approx \exp(-\frac{(\log N_i - \overline{\log N_i})^2}{2\sigma})$, as shown in Fig. 5.7. By measuring the variance $\overline{(N_i - \overline{N_i})^2}$ ($\overline{\cdots}$ denotes the average over time), it is shown that the variance in the core network is small, especially for the minority species (i.e., 13) (Kaneko, 2005a).

The origin of the log-normal distributions here can be understood by the argument already given in Sect. 5.1: for a replicating system, the growth of the molecule number $N_m$ of the species $m$ is given by $dN_m/dt = AN_m$, where $A$ is the average effect of all the molecules that catalyze $m$. We can then obtain the estimate $d \log N_m/dt = \overline{a} + \eta(t)$. By replacing $A$ with its temporal average $\overline{a}$ plus fluctuations $\eta(t)$ around it. If $\eta(t)$ is approximated by a Gaussian noise, $P(N_m)$ follows a log-normal distribution (this argument is valid if $\overline{a} > 0$). For the fast-switching state the growth of each molecule species is close to zero on the average and in this case, by considering the Langevin equation with boundary conditions, the power law is obtained as discussed in (Kaneko, 2005a).

By studying a variety of networks, the observed distributions of the molecule numbers can be generally summarized as (1) Distribution close to Gaussian form, with relatively small variances in the core (hypercycle) of the network such as (A,B,C) of Fig. 5.6b. (2) Distribution close to log-normal, with larger fluctuations for a peripheral part of the network with others; (3) Power-law distributions for parasitic molecules that appear intermittently.

## 5.5 Experiment

Let us discuss the synthesis of a cell that reproduces itself. To construct such a cell, we need the following steps:

(1) a system consisting of chemicals (polymers) with some catalytic activity, which reproduces itself as a set, even though the reproduction may not be precise.
(2) a membrane that splits the cell's internal system from outside grows through chemical reactions, so that it divides when its size is large enough.
(3) within the membrane the reaction system (1) works, while the synthesis of the membrane is coupled to this internal reaction system.
(4) the internal reaction process and the synthesis of membrane work in some synchrony, so that a system with membrane and internal chemicals is reproduced recursively.

Here we choose liposome (vesicle) as a "membrane" for a self-replicating cell, while the constituents include protein (enzyme) and nucleotides (RNA and DNA). If the synthesis of membrane and other internal reactions (that are also involved for the former) sustain some synchrony, we can construct a protocell (or protolife) artificially.

The experimental study at the step (1) is already discussed in Chap. 4. For the step (2), replication of liposome has been successful, while for the step (3), transcription of RNA, and protein synthesis in a liposome have been achieved separately. Here we discuss ongoing experiments on steps (2) and (3).

### 5.5.1 Replicating Liposome

All existing cells are surrounded by a lipid bilayer. Their in- and out-side are separated by this membrane, within which the membrane catalytic reactions progress. The synthesis of such a cell structure is an important problem (Bachman et al., 1992; Szostak et al., 2001; Hanczyc et al., 2003).

In general, oil molecules have a part that avoids the contact with water molecules. Hence they often form a bilayer membrane with this hydrophobic part inside the membrane. This membrane often forms a closed spherical structure. A closed unit with lipid-bilayer structure is called a liposome or a vesicle.[6] If the constituent molecules are supplied, the membrane surface increases. Because of the balance with the surface tension, this growth cannot continue forever, and a large membrane becomes unstable. In some case, this results in division of the vesicle. Indeed Luisi and others succeeded in observing such division process of liposomes (Bachmann et al., 1992).

Quite recently, Sugawara and collaborators succeeded in synthesizing a system with stable replication of liposome that recursively replicates over several

[6] The terms *vesicle* and *liposome* are used interchangeably, mostly according to the field of research.

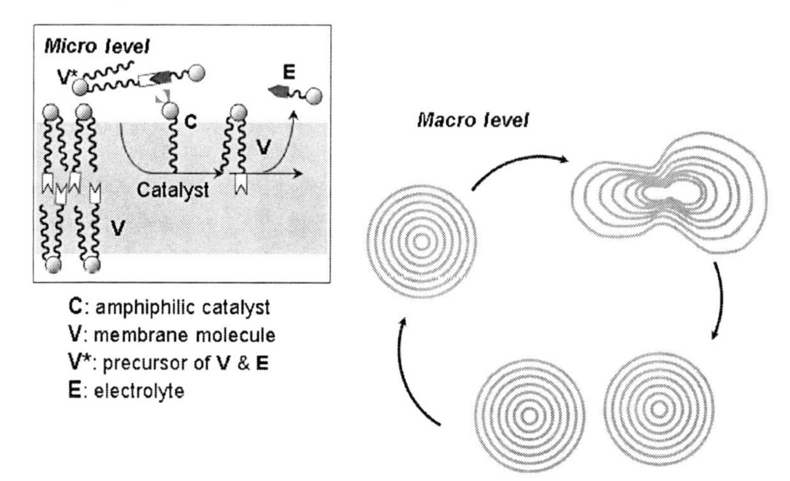


**Fig. 5.8.** Schematic representation of the replication process of a liposome. Drawings reproduced under the courtesy of Takakura, Shoda, Toyota, and Sugawara

generations (Takakura et al., 2003; Takakura & Sugawara, 2004). Schematic illustration of their self-replicating membrane system is shown in Fig. 5.8. The original liposome is composed of amphiphile V. Now for the growth of liposome, some nutrients in the solution have to be transformed into the amphihile V. Here, a precursor molecule $V'$ is put in the solution, which flows into the liposome, while within the liposome, an enzyme E to transform $V'$ to V exists. After the synthesis of V within the liposome, it fuses into the membrane. With this process the liposome increases its size, as displayed in Fig. 5.9. As it grows, there is "budding" of liposome as shown, and when the budded part is sufficiently large, the liposome splits. This growth process continues in the daughter liposome so that the next generation of liposome splits from it. As shown in Fig. 5.9, the process from mother to daughter and then to granddaughter liposome is observed, while the succession of this process is statistically confirmed by using the flow cytometry.

### 5.5.2 Protein Synthesis within a Liposome

The second necessary step to construct an artificial replication cell is to achieve synthesis of nucleic acids (such as RNA or DNA) as well as proteins within the liposome. This is a difficult task, since the environment within the liposome is very much "oily."

Quite recently the process of DNA→mRNA and separately the process of mRNA→ protein were realized by Yomo's group as schematically shown in Fig. 5.10 (Yu et al., 2001). (See also (Noireaux and Libchaber, 2004) for cell-free expression system encapsulated in a liposome). Within a liposome of diameter 1 $\mu m$ there is an autonomous replicating system consisting of RNA that produces its RNA syntheses. Here RNA and the enzyme mutually

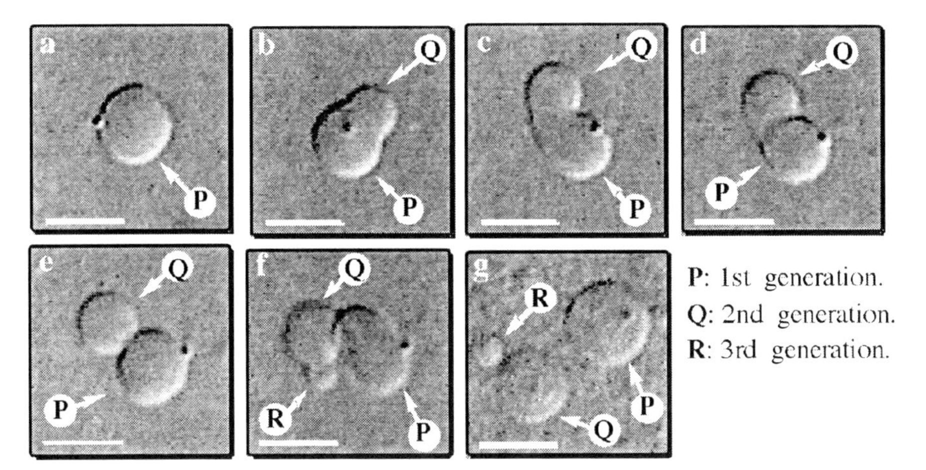


**Fig. 5.9.** Division process of liposome measured by differential interference contrast optical microscopy. Morphological changes in giant vesicle (liposome) are displayed. The white bars show 10 μm. For details of the experimental condition, see Takakura and Sugawara (2004). Reproduced with the courtesy of Takakura and Sugawara

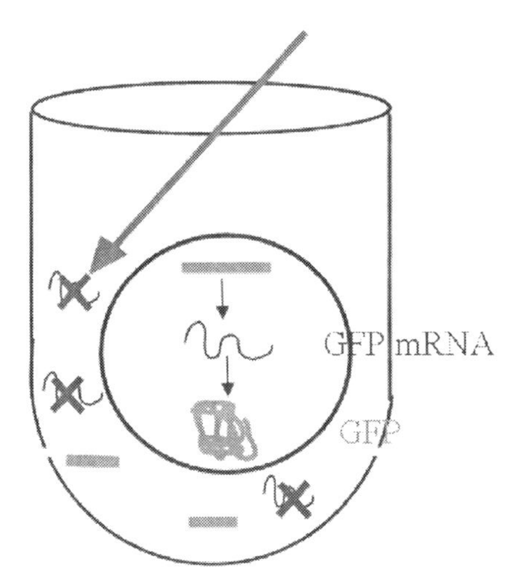


**Fig. 5.10.** Schematic representation of RNA and protein synthesis in a liposome. Drawings by Sato et al., (reproduced under their courtesy). For details see (Sato et al. 2005)

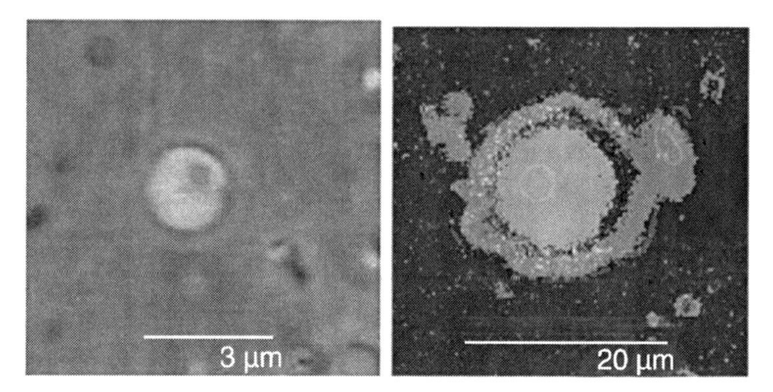


**Fig. 5.11.** Synthesis of fluorescent proteins in a liposome, detected by fluorescence microscopy. The left figure shows the existence of liposome. The figure on the left-hand side shows a liposome. The figure on the right-hand side is an image by fluorescence microscopy, where fluorescent proteins are clearly detected within the liposome. Based on the result by Sato et al. (2005). Reproduced by the courtesy of Sato et al.

synthesize each other as in the experiment of Sect. 4.5. This was possible by modifying genes so that a higher rate of synthesis of enzymes (that catalyze the RNA synthesis) is possible.

Furthermore, Sunami et al. (2005) have also succeeded in synthesizing proteins from the mRNA. By making the corresponding protein fluorescent, they have confirmed that the synthesis indeed occurs within the liposome by the measurement of fluorescence as shown in Fig. 5.11. This rate of protein synthesis is so high that the fluorescence of protein is measured by flow cytometry. In Fig. 5.12, we show the two-dimensional distribution of the size of liposome (longitudinal axis) and fluorescence intensity (horizontal axis). The liposomes at the left region correspond to those that could not synthesize proteins well. Hence, by selecting liposomes by using the cell sorter, one can select "good" RNA to have a higher protein synthesis rate within the liposome. Now the evolution to a better RNA is possible. (Sunami et al., 2005; Sato et al., 2005).

This selection process can also yield a better "oil" molecule for the liposome. Then evolution will ensure that the liposome is well suited to the protein synthesis within. By repeating the same process the third step (of Sect. 5.1) leading to the realization of an artificial cell will be achieved.

Since the protein synthesis efficiency in liposomes in this system has reached $10^3$ molecules per gene, which is comparable to that in a living cell, more complex genetic reactions can occur. Indeed they have succeeded in carrying out a cascading genetic network in liposomes (Ishikawa et al., 2004). Indeed, they have encapsulated a plasmid encoding the gene network "SP6 promoter" $\rightarrow$ "T7 RNA polymerase" $\rightarrow$ "T7 promoter" $\rightarrow$ GFP gene into liposomes, to detect the synthesis of GFP. As long as the networks are not

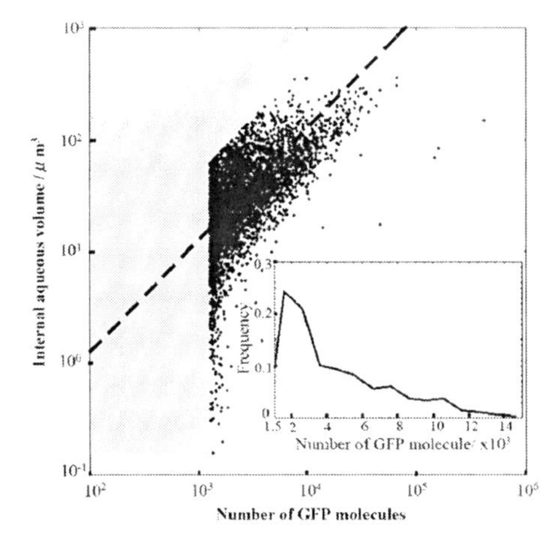


**Fig. 5.12.** The distribution of fluorescence (corresponding roughly to the abundance of the GFP protein) versus the size of the liposome, obtained from the same experiment as in Fig. 5.11. The number of GFP molecules synthesized and internal aqueous volume for each liposome are estimated by flow cytometry (*dots*). The internal aqueous volume is estimated on the basis of the red fluorescent protein, which was added to the solution before encapsulation. Gray dots represent background noise, which is detected from protein synthesis in liposome without GFP gene. The inset shows the distribution of GFP molecules among liposomes for the internal aqueous volume from 50 $\mu m^3$ to 100 $\mu m^3$. Reproduced from Sunami et al. (2005), under the courtesy of the authors

too complex, genetic networks can proceed with the current level of protein synthesis.

Finally, we discuss the distribution of protein abundances. Since the fluorescent protein is synthesized with the aid of other molecules (such as RNA), the fluctuations should be large, and the distribution will be close to log normal, if the argument of Sect. 5.2 is valid. The distribution obtained from the experiment is given in Fig. 5.12. Although it is not easy to confirm if the size distribution is log normal as yet, at least it is sure that the distribution is far from Gaussian, and has a large tail for the abundant size. Indeed by taking the logarithmic scale for the abundance, the distribution is close to be symmetric, which suggests a log-normal-type distribution.

The above result was obtained for protein abundances within a liposome synthesized artificially. In a living cell, do fluctuations obey a Gaussian distribution? Or, is the distribution still log normal? Indeed the data suggest the latter answer, as we shall discuss in the next chapter.

### 5.5.3 For the Synthesis of Artificial Replicating Cell

The last step leading to the synthesis of an artificial replicating cell is to combine the replication of liposome with that of the proteins it contains. Here it is important to balance the two processes to obtain recursive production. If the replication of the membrane is faster, then the inside ingredients will be sparse, while if the protein synthesis is faster the density of molecules will be too high to destroy the liposome. To achieve this balanced replication, some link between the liposome growth and the internal reaction will be required. Unfortunately, this step has not yet been completed.

One might think that there will be a long way to achieve the autonomous replication of artificial cell. However, once a minimal replication process with any loose reproduction is realized, then such "cell" can be an object for Darwinian selection process. By selecting a relatively "better" reproducing cell through cell sorter, a more reliable replication system will be obtained.

## 5.6 Relevance to Biology

Understanding how recursive production of cells arises is important not only with respect to the question of the origin of life, but also in order to understand the characteristics of existing cells. This second aspect will be discussed in the next chapter (Sect. 6.5).

Another important question is with regards to that of evolvability. How does the primitive cell with loose recursive production evolve into the present finely tuned cell? Is the minority control important as discussed in Chap. 4? We have recently carried out several simulations to discuss the evolution of the network (Kaneko, 2005a). For example, we set the catalytic activity as $c(i) = i/k$ so that the activity is monotonically increasing with the chemical species index, and then instead of global change to any molecule species by replication error, we modify the rule so that the change occurs only within a given range $i_0 (\ll k)$, in order to prohibit sudden change to molecules with very high activity and only to allow for gradual changes. In other words, when the molecule species $j$ is synthesized, with the error rate $\mu$, the molecule $j + j'$ with $j'$ a random number over $[-i_0, i_0]$ is synthesized.

The network is fixed and is not changed through the simulation. However, by local change of structural error, the range of species evolves over generations. Here we take species only with $i < i_{\mathrm{ini}}$ in the initial condition, and examine whether the evolution to a network with higher catalytic activities (i.e., with much larger $i$) progresses or not. In other words, we examine whether the indices $i$ in the network increase successively or not. An example of the evolution is shown in Fig. 5.13, where the catalytic activity increases through successively switching from one (quasi-)recursive state (consisting of species within the width of the order $2i_0$) to another.

Here the switching occurs as in phase (3). With the pressure for selection of the protocells, cells with a new (quasi-)recursive state are selected, which consist of molecules with higher catalytic activities (i.e., with larger indices of species). Again each recursive state consists of IHN, and the species with the highest catalytic activity in the core hypercycle has the lowest population

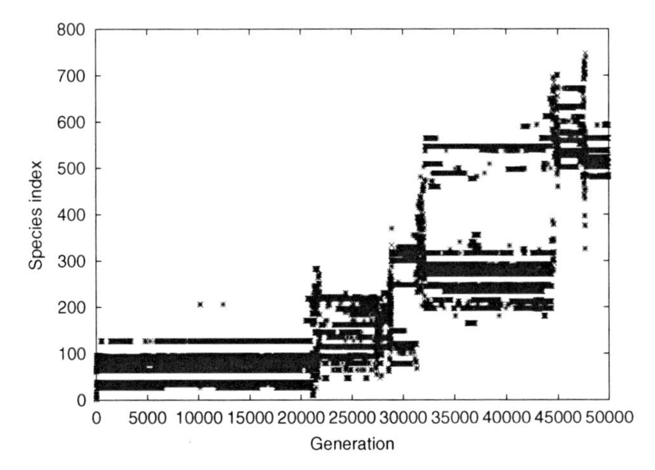


**Fig. 5.13.** Evolution of species in a cell: Those species $i$ with $N(i) > 100$ are plotted with the vertical axis as the species index $i$, and the longitudinal axis as the generation. The total number of species $k$ is 5000, where $c(i)$ is chosen as $c(i) = i/k$, so that it ranges from 0.0002 to 1.0 equally distributed. The number of molecules in a cell is set at 8000, so that the cell divided when the total molecule number is 16,000. The path rate is set at $\rho = 0.1$. The replication error for the species occurs within the range of species $[i - 100, i + 100]$, instead of global selection from all species. Totally there are $M_{\rm tot} = 10$ cells so that one of 10 cells is eliminated when one cell is divided into two

(minority species). Once the population of such species is decreased because of fluctuation, there occurs a switch to a new state that has higher catalytic activities, and the species indices successively increase. Hence, evolution from a rather primitive cell consisting of low catalytic activities to that with higher activities is possible, by taking advantage of minority molecules.

Now the itinerancy over several recursive states leads to evolvability. Here the fluctuation in the minority molecules is a trigger to the instability of the recursive production state. Such relationship between fluctuation and evolvability will be discussed again in Chap. 10, together with the results of laboratory experiments.

Note that this switching cannot occur if the total number of molecules $N$ is small. When the number is too small, the mutation of paths to destroy the recursive state is very unlikely to occur. On the other hand, if the total number of molecules is too large, it is harder to establish a recursive state, because of a larger possibility to change the network. Hence, the realization of both recursive production and evolution require an optimal value of the number of molecules in a protocell. Similarly, it will be important to determine what is the minimal (or optimal) size of cells, or minimal number of chemicals (or genes) to realize both sustainability and evolvability of a cell.

# 6

# Universal Statistics of a Cell with Recursive Growth

## 6.1 Question to Be Addressed

**Question: A cell takes nutrient chemicals from the outside and transforms them into other chemicals, among which catalysts to further these reactions are synthesized, while a membrane that roughly separates a cell from the outside is also to be synthesized. These syntheses of chemicals need to respect some degree of synchronization. For example, for a higher growth rate of a cell, it is better to take in more nutrients, while the catalysts for other reactions have to follow up. Both catalysts and nutrient chemicals exist in some balance to maintain recursive production. How is such efficient recursive production possible while keeping the diversity of other chemicals including catalysts? Is there some universal statistics in abundances of chemicals and in the network for a cell with steady recursive growth?**

In the last chapter, we have discussed the problem of recursive production, by using a unit of replicating molecules (Level II model in Sect. 5.2). In reality, synthesis of each molecule requires some other molecules. Instead of the simple direct autocatalytic production process adopted in Chap. 5, biochemical reaction consists of a huge number of catalytic reaction process, where a huge number of molecule species participate. Then how is such balanced replication possible? Is there some universal statistical behavior inherent in a system that can grow with recursive production as a collection of catalytic reactions?

Currently this set of questions has been addressed in two ways: (1) what is the topology of the reaction network structure (static aspect) and (2) what is the number distribution of chemical species and their dynamics? Of course, we need combine the two aspects to fully understand the condition for recursive production of a cell.

There is currently a very strong interest concerning the reaction network structure. For example, Barabasi et al studied the metabolic reaction network, without going into details of the topology (Barabasi & Albert, 1999; Jeong et al., 2000). Write down all (known) metabolic reaction equations. Here, the rate of reactions is neglected: only the presence or absence of each reaction equation is considered. Then compute how many times a specific molecule species appears in such reaction equations. If this number is large, the molecule species is related with many biochemical reactions. For example, $H_2O$ appears in a large number of reactions in either the left-handside or the right-hand side of the equation. Now the number of occurrence is treated as the number of connection paths in the reaction network. $CO_2$ must have a high number also. Among more complex molecules ATP must have a relatively high number. From these data they obtained the histogram $P(n)$ of the number of molecules species that appears $n$ times in the equations. Of course, this distribution gradually decreases with the increase of $n$. From the data they obtained $P(n) \approx n^{-\alpha}$ with $\alpha \approx (2.3 \sim 2.7)$ (Jeong et al., 2000).

If a network has a specific scale for the size of connection paths, this distribution decays exponentially as $P(n) \propto \exp(-n/n_0)$. The observed distribution decays more slowly with $n$ than this exponential decay. The observed power law distribution suggests that there is a relatively large number of molecule species that are involved in many reactions. Since the power law distribution does not have a specific scale of the size of connection in the network, such a network is also called a scale-free network (Barabási & Albert, 1999). In statistical physics, the power law distribution without specific scale is often observed when there is a hierarchical structure.

In other words, the scale-free network may suggest the following structure. In a network, there is a hub part that is connected with many other nodes, and then there is a little weaker hub that is connected to a smaller range of nodes, and so forth. This kind of hierarchical network structure is also observed in the link of Web, Internet connection. In these examples, the network structure is understood by assuming that the network is developed so that a new link is likely to be added to a node that is already well connected. If a similar argument is also applicable to the metabolic network, this may give some new ideas on the evolution path of the network until its present form.

Of course, these distributions on the connectivities in the reaction path are just one aspect of the network structure. For example, we detect larger and smaller loops in the metabolic network, while some linear chains of reactions are detected. Some kind of modular structure may exist. Distribution of paths is not sufficient to represent such structures. Furthermore, how such network structure has evolved is not yet fully explored.

To this day, most studies of the metabolic network focused on its static topological structure. In the reaction network dynamics, however, the number of molecules is different for each species. Hence on each "node" of the network, one should assign the abundances of the corresponding molecule species. Accordingly, some reactions occur much more frequently than others. In other words, some paths in the network are "thicker." Besides the static network structure, the statistics of abundances as well as the possible relationship between abundance statistics and static network structure also need to be studied.

In this chapter, we first investigate the statistics of abundances of the chemical components of a cell. The discovery of universal statistical laws leads to that of a hierarchical structure of reaction dynamics. We discuss how recursive production of a cell is possible amidst large fluctuations arising from the finiteness of the number of molecules.

## 6.2 Logic

Following the study of the last chapter, we study the statistical characteristics that a cell with recursive growth has to satisfy. In contrast to the last chapter, however, we do not start from a replicating unit, since the replication of a molecule appears as a result of several catalytic reactions.

Through the membrane, some nutrient chemicals flow in, which are successively transformed to other chemicals through this catalytic reaction network. For a cell to grow recursively, a set of chemicals has to be synthesized for the next generation. Each chemical for its synthesis requires some other chemicals as a catalyst. Then, generally speaking, there should exist some mutual relationship among molecules to catalyze each other.

The number of molecule species is huge in a cell, while the number of each molecule species is not necessarily large. Then fluctuations in molecule number are inevitable, since reactions occur through stochastic collision. The number of some molecules may go to zero, which sometimes may be dangerous, since the molecule may be essential as a catalyst to the synthesis of some molecules. Then, it is not trivial how recursive production of molecules in a cell is sustained.

Requiring that all the molecule species within a cell keep an identical number after recursive production would be too demanding. The chemical compositions are expected to vary between cells obtained by division, where some fluctuations always occur. Even if the reproduction of a cell is not faithful, the catalytic activities should be sustained to keep reproduction of cells. In this sense we need to understand how an ensemble of molecules keep catalytic activity and loose reproduction amid large fluctuations in the chemical compositions.

The biochemical reactions for metabolism, synthesis of membrane, and nucleic acid keep some synchrony. In a cell, some transported nutrients are successively transformed to some other chemicals that include catalysts for other reactions. If the transport of nutrients is faster, growth will also be faster. However, with too many nutrients, catalysts may become too few, to the point where the reaction may stop.

Let us assume that all chemicals have roughly the same concentration. Then the probability for each molecule to meet its catalysts will be lower, and the effective transformation of nutrients may become impossible. Furthermore, the reaction events will progress in a completely random fashion since there are many molecule species that are low in concentration. As a result, chemical compositions will vary substantially between generations.

We conclude that some structuring within the abundances of chemicals should improve the effective transformation of nutrients. Indeed, in catalytic reactions, there is a successive structure, as indicated by the existence of catalysts for the reactions to transform nutrients, then catalysts for such catalysts for nutrients, and then catalysts for catalysts for catalysts for nutrients, and so on, as this cascade continues. It would then be expected that these levels of catalysts do exist in different levels of abundances. On the other hand, if such biased distribution in abundances of chemicals exist, the reaction probability is also not homogeneous for each reaction. This will lead to decrease the random change of concentrations, as compared with the case of almost equal distribution in numbers.

Hence, for an effective use, some hierarchical structure among catalysts is expected, with regards to the abundances. Indeed, by taking a specific cell model, we explicitly demonstrate the existence of such cascade structure, when recursive production of chemicals is sustained. When a hierarchical structure exists, there often appears some power law in the statistics of abundances over all chemical species. The probability $\rho(x)dx$ that the abundance of a chemical is between $x$ and $x+dx$ then gives some characteristics of the possible structure in the reaction dynamics, and it is expected that a power law in abundance statistics may appear. Indeed, from several model simulations, we find universal power law statistics for a cell that grows efficiently and recursively.

The next question concerns fluctuations. As discussed in the last chapter, the chemical abundance of each molecule is a fluctuating quantity, since the collision process of molecules is basically stochastic. Still, the argument we used in the last chapter to derive the log-normal distribution cannot be directly applied, since we assumed then the existence of a replication unit. However, the "multiplicative" nature of catalytic reaction as mentioned in Sect. 5.2 is general.

Now recalling the above hierarchical structure, consider the successive catalytic process as a chemical in $i$-th group is catalyzed by $(i+1)$-th, and that in the $(i+1)$-th group is catalyzed by $(i+2)$-th, and so forth. In other words, let us assume that a "modular structure" with groups of successive catalytic reactions is self-organized. With this cascade of catalytic reactions, the fluctuations are propagated "multiplicatively"; for example, the concentration fluctuation of a chemical in the $(i+2)$-th group influences multiplicatively on that of $(i+1)$-th, which then influences multiplicatively on that of the $i$-th, and so forth. By taking the logarithm of concentrations (i.e., $\log x_m$), these successive multiplications become successive additions of fluctuations. Considering the central limit theorem, the additive fluctuations for $\log x_m$ implies that the dynamics of $\log x_m$ is driven by Gaussian noise, resulting in the log-normal distribution of $x_m$.

This argument is rather primitive, and we need to check whether it really works in a model and in experiments. In the following, we will investigate these statistical laws by considering a class of catalytic reaction network models and compare the results with experimental data measured in natural cells.

## 6.3 Model

To investigate the above questions, we adopt a simple model of cellular dynamics that captures only its basic features. It consists of intracellular catalytic reaction networks that transform nutrient chemicals into proteins. By studying a class of simple models with these features, we demonstrate the existence of universal statistical laws on the abundances of chemicals for a cell that grows recursively (Furusawa & Kaneko, 2003a).

Of course, real intracellular processes are much more complicated, but if the statistical laws to be presented are universal, they should be valid, regardless of how complicated the actual processes are. Hence it is relevant to study a model as simple as possible, when trying to find such laws and compare with real data.

Consider a cell consisting of a variety of chemicals. The internal state of the cell can be represented by a set of numbers $(n_1, n_2, \ldots, n_k)$, where $n_i$ is the number of molecules of the chemical species $i$ with $i$ ranging from $i = 1$ to $k$. For the internal chemical reaction dynamics, we chose a catalytic network among these $k$ chemical species, where each reaction from some chemical $i$ to some other chemical $j$ is assumed to be catalyzed by a third chemical $\ell$, that is, $(i + \ell \rightarrow j + \ell)$. The rate of increase of $n_j$ (and decrease of $n_i$) through this reaction is given by $\epsilon n_i n_\ell / N^2$, where $\epsilon$ is the coefficient for the chemical reaction. For simplicity all the reaction coefficients were chosen to be equal,[1] and the connection paths of this catalytic network were chosen randomly such that the probability of any two chemicals $i$ and $j$ to be connected is given by the connection rate $\rho$ (Kaneko & Yomo, 1997).

Some nutrients are supplied from the environment by diffusion through the membrane (with a diffusion coefficient $D$), to ensure the growth of a cell. Through the catalytic reactions, these nutrients are transformed into other chemicals. (Note that the nutrient chemicals have no catalytic activity, since they are not products by intracellular reactions. Indeed, catalytic reactions do not occur in the environment.) Besides these nutrients, some of these chemicals may penetrate[2] the membrane and diffuse out while others will not. With the synthesis of the impenetrable chemicals that do not diffuse out, the total number of chemicals $N = \sum_i n_i$ in a cell can increase, and accordingly the cell

[1] We have examined the case where the coefficients are distributed, but the results to be discussed are not altered.

[2] Even if the reaction coefficient and diffusion coefficient of penetrating chemicals are not identical but distributed, identical results are obtained.

volume will increase. We study how this cell growth is sustained by dividing a cell into two when the volume is larger than some threshold. For simplicity the division is assumed to occur when the total number of molecules $N = \sum_i n_i$ in a cell exceeds a given threshold $N_{max}$. Each daughter cell inherits one half of the mother cell's molecules, chosen randomly.

In the case with $N \gg k$ (i.e., continuous limit), the reaction dynamics is represented by the following rate equation:

$$dn_i/dt = \sum_{j,\ell} Con(j,i,\ell)\ \epsilon\ n_j\ n_\ell/N^2 - \sum_{j',\ell'} Con(i,j',\ell')\ \epsilon\ n_i\ n_{\ell'}/N^2 + D\sigma_i(\overline{n_i}/V - n_i/N)\ , \quad (6.1)$$

where $Con(i,j,\ell)$ is 1 if there is a reaction $i + \ell \to j + \ell$, and 0 otherwise, whereas $\sigma_i$ is set to 1 if the chemical $i$ can penetrate the membrane, and 0 otherwise. The third term describes the transport of chemicals through the membrane, where $\overline{n_i}$ is a constant, representing the number of the $i$-th chemical species in the environment and $V$ denotes the volume of the environment in units of the initial cell size. The number $\overline{n_i}$ is nonzero only for the nutrient chemicals.

Here, however, we are also interested in the fluctuations of chemicals because of the finiteness of $N$. Hence, we adopt a stochastic simulation as in Chaps. 4 and 5. In our numerical simulations, we randomly pick up a pair of molecules in a cell, and transform them according to the reaction network. In the same way, diffusion through the membrane is also computed by randomly choosing molecules inside the cell and nutrients in the environment.

## 6.4 Result

If the total number of molecules $N_{max}$ is larger than the number of chemical species $k$, the population ratios $\{n_i/N\}$ are generally fixed, since the daughter cells inherit the chemical compositions of their mother cells. For $k > N_{max}$, the population ratios do not settle down and can change from generation to generation. In both cases, dependence of the intracellular reaction dynamics on the membrane diffusion coefficient $D$ is similar, though.[3]

As $D$ is increased, the growth speed of a cell is increased as shown in Fig. 6.1, since the intake of nutrients is hastened. The increase of the growth speed with $D$ is quite natural. However, we found that there is a critical value $D = D_c$ beyond which the cell cannot grow. For $D > D_c$, a cell cannot grow continuously. Even if it can grow and divide a few times, it stops

[3] Instead of changing $D$, the same change of behaviors is observed when changing the connection rate $\rho$. Instead of a critical $D$ (to be described), there is a critical value $\rho = \rho_c$, where in the case $\rho < \rho_c$ the cell stops growing. The power law distribution of chemical abundances with an exponent$-1$ found at $D_c$ appears at $\rho = \rho_c$ also.

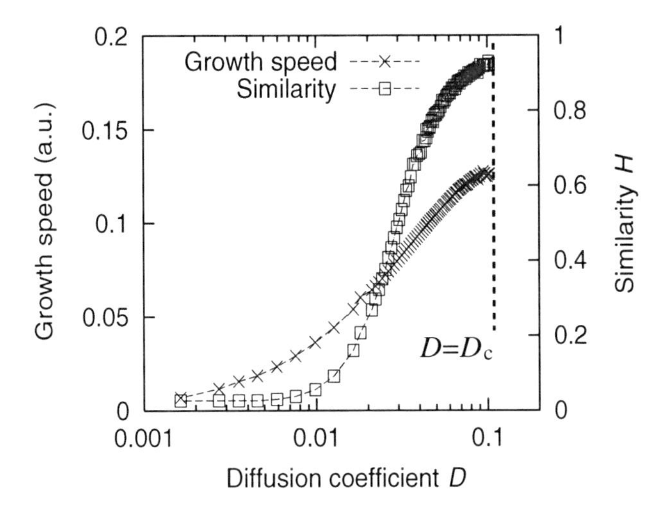


**Fig. 6.1.** The growth speed of a cell and the similarity between the chemical compositions of the mother and daughter cells, plotted as a function of the diffusion coefficient $D$. The growth speed is measured as the inverse of the time for a cell to divide. Following Sect. 5.4, the degree of similarity between two different states $m$ (mother) and $d$ (daughter) is measured as the scalar product of k-dimensional vectors $H^1(\mathbf{n}_m, \mathbf{n}_d) = (\mathbf{n}_m/|\mathbf{n}_m|) \cdot (\mathbf{n}_d/|\mathbf{n}_d|)$, where $\mathbf{n} = (n_1, n_2, \ldots, n_k)$ represents the chemical composition of a cell and $|\mathbf{n}|$ is the norm of $\mathbf{n}$, as defined in Sect. 5.3 (see Fig. 5.4). Both the growth speed and the similarity are averaged over 500 cell divisions. Note that the case $H^1 = 1$ indicates an identical chemical composition between the mother and daughter cells

growing then. This is explained as follows. When $D > D_c$, the flow of nutrients from the environment is so fast that the internal reactions transforming them into chemicals sustaining "metabolism" cannot keep up. In this case all the molecules in the cell will finally be substituted by the nutrient chemicals and the cell stops growing since the nutrients alone cannot catalyze any reactions to generate impenetrable chemicals, while the diffusive transport stops as the nutrient concentration in a cell reaches the level at the environment. Hence, continuous cellular growth and successive divisions are possible only for $D \leq D_c$. As shown in Fig. 6.1, the growth speed of a cell is maximal at $D = D_c$. This suggests that a cell whose reaction dynamics are in the critical state should be selected by natural selection.

Second, we discuss the similarity of chemical compositions between the mother cell and the daughter cell. To characterize the similarity of cell compositions by divisions, we again use the quantity introduced in Sect. 5.3. As plotted in Fig. 6.1, the similarity is maximal around $D \sim D_c$.

When the diffusion coefficient $D$ is sufficiently small, the internal reactions progress faster than the flow of nutrients from the environment, and all the existing chemical species have small numbers of approximately the same level.

There, the similarity between the mother and daughter cells is low, because of the stochastic change of chemical abundances. As $D$ approaches $D_c$, the abundance of nutrients increases, and the abundance of those that are synthesized effectively from the nutrients increases. Such inhomogeneous abundances by chemical species is preserved over generations, leading to a higher similarity as given in Fig. 6.1.

For $k > N$, the chemical compositions differ significantly from generation to generation when $D \ll D_c$. When $D \approx D_c$, several semistable states with distinct chemical compositions appear. Daughter cells in the semistable states inherit chemical compositions that are nearly identical to their mother cells over many generations, until fluctuations in molecule numbers induce a transition to another semistable state.

To sum up, the most faithful transfer of the information determining a cell's intracellular state is at the critical state $D \sim D_c$. In this state, chemical compositions of cells are well transferred to offspring through divisions, amid fluctuations, and furthermore the growth of a cell (division speed) is maximal. For these reasons it is natural to conclude that evolution favors a critical state for the reaction dynamics.

Accordingly, we study the statistics of the abundances of chemicals focusing on the case with $D \sim D_c$. Here we are interested in the inhomogeneity of abundances by chemical species. The number of some molecule species is quite high, while some others are not, at around $D \sim D_c$. As a measure of such inhomogeneity in abundances, we have computed the rank-ordered number distributions of chemical species. They are plotted in Fig. 6.2, where the ordinate indicates the number of molecules $n_i$ and abscissa shows the rank determined by $n_i$. As shown in the figure, the slope in the rank-ordered number distribution increases with an increase of the diffusion coefficient $D$. We found that at the critical point $D = D_c$, the distribution converges to a power law with an exponent $-1$. This power law with the exponent $-1$ is observed independently of the choice of the network, as long as $D \sim D_c$ (Furusawa & Kaneko, 2003a). The power law in the abundance–rank relationship was first studied in linguistics as the frequency of appearance of words by Zipf (1949), and is called Zipf's law.

The power-law distribution at this critical point is maintained by a hierarchical organization of catalytic reactions, where the synthesis of higher ranking chemicals is catalyzed by lower ranking chemicals. For example, major chemical species (with, e.g., $n_i > 1000$) are directly synthesized from nutrients and catalyzed by chemicals that are slightly less abundant (e.g., $n_i \sim 200$). The latter chemicals are mostly synthesized from nutrients (or other major chemicals), and catalyzed by chemicals that are much less abundant. In turn, these chemicals are catalyzed by chemicals that are even less abundant, and this hierarchy of catalytic reactions continues until it reaches the minor chemical species (with, e.g., $n_i < 5$). In fact, in the case depicted in Fig. 6.3, a hierarchical organization of catalytic reactions with $5 \sim 6$ layers is observed at the critical point.

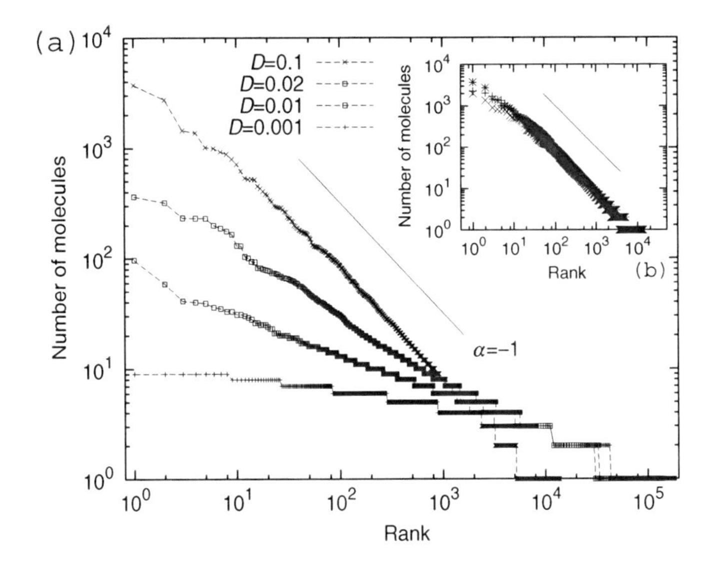


**Fig. 6.2.** Rank-ordered number distributions of chemical species. (**a**) Distributions with different diffusion coefficients $D$ are overlaid. The parameters were set as $k = 5 \times 10^6$, $N_{max} = 5 \times 10^5$, and $\rho = 0.022$. Thirty percent of chemical species are penetrating the membrane, and others are not. Within the former ones, chemical species are continuously supplied to the environment, as nutrients. In this figure, the numbers of nutrient chemicals in a cell are not plotted. With these parameters, $D_c$ is approximately 0.1. (**b**) Inset: Distributions at the critical points with different total number of chemicals $k$ are overlaid. The numbers of chemicals were set as $k = 5 \times 10^4$, $k = 5 \times 10^5$, and $k = 5 \times 10^6$, respectively. Other parameters were set the same as those in (a) (Reproduced from Furusawa and Kaneko (2003a)

On the basis of this catalytic hierarchy, the observed exponent $-1$ can be explained using a mean field approximation. First, we replace the concentration $n_i/N$ of each chemical $i$, except the nutrient chemicals, by a single average concentration (mean field) $x$, while the concentration of nutrient chemicals $S$ is given by the average concentration $S = 1 - k^*x$, where $k^*$ is the number of nonnutrient chemical species. From this mean field equation, we obtain $S = \frac{DS_0}{D+\epsilon\rho}$ with $S_0 = \sum_j \overline{n_j}/V$. With linear stability analysis, the solution with $S \neq 1$ is stable if $D < \frac{\epsilon\rho}{S_0-1} \equiv D_c$. Indeed, this critical value does not differ much from numerical observation.

Next, we study how the concentrations of nonnutrient chemicals differentiate. Suppose that chemicals $\{i_0\}$ are synthesized directly from nutrients through catalysis by chemicals $j$. As the next step of the mean-field approximation, we assume that the concentrations of the chemicals $\{i_0\}$ are larger than the others. Now we represent the dynamics by two mean-field concentrations: the concentration of $\{i_0\}$ chemicals, $x_0$, and the concentration of the others, $x_1$. The solution with $x_0 \neq x_1$ satisfies $x_0 \approx x_1/\rho$ at the critical point

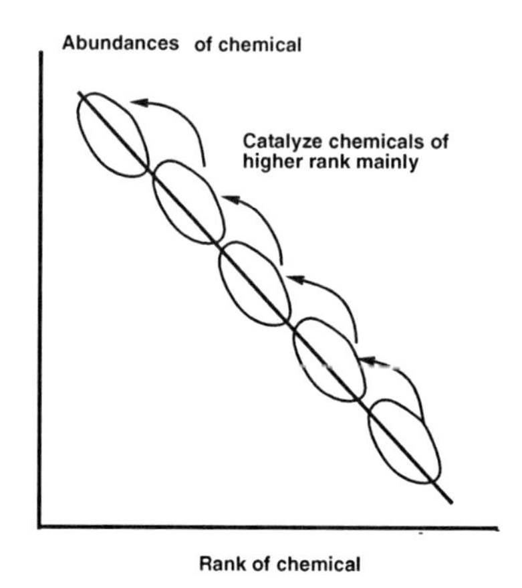


**Fig. 6.3.** Schematic representation of the catalytic cascade. Higher rank chemicals are mainly catalyzed by lower level ones. This is a schematic representation showing a rough structure and indeed there are several other cascade paths also

$D_c$. Since the fraction of the $\{i_0\}$ chemicals among the non-nutrient chemicals is $\rho$, the relative abundance of the chemicals $\{i_0\}$ is inversely proportional to this fraction. Similarly, one can compute the relative abundances of the chemicals of the next layer synthesized from $i_0$. At $D \approx D_c$, this hierarchy of the catalytic network is continued. In general, a given layer of the hierarchy is defined by the chemicals whose synthesis from the nutrients is catalyzed by the layer one step down in the hierarchy. The abundance of chemical species in a given layer is $1/\rho$ times larger than chemicals in the layer one step down. Then, in the same way as this hierarchical organization of chemicals, the increase of chemical abundances and the decrease of number of chemical species are given by factors of $1/\rho$ and $\rho$, respectively. This is the reason for the emergence of power-law with an exponent $-1$ in the rank-ordered distribution. Within a given layer, a further hierarchy exists, which again leads to the Zipf rank distribution.

In general, as the flow of nutrients from the environment increases, the hierarchical catalyzation network pops up from random reaction networks. This hierarchy continues until it covers all chemicals, at $D \to D_c - 0$.[4] Hence, the emergence of a power-law distribution of chemical abundances near the critical point is quite general, and does not rely on the details of our model, such as the network configuration or the kinetic rules of the reactions. Instead it is a universal property of a cell with an intracellular reaction network to grow, by taking in nutrients, at the critical state, as has been confirmed from simulations of a variety of models.

[4] $D_c - 0$ here means that $D$ approaches $D_c$ from the side with $D < D_c$. Since the cell cannot grow for $D > D_c$, the behavior at the limit of $D_c + 0$, that is, from the side with $D > D_c$ is different.

We have checked the universality of Zipf's law by adopting distributed network connectivity, distributed parameters, and so forth. Our result is robust against these changes of the model. This power law in abundances seems to be a quite universal law as long as a cell achieves a recursive growth (i.e., relatively faithful reproduction) and efficient growth. Since the current cell also satisfies relatively faithful reproduction (as well as effective growth for a suitable condition), the above behavior at $D \sim D_c$ may be expected to be true also in a real cell. Then, does Zipf's law also hold in a natural cell? To investigate possible universal properties of the reaction dynamics, we examine the distributions of the abundances of expressed genes (that are the abundances of the mRNA that produce corresponding proteins). As will be discussed in Sect. 6.4, the data from several tissues suggest the validity of the present law.

*Remark:* Equivalently with the Zipf's law on the abundance-rank relation, the distribution $p(x)$ of the chemical species with abundance $x$ is proportional to $x^{-2}$. This is easily obtained by noting that the rank distribution, that is, the abundances $x$ plotted by rank $n$, can be transformed into the density distribution function $p(x)$, which is the probability that the abundance is between $x$ and $x + dx$. Recalling that $dx = dx/dn \times dn$, there are $-(dx/dn)^{-1}$ chemical species between $x$ and $x + dx$ (recall $dx/dn < 0$ since the rank is in descending order). Thus, if the abundance–rank relation is given by a power law with exponent $-1, p(x) = -(dx/dn)^{-1} \propto n^2 \propto x^{-2}$.

### 6.4.1 Log-Normal Distribution

In the last subsection, we discussed the average abundances of each chemical, and studied the universal statistic over molecule species (Furusawa et al., 2005). Since the chemical reaction process is stochastic, each number of molecule differs by cells. For example, if $x_3$ in Sect. 6.2.1 is 3000, it can be 2831, 3203, 2903, 3111,... for each cell. Now we study the distribution of each molecule number, sampled over cells. In Fig. 6.4, we plotted the number distributions of several chemicals in the condition $D \sim D_c$. We measure the number of chemicals when a cell is divided into two. Since the total number of chemicals is constant when the cell is divided, the abundances of each chemical are just proportional to the concentration of each. As shown in Fig. 6.4, the distribution is fitted quite well by the log-normal distribution, that is,

$$P(n_i) \propto \left(\frac{1}{n_i}\right) \exp\left(-\frac{(\log n_i - \log \overline{n_i})^2}{2\sigma}\right) , \tag{6.2}$$

where $\overline{n_i}$ indicates the average of $n_i$ over cells.

This log-normal distribution holds for the abundances of all chemicals, except for a few chemicals that are supplied externally to a cell as nutrients, which obey the standard Gaussian distribution. In other words, those molecules that are reproduced in a cell obey the log-normal distributions, while those that are just transported from the outside of a cell follow the normal distribution.

Why does the log-normal distribution law generally hold, in spite of the threat by the central limit because of addition of several fluctuation terms? We have discussed already the possible mechanism for the log-normal distribution in the last chapter. However, there is slight difference here. In the discussion of the last chapter, there is autocatalytic reaction process, as given by $dx_i/dt = cx_jx_i$, which leads to multiplicative stochastic process. In the present example, the reaction process is given by $dx_i/dt = cx_jx_\ell$. Thus, the discussion of the last chapter is not directly applicable.

Still, a multiplicative reaction process is also involved here. Furthermore, as discussed in the last subsection, there is a cascade reaction process, to support the recursive production around the critical state $D \sim D_c$. There, only a fraction of all possible reaction pathways is used dominantly to organize a cascade of catalytic reactions so that a chemical in the $i$-th group is catalyzed by a chemical in the $(i+1)$-th, and one in the $(i+1)$-th group is catalyzed by one in the $(i+2)$-th group. As expected in Sect. 6.2, a "modular structure" with groups of successive catalytic reactions is self-organized in the network. Then the fluctuations are amplified multiplicatively through the cascade. By taking the logarithm of concentrations (i.e., $\log x_m$), these successive multiplications are transformed to successive additions. Now, if many numbers of random numbers are added, its distribution approaches the Gaussian distribution, as given by the central limit theorem. The distribution of $\log x_m$ is then expected to obey the normal distribution. Hence, the log-normal distribution of $x_m$ is derived. Note that at the critical state we study here, this cascade of catalytic reaction continues for all chemical species that are reproduced, and the log-normal distribution holds clearly.

Next we discuss how the magnitude of fluctuations depends on each chemical species. As shown in Fig. 6.4, the width of the distribution hardly changes with the average abundance, itself plotted with a logarithmic scale. This suggests that some relationship holds between the fluctuation and the average for all chemicals. We have thus plotted the standard deviation of each chemical $\sqrt{\overline{(n_i - \overline{n_i})^2}}$ as a function of the average. As shown in Fig. 6.5, the standard deviation (*not* the variance) increases linearly with the average.

To discuss the relationship between the mean and the standard deviation, one should recall the steady growth and cascade structure in catalytic reactions. Consider two chemicals $i$ and $j$, one of which ($j$) belongs to a group of catalyzing molecules for the other. Then the balance between the synthesis and the conversion implies $x_i \times A - x_j \times B = 0$, where $A$ and $B$ are average concentrations of other chemicals involved in the catalytic reaction. Assuming the steady growth of a cell, the average concentration satisfies $\overline{x_i}/\overline{x_j} = A/B$. If we further assume the steady state condition for the fluctuations also, then $\langle(\delta x_i)^2\rangle/\langle(\delta x_j)^2\rangle = (A/B)^2 = \overline{x_i}^2/\overline{x_j}^2$. Hence with this rough argument, the

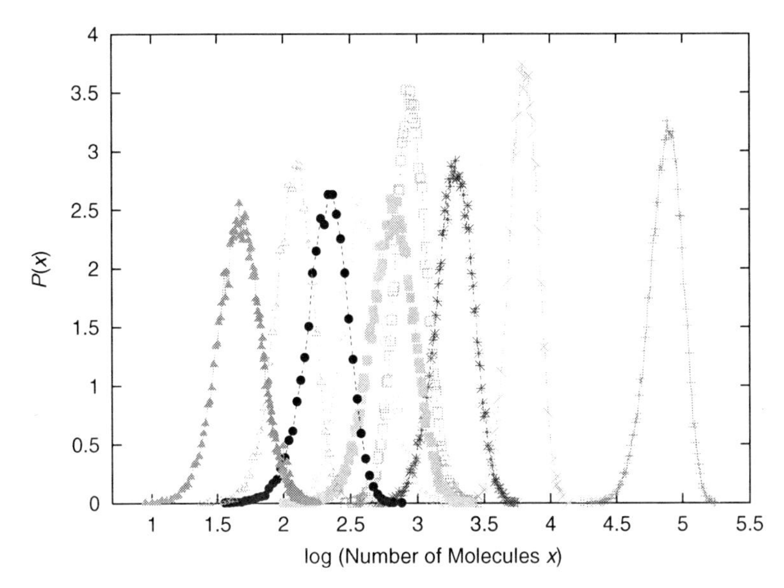


**Fig. 6.4.** The number distribution of the molecules of chemical abundances. Distributions are plotted for several chemical species with different average molecule numbers. The data were obtained by observing 178,800 cell divisions. The parameters were set as $k = 5 \times 10^3$, $N_{max} = 10^6$, and the connectivity to $\rho = 0.02$. Thirty percent of the chemical species are penetrating the membrane, and the others are not. Within the penetrable chemicals, two chemical species are continuously supplied from the environment as nutrients. The diffusion coefficient $D$ of the membrane was set as 0.04, which is close to the critical value $D_c$ (Reproduced from Furusawa et al. 2005)

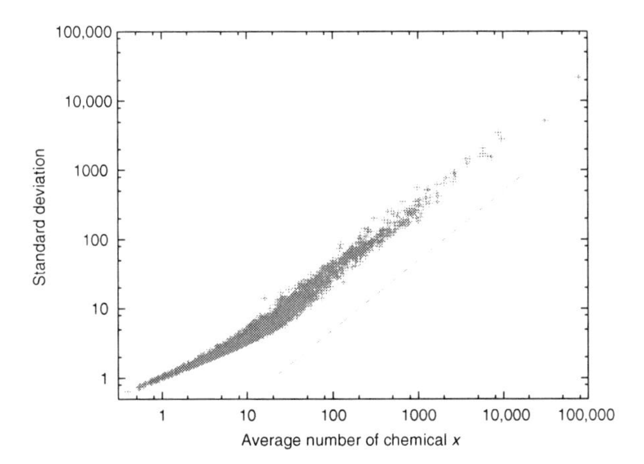


**Fig. 6.5.** Standard deviation versus average number of molecules. Using the same data set and parameters as for Fig 6.4, the relationship between the average and standard deviation is plotted for all chemical species. The broken line is for reference (Reproduced from Furusawa et al. 2005)

variance is expected to be proportional to the square of the mean, leading to the linear relationship between mean and standard deviation.

The linear relationship is also found with regards to the variation of chemical abundances by the change of external conditions. For example, we have computed the change from $\overline{x_i}$ to $\overline{x_i'}$ by changing the concentrations of supplied nutrients. The variation $|\overline{x_i'} - \overline{x_i}|$ is again found to be proportional to $\overline{x_i}$ for each chemical $i$, similarly as the data plotted in Fig. 6.5.

The discovered laws on the distribution and the linear relationship between the average and fluctuation are universally observed, near the critical point with the largest reproduction speed, hold generally and do not rely on the details of the model, such as the network configuration of the kinetic rules of the reactions, as has been confirmed from simulations of a variety of models. Instead it is a universal property of replicating cellular dynamics near the critical state $D \sim D_c$, giving a recursive and optimal reproduction of cells.

Note, however, that the above arguments for the two laws are based on the steady growth of a cell with catalytic cascade process, realized at $D \sim D_c$. Indeed, as the parameter $D$ is much smaller, all possible reaction pathways are used with a similar weight, where the cascades of catalytic reactions are replaced by random reaction network. In this case, the fluctuations of each molecule number are highly suppressed, and the distribution is close to normal Gaussian. The variance (not the standard deviation) increases linearly with the average concentrations. In other words, the behavior is "normal" as expected from the central limit theorem.

## 6.5 Experiment

### 6.5.1 Confirmation of Zipf's Law

First we discuss the validity of Zipf's law with regards to the abundances of many chemicals within the cell. Here we study this problem in a real cell, since an artificial cell with a large variety of chemicals has not yet been created. To investigate possible universal properties of the reaction dynamics, examined are the distributions of the abundances of expressed genes, that are nothing but the abundances of the mRNA that produce the corresponding proteins in a variety of organisms and a variety of tissues. In Furusawa and Kaneko, (2003a), we used the data publicly available from SAGE (Serial Analysis of Gene Expression) databases (Lash et al., 2000; Velculescu et al., 1997; Jones et al., 2001) over 6 organisms and more than 40 tissues. SAGE allows the number of copies of any given mRNA to be quantitatively evaluated by determining the abundances of the short sequence tags that uniquely identify it (Velculescu et al., 1995).

In Fig. 6.6, we show the rank-ordered frequency distributions of the expressed genes, where the ordinate indicates the frequency of the observed sequence tags (i.e., the population ratio of the corresponding mRNA to the

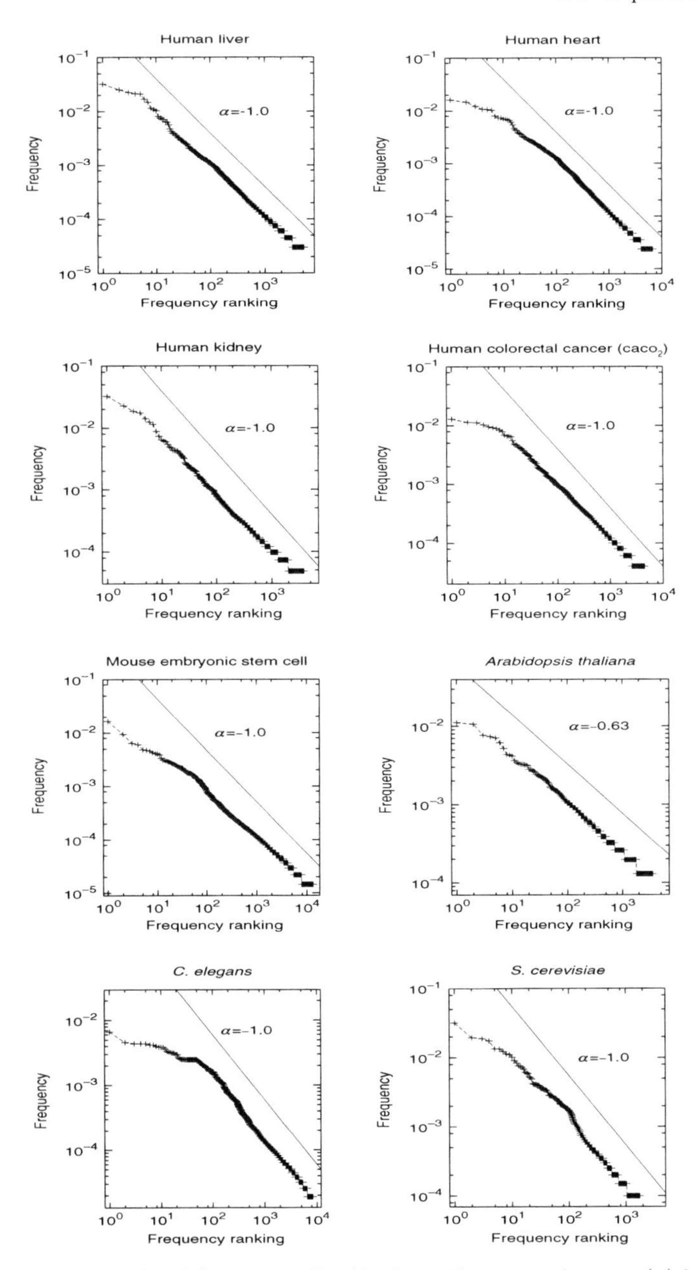


**Fig. 6.6.** Rank-ordered frequency distributions of expressed genes. (**a**) human liver; (**b**) kidney; (**c**) human colorectal cancer ($CACO_2$); (**d**) mouse embryonic stem cells; and (**e**) *C. elegans*; (**f**) yeast (*Saccharomyces cerevisiae*). The exponent $\alpha$ of the power law is in the range from $-1$ to $-0.86$ for all the samples inspected, except for two plant data (seedlings of *Arabidopsis thaliana* and the trunk of *Pinus taeda*), whose exponents are approximately $-0.63$ (Reproduced from Furusawa & Kaneko 2003a)

total mRNA), and the abscissa shows the rank determined from this frequency. As shown, the distributions follow a power-law with an exponent close to $-1$ (Zipf's law). We observed this power-law distribution for all the available samples, including 18 human normal tissues, human cancer tissues, mouse (including embryonic stem cells), rat, nematode (*C. elegans*), and yeast (*S. cerevisiae*) cells. All the data more than 40 samples (except for two plant seedling data) show a power-law distribution with an exponent in the range from $-1$ to $-0.86$. Even though there are some factors that may bias the results of the SAGE experiments, such as sequencing errors and nonuniqueness of tag sequences, it seems rather unlikely that the distribution is an artifact of the experimental procedure. Indeed, by using gene-chip (microarray analysis), Zipf's law is also observed (Kuznetsov et al., 2002), and recently for more than 100 examples ranging from bacteria to human (Ueda et al., 2004).

### 6.5.2 Confirmation of Laws on Fluctuations

Now, we report experimental confirmations of the law on the distributions of abundances (Furusawa et al., 2005). Recalling that the laws are expected to hold for the abundances of a protein synthesized within cells with recursive (steady) growth, we measured the distribution of the protein abundances in *Escherichia coli* that are in the log phase growth, that is, in a stage of steady growth. To obtain the distribution of the protein abundances, we introduced the fluorescent proteins with appropriate promoters into the cells and measured the fluorescence intensity by flow cytometry. To demonstrate the universality of the laws, we have carried out several sets of experiments by using a variety of promoters and also by changing locations where the reporter genes are introduced (i.e., on the plasmid and on the genome).

In Fig. 6.7, we have plotted the distributions of the emitted fluorescence intensity from *E. coli* cells with the reporter plasmids containing either EGFP under the control of the tetA promoter without repression, or DsRed[5] under the control of the trc promoter with and without IPTG[6] induction. In general, the fluorescence intensity (the abundance of the protein) increases with the cell size. To avoid the effect of variation of cell size, which may also obey log-normal distribution, we normalized the fluorescence intensity by the volume of each cell. Here we adopt the forward-scatter (FS) signal from the flow cytometry, to estimate the cell volume. Indeed, by plotting data of the fluorescence intensity versus FS signal, the two are proportional on the average. (The data points are distributed around the proportionality line between the two, as are generally observed for the plot of fluorescence intensity by flow cytometry). Hence, we normalized the fluorescence intensity by dividing the FS

[5] Both EGFP (enhanced green fluorescent protein) and DsRed are some fluorescent proteins.

[6] Isopropyl-$\beta$-D-thiogalactopyranosid, which induces synthesis of $\beta$-galactosidase, an enzyme that promotes lactose use.

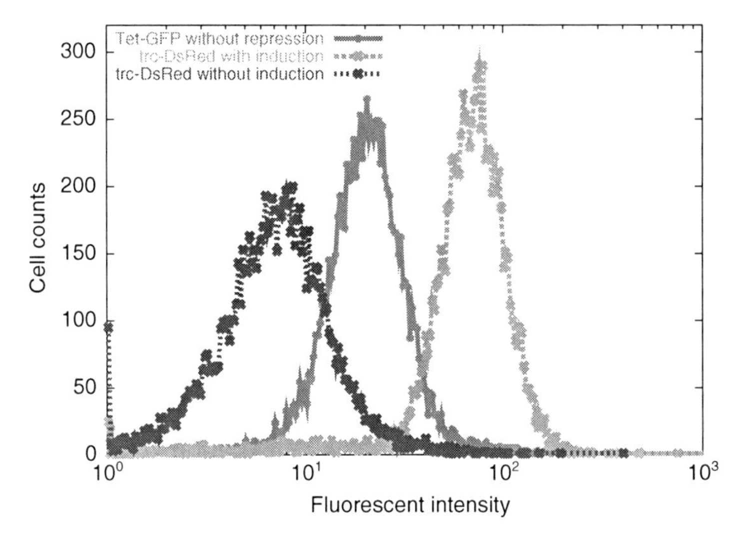


**Fig. 6.7.** The number distribution of the proteins measured by fluorescence intensity, normalized by the cell volume. Distributions are obtained from three *Escherichia coli* cell populations containing different reporter plasmids (see text). Note that, although the IPTG induction changes the average fluorescence intensity, both the distributions (with and without the induction) can be fitted by log-normal distributions well

signal. Figure 6.7 is the distribution of this normalized fluorescence intensity. Note that all these data are fitted well by log-normal, rather than Gaussian, distributions, even though each of the expressions is controlled by a different condition of the promoter.

The abundances of fluorescent proteins expressed from the chromosome are also found to obey the log-normal distribution, as shown in Fig. 6.8. Here, the data are obtained by *Escherichia coli* cells with an expression of glutamine synthetase (GS) fused to GFP (Green Fluorescent protein) in the chromosome, whose expression is controlled by the upstream tetA promoter. In Fig. 6.8, plotted are the distribution of fluorescence intensity again normalized by the cell volume. As can be seen, when using the logarithmic scale (a), the distribution is roughly symmetric and close to Gaussian, while when using the normal scale (b), the distribution has a larger tail on the side of greater abundances. We have also examined several other cases, using different reporter genes both on the plasmids and on the genome, and obtained similar results supporting the universality of the log-normal distributions. It is furthermore interesting to note that the abundances of the fluorescent proteins, reported in the literature so far, are often plotted with a logarithmic scale (Hasty et al., 2000). (See also Banerjee et al., [2004] for a report of log-normal distribution of gene expression, and Krishna et al., [2005] for a theoretical arguement.)

It should be noted that the log-normal distribution of protein abundances is observed when the *Escherichia coli* are in the log phase of growth, that is,

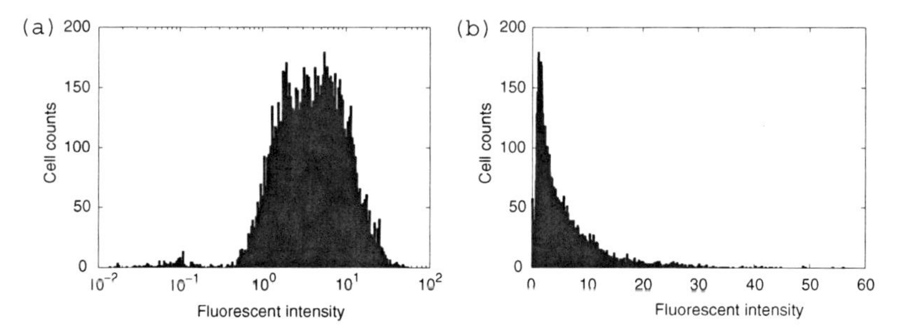


**Fig. 6.8.** The distribution of the fluorescence intensity normalized by the cell volume, plotted with a logarithmic scale and with a normal scale. Data are obtained from a population of isogenic bacterial cells with an expression of GFP-GS fusion protein in the chromosome. It is clear that the distribution with the logarithmic scale is symmetric and close to a Gaussian form (Reproduced from Furusawa et al. 2005)

when the bacteria are in steady growth. For other phases of growth, without steady growth, the distribution can deviate from the log-normal distribution and sometimes possesses two peaks, as will be reported in the future. Note that the theory also supports the log-normal distribution for the steady growth case only.

## 6.6 Relevance to Biology

### 6.6.1 Cluster Analysis of Gene Expression

First, we discuss the possible biological significance of Zipf's law in abundances. This law theoretically predicted agrees well with experimental data. This agreement is universal over 50 cell types we have examined. On the other hand, it is too universal to lead to some specific biological meaning. It would, however, be possible to deduce such a meaning if a deviation from this law were observed under some condition.

Theoretically, the validation of Zipf's law is based on recursive production. If a cell's state is not in the steady recursive growth, there could be deviation from it. In this respect, further studies on embrionic stem (ES) cells should be important. As can be seen in Fig. 6.6, there might be some deviation from the power law at abundant mRNAS for ES cells. However, at the current accuracy of microarray data, it would be difficult to make conclusive statement.[7]

Note that in the observed Zipf's law, the ranking can change in time, or depending on the environment. In a theoretical model, we can see some

[7] In some data in the seedlings of a plant, the power seems to be than 1. Since the state of seedlings is a transient from the seed to the steady tissue, this deviation if confirmed, is consistent with the theoretical argument.

cascade structure with regards to the catalytic relationship. Then, it should be important to study from microarray data how the abundances and the relationship within the network change in time.

### 6.6.2 Remark on Universal Log-Normal Statistics

In this chapter we have observed ubiquity of log-normal distribution, in several models.[8] The fluctuations in such distribution are generally very large. This is in contrast to our naive impression that a process in a cell system must be well controlled.

Then, is there some relevance of such large fluctuations to biology? Quite recently, we have extended the idea of fluctuation–dissipation theorem in statistical physics to evolution and proposed a linear relationship (or high correlation) between (genetic) evolution speed and (phenotypic) fluctuations. This proposition turns out to be supported by experimental data on the evolution of bacteria to enhance the fluorescence in its proteins, as will be discussed in Chap. 10, while relevance of fluctuation to adaptation is recently discussed by Kashiwagi et al. (2005). Hence the fluctuations are quite important biologically.

The log-normal distribution is also rather universal in the present cell, as demonstrated in the distribution of some proteins, measured by the degree of fluorescence. We have to be cautious here, since too universal laws may not be so relevant to biological function. In fact, chemicals that obey the log-normal distribution may have too large fluctuations to control some function. For example, the abundances of DNA should deviate from the log-normal distribution. Some other mechanism to suppress the fluctuation may work in a cell.[9]

Indeed, the minority control suggests the possibility of such control to suppress the fluctuation, as discussed in Sect. 4.5. For a recursive production system, some mechanism to decrease the fluctuation in minority molecule may be evolved. As discussed in Chap. 5, there can be two possibilities to decrease the fluctuation leading to deviation from log-normal distribution.

The first one is some negative feedback process. In general, the negative feedback can suppress the response as well as the fluctuation. Still, it is not a trivial question how chemical reactions can suppress fluctuations: the production of some molecules, which may further contribute to fluctuations, is indeed necessary to realize such a negative feedback.

---

[8] Lognormal distribution is also discussed in ecology (see e.g., [Preston 1962, May 1999, Hubbel 2001]. However, the log-normal distribution in ecosystem is observed with regards to the number distribution of species that have population $x$. In the term of cell, it corresponds to the distribution of chemical species (proteins) that have abundances $x$, which is $p(x)$ of Sect. 6.4 (page 145), and is estimated not as lognormal but as $x^{-2}$, corresponding to Zipf's law.

[9] It is interesting to note that the weight distribution of adult humans obeys the log-normal distribution while the height distribution obeys the Gaussian distribution.

The second possible mechanism is the use of multiple parallel reaction paths. If several processes work sequentially, the fluctuations would generally be increased. When reaction processes work in parallel for some species, the population change of such molecule is influenced by several fluctuation terms added in parallel. If a synthesis (or decomposition) of some chemical species is a result of the average of these processes working in parallel, the fluctuation around this average can be decreased by the law of large numbers. Indeed, the minority in the core network that has higher reaction paths has relatively lower fluctuation as discussed in Sect. 5.3. Suppression of fluctuation by multiple parallel paths may be a strategy adopted in a cell. Note that this is also consistent with the scenario that more and more molecules are related with the minority species as discussed in Sect. 4.5. With the increase of the paths connected with the minority molecules, the fluctuation of minority molecules is reduced, which further reinforces the minority control mechanism. Hence the increase of the reaction paths connected with the minority molecule species through evolution, decrease of the fluctuation in the population of minority molecules, and enhancement of minority control reinforce each other. In the respect, the search for molecules that deviate from log-normal distribution should be important, in future. Here it is important to measure the distribution of chemicals in relationship with its characteristics (such as the connectivity) in the reaction network.

In physics, we are often interested in some quantities that deviate from Gaussian (normal) distribution, since the deviation is exceptional. Indeed, in physics, search for power-law distribution or log-normal distributions has been popular over a few decades. On the other hand, a biological unit can grow and reproduce, to increase the number. For such system, the components within have to be synthesized, so that amplification process is common. Then, the fluctuation is also amplified. In such system, the power-law or log-normal distributions are quite common, as already discussed here. In this case, the Gaussian (normal) distribution is not so common (normal). Then exceptional molecules that obey the normal distribution with regards to their concentration may be more important.

Also, the ubiquity of log-normal distribution we found is true for a state with recursive production. If a cell is not in a stationary growth state but in a transient process switching from one steady state to another, the universal statistics can be violated. Search for such violation will be important both experimentally and theoretically.

### 6.6.3 Relationship Between Abundance Statistics and Network Topology

In the beginning of this chapter, we referred to some universal statistics in network topology, in particular power-law distribution in the connectivity in the network (Jeong et al., 2000; Almaas et al., 2004). On the other hand, the power law in abundances we discussed here holds independently of the network structure, as long as the cell satisfies efficient and recursive growth. Then is there some relationship between the two power laws?

Of course, how these reactions progress depends on the intracellular reaction network. Still we can set the parameter values for each reaction network dynamics so that the cell state is at a critical point satisfying recursive and efficient growth, where the power law in abundances is universally observed. Still, the growth speed at each critical point can differ by each network. Hence it will be possible to study the network structure in relationship with the abundance distribution by taking into account of the evolutionary process in the network structure to select cells with higher growth.

To discuss this problem, we studied the evolution of the network, by using the model in Sect. 6.3 (Furusawa & Kaneko, 2005a). We generate slightly modified networks and then select those that grow faster. We repeat this procedure. To be specific, $n$ cells are first generated, having a randomly connected catalytic network with a given initial path number. Then, from each of $n$ mother cells having a randomly connected catalytic network, $m$ mutant cells are generated by randomly adding one reaction path to the reaction network of the parent cell. Then, for each of the $n \times m$ cells, reaction dynamics are simulated, to obtain the growth speed of each cell. Among the cell population, $n$ cells with higher growth speeds are selected as the mother cells of the next generation, and $m$ mutant cells are generated again from each mother cell in the same manner. We have carried out a variety of these network evolution simulations by using several different initial networks, different parameters, and variations of settings.

Recall that the critical $D$ value in Sect. 6.4 depends on the network. Initially, the system is not near the critical point, and hence the abundance distribution does not follow the power law. Through the evolutionary process to select cells with a higher growth, however, the Zipf law in abundances emerges within 10 generations or so. Indeed, the emergence of such power law by selecting cells with higher growth speeds is a natural consequence of the result in Sect. 6.4. To increase the growth speed of cells, change in the network that enhances the uptake of nutrients from the environment is favored. This nutrient uptake is facilitated by increasing the concentrations of some catalysts, while if the uptake of nutrient is too large, the cell can no longer grow continuously, since there is no room to synthesize the catalysts to "metabolize" the nutrient, as discussed in Sect. 6.4. Hence with the change in the network, a critical value for the nutrient uptake is approached, where the power-law distribution of chemical abundance appears in the intracellular dynamics.

So far the network structure does not differ much from the initial networks that are generated randomly. Next, we investigate topological properties of the reaction networks. Although the additions of reaction paths are chosen randomly, topological properties of networks after this evolutionary process show a significant deviation from those expected from random networks. In Fig. 6.9, the connectivity distributions $P(k)$ of chemicals obtained from the network

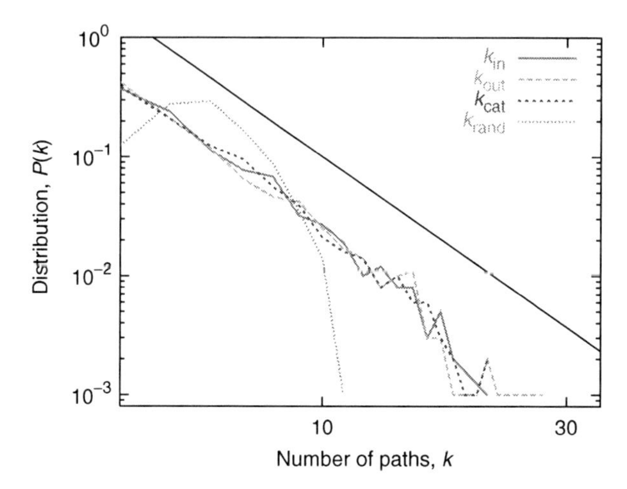


**Fig. 6.9.** Connectivity distribution $P(k)$ of chemical species obtained from the network of the 1000th generation. The model given in Sect. 6.3 is adopted. At each generation, networks giving a higher growth speed are selected. In the figure, $k^{in}$, $k^{out}$, and $k^{cat}$ indicate the number of incoming, outgoing, and catalyzing paths of chemicals, respectively. The *solid line* indicates the power law $P(k) \propto k^{-3}$. In comparison, we show the distribution of $k^{rand}$, obtained by a randomly generated reaction network having the same number of paths with the network of 1000th generation (For details, see Furusawa and Kaneko [2006a] from which the figure is reproduced)

of the 1000th generation are plotted, where $k^{in}$, $k^{out}$, and $k^{cat}$ indicate the number of incoming, outgoing, and catalyzing paths of chemicals, respectively, and $P(k)$ is the frequency of nodes with the connection number $k$. As shown, these distributions are fitted by power laws with an exponent close to $-3$. The evolved reaction network here forms a cascade structure, in which each chemical species is mainly synthesized from more abundant species.

The reason why the scale-free-type connectivity distribution emerges in this evolution is the following: attachment of paths to the chemicals with larger abundances is preferred as a result of selection process. Note that the power-law distribution of chemical abundances has already been established through evolution. Now, a connection of a reaction path to a more abundant chemical is more effective to increase the growth speed of a cell. For example, it is natural to assume that change of the growth speed by the addition of a path outgoing from a chemical increases with its abundance $x$, since the degree of influence on the cellular state is generally proportional to the flux of the reaction path added to the network, that is the product of substrate and catalyst abundances. Then the probability $q_{out}(x)$ to have such outgoing path *after selection* will increase with $x$, even though the network change itself is random. If such probability linearly increases with $x$, then the abundance power law ($p(x) \propto x^{-2}$) is transformed to the connectivity power law as $P(k_{out}) \propto {k_{out}}^{-2}$, as the abundance $x$ is mapped to the path number $k$ according to the probability. Then, the scale-free network with the exponent $-2$ should be evolved.

Numerically, we found that the probabilities $q_{out}(x)$ and $q_{cat}(x)$ are fitted by $q(x) \propto x^{\alpha}$ with $\alpha \approx 1/2$. In this case, the connectivity distribution is given by $k^{-(\alpha+1)/\alpha} = k^{-3}$. (See Furusawa and Kaneko [2006a] for detailed analysis.)

As for the evolution of the reaction network, preferential attachment to more connected nodes is often discussed (Simon 1955, Barabasi & Albert, 1999; Jeong et al., 2000). The present mechanism is related with this preferential attachment, but here the selection depending only on the cellular growth speed results in such preference, even though the attachment itself is random. Chemical species with a higher abundance acquire more reaction links, since attachment of new links to such chemicals gives larger influence to the cellular state and higher probability to be selected. With this mechanism, the power law in abundance is naturally embedded into the intracellular reaction network structure through evolution, that is simply the process to select cells with higher growth speed.

The emergence of the two statistical features is quite general and does not rely on the details of our model, such as the kinetic rules of the reactions and the parameters. We have studied various extensions of the present model, but the power laws both in abundances (through reaction dynamics) and in connectivity (in network structure) are universally observed as a result of evolution.

Here it is interesting to note that first the power law in abundances appears, and later through the evolution it is embedded into the power law in the network connectivity. Here the power law abundance is that of proteins (or other metabolites), while the paths in the network is given by genes, that ultimately determine whether a particular enzyme species that catalyzes a given reaction is present or not. When genes mutate, the network path is changed accordingly. In this sense, what we have observed here can be rephrased as "metabolic process (with proteins) first, and genes later." The power law in the abundance in the former is later "assimilated" by the gene, as network structure. We will discuss again such genetic assimilation (Waddington, 1957), in Chap. 10.

# 7

# Cell Differentiation and Development

## 7.1 Question to Be Addressed: Stability of Development

**Question: How does an identical cell diversify into a discrete set of types through development? Together with the diversification, how is recursive production of cells sustained? How is each character of each cell type stable, and how is the number distribution of cell types stably maintained?**

The process whereby just a single fertilized egg gives rise to a body consisting of many different cell types and showing distinctive patterns looks quite mysterious. Through a complicated process, a complex adult body is formed, which takes almost identical patterns – this looks like a miracle.

Each cell has at least the potentiality to process almost identical biochemical reactions. Through the developmental process, the character of cells starts to be differentiated. Organisms of the same species have identical sets of cell types, which form organized patterns that are again almost identical.

Studies to attribute a cause for differentiation to specific molecules have been pursued over several decades. Indeed, at early stages of development, a gradient of specific signaling molecules, called morphogen, is formed in a fertilized egg. This gradient brings about asymmetric cell divisions, which in turn lead to spatial pattern such as anteroposterior and dorsoventral axes. The spatial asymmetries formed by cells then lead to the creation of further asymmetry. As development proceeds in constructing more detailed structures, neighboring similar cells should be able to recognize small differences among them in order to differentiate.

Let us briefly summarize the standard picture of the cell differentiation process. Initially there is a gradient of some chemical, which works as a signaling molecule. Depending on the concentration of the signaling molecule, states of receptors, which are located at the cell membrane and influenced by the signaling molecules, are changed. Then this change is transferred into a cell state, and finally the expression of genes of the cell (or, in other words, composition of chemicals in the cell) is changed. In this way, expressions of genes in a cell are changed according to the concentration of a morphogen.

In general, small differences among cells brought about by their surroundings or by physiological conditions can be amplified if a given threshold of some biochemical is within the range of the difference. That is, if, within one cell, the concentration of a certain chemical (for instance, a morphogen) exceeds a threshold value, the corresponding gene is switched on. Vice versa, if a neighboring cell has slightly lower concentration of the biochemical, it turns off the gene. (See Forgacs and Newman (2005) for physical approach to development).

In the traditional threshold mechanism, it is generally believed that genes switch on and off depending on the concentrations of signal molecules. Indeed, several researches have done good justice in showing how cells interact with each other through signals, and how such received signals are amplified to change the state of a cell.

Hence, the standard answer to the origin of differentiation is as follows: Assume the existence of a gradient of concentrations of some chemicals. Depending on the position of the cell, the chemical concentration changes. Then, assume that a gene is on or off depending on whether the chemical is larger than a given threshold or not. Then depending on the position of the cell, a certain gene is on or off, and cells differentiate accordingly. By differentiation of cells, then each cell starts to have a different chemical composition. Then, a new gradient of other chemicals may be formed, which may lead to a new differentiation.

Although this picture may be correct to explain each step of cell differentiation, the above-mentioned experimental results cast a question on the threshold mechanism when it comes to the robustness of development process. Since each signal involves stochastic error due to molecular fluctuations, the issue then is basically on how the matured body comes to have a well-developed form with all the accumulated uncertainties involved in the whole development (Kaneko & Yomo, 1999).

Recall that all the intracellular processes consist of molecular reactions. The number of molecules in a cell and the signalling molecules at each cell position fluctuate. Typically, the concentration fluctuation for a molecule with the number $N$ should be $1/\sqrt{N}$. Since the number is not so large, fluctuations of the concentration are not necessarily tiny. As long as the molecules show Brownian motion, these fluctuations are inevitable. When there are $N$ signal molecules per cell, there exists fluctuation of the order of $\sqrt{N}$ for "$N$" molecules as a result of Brownian motion. This stochasticity is a general consequence of probability theory and thermodynamics, and cannot be avoided. (Recall also the quantitative measurements of fluctuations presented in Chap. 6).

The condition that the concentration is larger than some threshold is given by the *number* of molecules per cell, that is not necessarily so large. Then, in the terms of concentrations, fluctuations of the order of $1/\sqrt{N}$ are inevitable.

Accordingly the probability that a gene is not switched on or off as designed is $1/\sqrt{N}$ (see Fig. 7.1).

For example, consider the differentiation process, schematically shown in Fig. 7.1, where a cell located beyond the threshold concentration becomes "white" type and otherwise "black" type. As a concentration crosses the threshold value between two neighboring cells, one becomes white and the other black, which leads to cell differentiation with white–black pattern. Still, with the above rate of error, the cell pattern with white–white or black–black may appear.

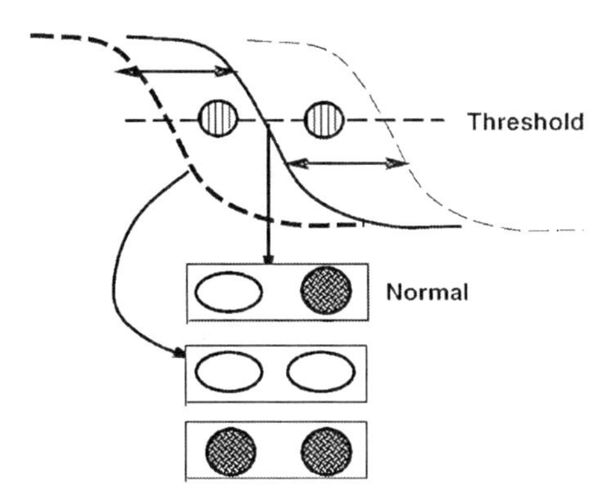


**Fig. 7.1.** The threshold mechanism and its robustness against fluctuations. The chemical concentrations show fluctuations as shown. Then the differentiation to distinct type of cells does not follow

Let us take an example from *Drosophila*, a fruit fly. In *Drosophila*, the developmental process is studied in depth. The process is believed to be rather well programmed among multicellular organisms. As shown in Fig. 7.2, recent experiments show that there are about 30% fluctuations in the Bicoid protein, that is believed to be an important signal molecule to determine the cell differentiation from the Drosophila egg. This is nothing but the fluctuation expected from the above argument (Houchmandzadeh et al., 2002). In spite of such errors, however, the morphogenesis proceeds without large errors. In fact, even by disturbing the Bicoid concentrations externally, the development was found to progress normally (Lacalli & Harrison, 1991).

This is an error for a single gene. Let us assume that there are $M$ genes, each of which is turned on or off in accordance to the signal with the threshold of $N$ molecules. Each on/off pattern of $M$ genes corresponds to a cell type. Then the probability that a "correct" cell type is formed as designed is $(1 - 1/\sqrt{N})^M \approx \exp(-M/\sqrt{N})$. This probability is so small (e.g., 0.04 for $N = 1000$ and $M = 100$) that the developmental process will scarcely proceed constitutively.

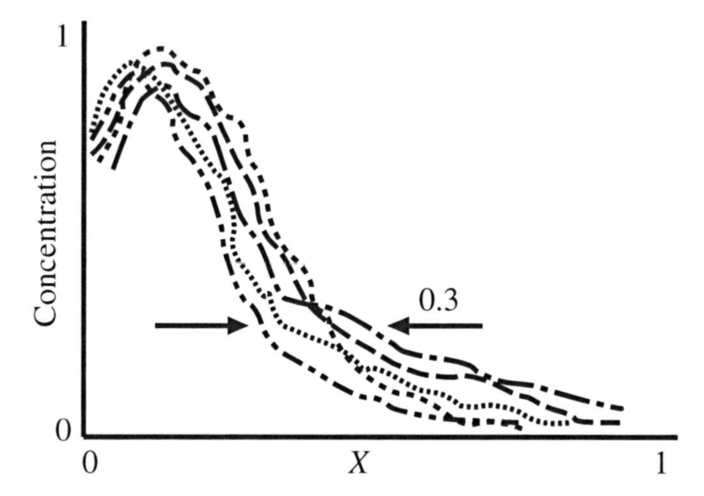


**Fig. 7.2.** Concentration of Bicoid along the axis of Drosophila. The $x$ axis denotes the position, normalized at one end as 0, and at the other end as 1. The concentrations of several eggs are plotted with this position $x$, which is scaled so that the maximum is 1. The range denoted by the arrows is roughly 0.3, which means about 30% error by each egg. Reproduced from the result of Houchmandzadeh et al. (2002)

One might argue that the error could be eliminated by proofreading mechanisms that exist in a natural cell to some extent, to support the threshold mechanism. However, such proofreading mechanisms will be triggered only when a certain condition is satisfied. Since the required condition is also error prone, then another molecular fluctuation sets in. In short, different mechanisms to correct errors in a cell give rise to a chain of fluctuations. Hence, there will always be fluctuations even if we consider all possible fine-tunings of the threshold value through evolution.

Put differently, a threshold mechanism is regarded as a process of cutting the state space into two, along a plane. The cellular state with a signal molecule larger than a threshold is given on one side of the state space separated by the surface. Then, several conditions for successive gene expressions correspond to successive cuts by surfaces, as schematically shown in Fig. 7.3. By several switches of on and off of genes, state space is separated into several areas corresponding to different cell types. Then assume each surface fluctuates by a few percent error width, as shown in the figure. By successive errors of these "blurred" surfaces of section, the separation into areas would be fused. As the number of surfaces increases, the probability of forming a "correct" section will go to zero.

Let us discuss this problem again from a different point of view, that is, in terms of programming. Indeed developmental process is often described in terms of "program" as in the discussion of body plan. Each threshold mechanism gives a condition like a program "**If** concentration of signal molecule is

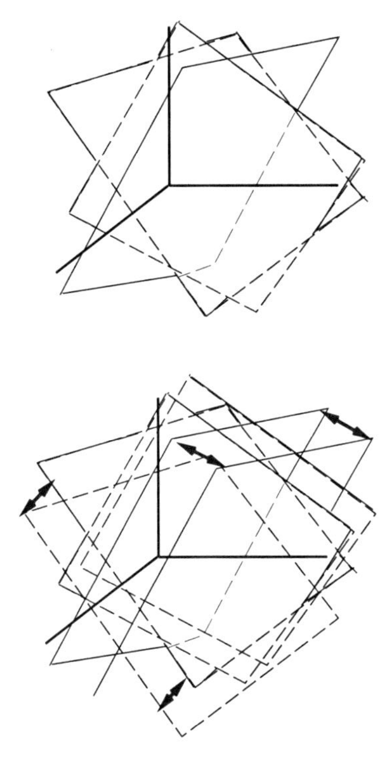

**Fig. 7.3.** Threshold condition for cellular state corresponds to splitting the phase space by the surface given by the threshold condition. This surface is blurred because of the fluctuations. As the number of conditions for split increases, the number of surfaces increases also. If each surface is blurred with some width, then with the increase of the number of conditions, the split by some condition will not work

larger than a threshold, **then**, a gene is on, ...." Combination of these if–then processes form a program of development. This looks like a process written by computer programming. One important problem, however, is that each step in this procedure involves some errors (say 3%). Then it is hard to believe that the total development process will work correctly as planned. In other words, even if one intends to write a "computer program" in terms of reactions with the molecule number of $O(1000)$, the probability that the program works "correctly" will go to zero. We need some viewpoint other than logical "if–then" operations in a computer programming.

In spite of these fluctuations in molecule numbers, however, the development is surprisingly robust. All organisms of the same species have almost identical spatial patterns consisting of the same set of cell types. Cells of the same cell type, although they differ from each other, adopt roughly the same intracellular composition of chemicals. How is such robustness in development possible?

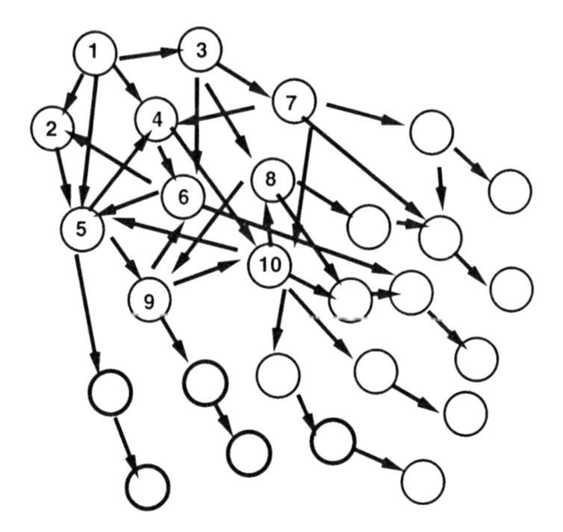


**Fig. 7.4.** Schematic picture of a cellular reaction network. Some chemicals form a complex network through catalytic reactions as discussed in the last chapter, while some other are synthesized passively. Here we concentrate on a part with positive feedback process through mutual catalytic process

### 7.1.1 Kauffman's Gene Network Model

One possibility to answer this "robustness" question is given by the use of the "attractor" picture. As mentioned already, metabolic and genetic reaction processes form mutually catalyzing reaction network, as schematically shown in Fig. 7.4. Then chemical concentrations of each cell (as well as gene expression) change in time as a result of temporal evolution. By adopting a dynamical systems viewpoint, cell state can be represented as a point of state space representing chemical concentrations of a cell. Then, each cell type may be regarded as an attractor of intracellular dynamics. If this picture is valid, the above error problem may be resolved: After some perturbation, the cell state represented by an attractor comes back to the original state through the temporal evolution of intracellular reaction dynamics.

Indeed, Kauffman (1969) pioneered this attractor picture of cell differentiation. In a cell, genes are expressed into mRNA and then into proteins including many enzymes, which may catalyze biochemical reactions, including metabolic reactions. These enzymes influence the expressions of genes. With these reaction processes, the concentrations of proteins are changed, which again activate or suppress gene expressions. These reactions are very much complicated and complex. Instead of considering reaction processes including metabolites, Kauffman focused only on the genetic expressions. Of course, these dynamics are not self-contained by the gene expressions only, but all other metabolites are involved. As a drastic simplification, however, Kauffman treated all states other than gene expressions as a black box. Through this black box, genes mutually activate or suppress their expressions. Furthermore, Kauffman represented the gene expressions by only two levels, that is, on or off. The feedback of the reaction process onto the gene expressions is represented by a transition rule, represented by a combination of Boolean operators. (Since each gene has just two states on–off, the influence of one gene upon another through the black box is given by a rule table of on–off patterns. For example, a rule "a specific gene is on, if and only if two genes are on" is represented by "AND" operation in Boolean algebra, and so forth.)

In this model there are $N$ genes with "on" or "off" states, a total of $2^N$ states. For example, take two genes. Depending on their state (on or off), another gene is on or off. This transition rule is then represented by Boolean algebra, say "AND," "OR," "exclusive-OR," and so forth. The connection among genes is chosen randomly and fixed. By again adopting a random table from all Boolean functions, one can set up a rule for the change of on–off states of each gene. Since each "on–off" is given by a randomly chosen network of Boolean functions, this model is called the random Boolean network model (Kauffman, 1969, 1993).

With this rule for the dynamics of switching genes, the on–off pattern of genes changes in time. Through the dynamics of the gene regulation network, only restricted combinations of on–off patterns of genes are allowed as attractors. In general, there exist multiple attractors, and in this case, depending on different initial genetic expression patterns, the state will be attracted to one of the attractors. Indeed, in this model, Kauffman has shown that many attractors coexist as a different pattern of gene expressions, each of which is regarded to correspond to a differentiated cell type.[1] Even if there are some fluctuations to bring about errors in the on/off of gene expression, such error can be removed to recover the original pattern of each cell type through the mutual interaction among genes, as each cell type is represented as an attractor. Hence, stability of each cell type could be expected.

As for differentiation from one type to another, Kauffman assumed that signal molecule may switch the expression of some gene between on and off, with which an attractor may be switched to a different attractor. Hence differentiation of cell type is represented as a switch among attractors by external inputs, while each cell type corresponds to a different attractor. Kauffman's work was pioneering and the picture he proposed is quite important. Still, some problems remain to be solved.

(1) Depending on the initial condition, an attractor is selected. However, how is the initial condition of a cellular state selected? In physics, the initial condition can be set by the experimentalist, but in the developmental biology, one must also understand how the initial conditions of cells are selected through the developmental process of an ensemble of cells.
(2) Once gene expression pattern falls on an attractor, the gene expression pattern will come back to the original pattern of the attractor even when some error leads to switch on or off a single gene. Thus, the attractor picture helps explain the robustness of a final cellular state. However, for cell differentiation in the developmental process, a switch from one state

[1] The number of attractors increases with some power of the number of genes.

to another has to occur, because of the external input supplied by signal molecules. Then, the original state has to be destabilized when some perturbation is applied. This, on the one hand, implies instability in the cellular state to specific perturbations. Now, there exists fluctuation in the concentrations. Hence, some fluctuation may determine whether the cell state is differentiated or to which attractor it is switched. Some error to change the gene expression at this event of differentiation process may alter the destination of the final cellular state (attractor). If error occurs during cell differentiation process, the cellular state, that is, the gene expression pattern after this differentiation would be affected. As dynamics for differentiation from one cellular state (attractor) to another generally passes through unstable orbit (imagine a path crossing the top of the hill), instability in the developmental path against molecular fluctuations is not resolved.

Summing up, although the stability of a given cell type may be explained by attractor picture, the stability of differentiation process is not answered yet. Stability of differentiation against molecular fluctuations is not clarified.

(3) In the attractor picture so far, only the stability of a single cell state is concerned. However, to discuss the stability of an ensemble of cells, we need to understand how the stability at a population distribution level is achieved – that is, the number distribution of existing cell types in a tissue, an ensemble of cells. This number distribution is selected through the course of developmental process. As the number increases, the cell differentiation proceeds, to form a tissue consisting of different cell types with given number distribution. We need understand the stability at this distribution level also, besides the stability of each cell state.

To discuss this problem, we need to seriously consider cell–cell interaction. The importance of cell–cell interaction to developmental robustness is also discussed as the community effect (Gurdon et al., 1993) from experimental observations, as well as from theoretical considerations (Kaneko & Yomo, 1999).

Considering the points (1)–(3), the attractor picture of a single cell state is not sufficient. To answer this question, we need to consider a system of interacting cells, as simple as possible, but not too simple. We need to take into account at least intracellular reaction network, cellular interactions through chemicals that serve as signals, and the increase of cell number by cell divisions.[2]

To close the present section, we repeat the question that is asked here: Each cell differentiation comprising the whole development is inevitably

[2] Attraction to different states is also studied by using a coupled system of Boolean-network-type differential equations (Glass & Kauffman, 1973), while a coupled-map-lattice corresponding to it is studied by Bignone (1993). Also, Mjoliness et al. (1991) studied a "realistic" genetic network model with cell-to-cell interaction.

accompanied by stochastic errors, since it is triggered by stochastically fluctuating events like diffusion of a molecule or binding to a certain receptor, which are no other than the so-called signals by most of the researches. Assuming that the embryo is a machine like a parallel processor, then all cell differentiations occurring in the development have to follow a strictly organized course. However, this does not hold true because of the uncertainty of each cell differentiation. Therefore, the whole development cannot proceed in a machine-like manner via only the threshold mechanism. Then, how is developmental stability with cell differentiations possible? How does the cell–cell interaction lead to cell differentiation to several cell types, and how are they stable? Besides the stability of each cell type, how is the realized number distribution of each cell type robust against molecular fluctuations?

## 7.2 Logic: Isologous Diversification

Here we will discuss a logic to answer the question raised in the last section. Indeed, this theory is derived from extensive simulations of models introduced in the next section. Here we sketch the logic, while detailed accounts are given in Sect. 7.4.

Let us start with a very primitive description of a cellular state, following the discussions in Chap. 6. In a cell, there are several chemical components. Then a cellular state depends on chemical compositions within a cell. These chemicals include proteins, RNA, membrane, and so forth. Through chemical reaction processes, the chemical compositions of a cell can change in time. Of course, a cell is more complicated, but this basic standpoint will at least be required to study a cell state.

With the intracellular reactions, some chemicals are converted to products, and at some stage, the cell becomes large enough to be divided into two. Without assuming any sophisticated programs, the two divided cells take almost identical chemical compositions, although there are some differences due to molecular fluctuations (Fig. 7.5).

Note that positive feedback process exists in the intracellular reactions. On the other hand, these cells interact with each other through diffusion of chemicals across the membrane. With nonlinear dynamics in the reaction, the fixed cell state may be unstable and change in time, also depending on the states of other cells. Furthermore, with the positive feedback process in the intracellular reaction, tiny differences between the two cells may be amplified. This amplification depends also on cell–cell interaction, including competition for nutrient chemicals for cell growth. As the cell number is increased, this cell–cell interaction is stronger, and the amplification of tiny difference can be stronger. At some stage of cell numbers, this amplification becomes large enough to make a macroscopic difference between the two cells. Now a cell ensemble with homogeneous states will be unstable.

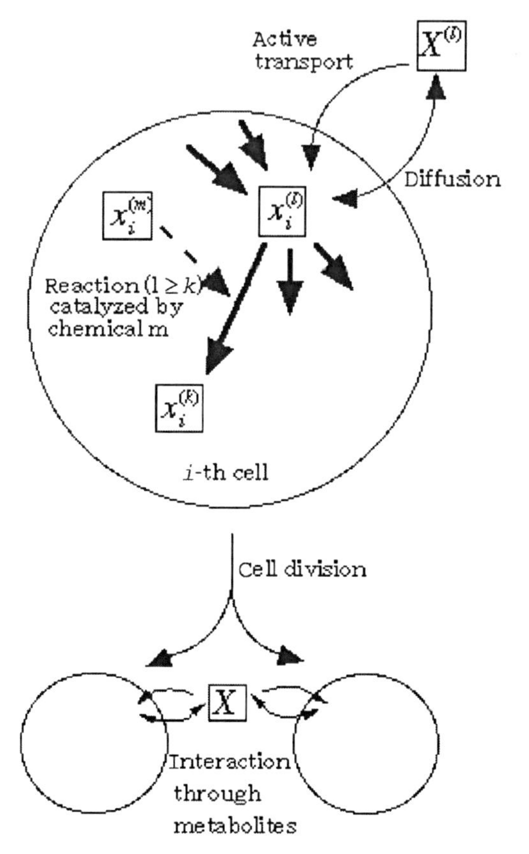


**Fig. 7.5.** Schematic representation of our logic. See text for details. Reproduced from Kaneko and Yomo (1999) with permission

Here, it is expected that cells with similar states compete strongly for nutrients for growth whereas cells taking different states do not compete with each other for the nutrients so much. It may then be expected that differentiation of cells comes to a stage that different types of cells mutually stabilize others' states.

A schematic example is shown in Fig. 7.6. Here two types of cells A and B exist, where existence of type B cells stabilizes the state of A cells through cell–cell interaction, and vice versa. In other words, the states A and B are "attractors" of intracellular dynamics, under the condition of cell–cell interaction with each other. Now, these states are stable under molecular fluctuations.

This mutual stabilization is possible under suitable cell–cell interactions. To achieve stability, the number distributions of cell types A and B should be under some range. If the distributions are biased, the states will be unstable. This is understood by taking an extreme limit. If only the type of A cells exists, the cell states are almost homogeneous. However, the type A cell state

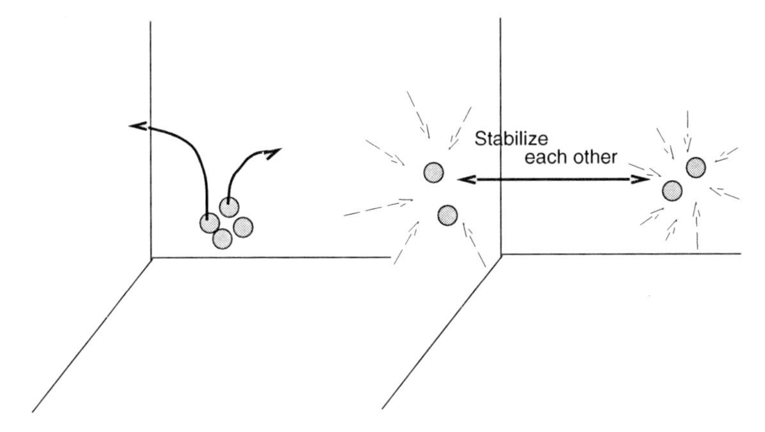


**Fig. 7.6.** Schematic representation of state space and differentiation. A homogeneous cell state is destabilized, and by taking distinct states, the cells mutually stabilize each other's state

is not an attractor of a single cell dynamics. Thus this homogeneous cell state becomes unstable, to make de-differentiation of type A cells, and some of them will be type B cells, so that the cell ensemble becomes stable. In other words, because of instability in the homogeneous cell state and cell–cell interactions, the differentiated cell types and the number distribution of cell types are stable.

This diversification of cell types into a discrete set of types from a single cell type is a general consequence of interacting cells with biochemical networks and cell divisions, as will be confirmed by several model simulations. According to the simulation results, differentiation proceeds first by loss of synchrony of intracellular oscillations of chemical concentration, as the number of cells increases. Then the chemical composition of the cells is differentiated. The differentiated compositions become inherited by the next generation and lead to determined cell types. As a result of successive occurrence of the cell differentiation, the cell society will be composed of different cell types. Here a threshold mechanism is not implemented in advance, but a threshold-type behavior emerges as an outcome, accompanied with robustness in the development process.

The present mechanism leading to differentiation is called isologous diversification, since it shows a mechanism how identical (cell) state can be diversified through the interplay between internal (reaction) dynamics and cell–cell interaction. With this mechanism, cell differentiation is shown to be stable against molecular and other external fluctuations, where amplification of noise-induced slight difference between cells leads to a noise-tolerant society with differentiated cell types.

## 7.3 Model

Following the prescription discussed so far, we adopt some simple models consisting only of basic features of cells. Some common features are extracted among the results from the computer simulations of several models, although, of course, there remain specific phenomena depending on each model and the parameter values adopted. A run-through on the common characteristics brings us essential features and mechanisms in cell differentiation.

The environment of a cell culture is always taken as one of the essential factors for cell differentiation. If there is spatial bias among the cells, for example, a gradient of an activator protein, the cells may easily undergo differentiation. However, experimental studies on some bacterial cultures (Ko et al., 1994) showed that even in a homogeneous environment, cells can differentiate in due time. In other words, spatial bias is, in fact, not required for cell differentiation. Hence, the cells in our models are grown in a homogeneous environment where there is no spatial bias in chemical concentration. (In fact, by further allowing for spatial bias in our model, the cells in our model show cell differentiation and morphogenesis, as will be shown in Chap. 9.)

Our model consists of intracellular biochemical reaction dynamics, cell–cell interaction through medium, molecular fluctuation, and cell division (see Fig. 7.7 for schematic diagram). Let us describe each process.

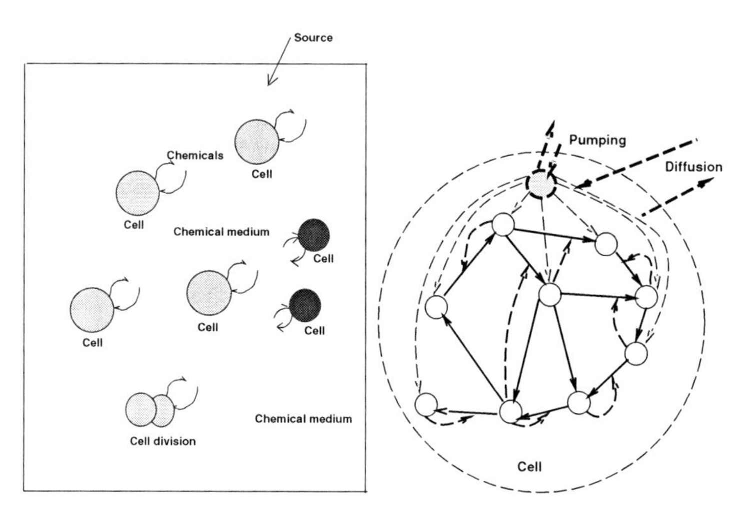


**Fig. 7.7.** Schematic representation of our model: reproduced from Kaneko and Yomo (1999) with permission. The cells are in a well stirred medium to which nutrition components are supplied. A cell takes the nutrition by inflow through the membrane, and there is a mutually catalytic reaction network in each cell, identical for every cell

### 7.3.1 Internal Biochemical Reaction Network

The reaction network in a cell consists of numerous biochemicals. A protein activates a certain gene, whose product (e.g., enzymes) catalyzes a certain metabolic reaction. The product of metabolic reaction plays the role of an activator to the previous activator protein or to an enzyme in the intermediate reaction. These chain reactions or autocatalytic reactions in a cell (or simply the nature of catalytic reaction with activation or inhibition) constitutes a nonlinear reaction network.

To be specific, we choose the concentration $x_i^\ell(t)$ of the chemical $\ell$ of the cell $i$ as dynamic variable, and study its evolution. With the internal reaction dynamics from the chemical $m$ to $\ell$ catalyzed by the chemical $j$, $\frac{dx_i^m(t)}{dt}$ includes the Michaelis–Mentens term $e_1 x_i^{(j)}(t) x_i^{(m)}(t)/(a_M + x_i^{(m)}(t))$. The total intracellular dynamics consists of a network of reactions. This network is chosen randomly and fixed throughout the simulation. When the reaction is autocatalytic, $j$ agrees with $m$, in the above term. Such autocatalytic reaction often leads to oscillatory behavior in chemical concentrations (Fig. 7.8).

As shown by Hess and Boiteux (1971), concentrations of metabolites involved in glycolysis oscillate the nonlinear reactions catalyzed by some enzymes involved in the reaction network. In general, the biochemical reaction network of a cell is composed of highly complex nonlinear reactions. Hence, we choose such reaction network allowing for oscillatory dynamics.[3] The concentrations of the biochemicals may be regarded as that of metabolites or expression level of a gene or of a gene network. The importance of temporal oscillations in cellular dynamics was studied in pioneering work by Goodwin (1963). Recently, the existence of oscillatory dynamics for cell division processes has been discussed both experimentally and theoretically in cycline and M-phase-promoting factor (Tyson et al., 1996).

### 7.3.2 Interaction Among the Cells

There are many types of cellular interaction ranging from the diffusion of morphogen to simple competition for nutrients among cells. Here, it will be relevant to choose the most fundamental type of intercellular interaction. Let us consider some of the simplest organisms like *Anabaena*, a cyanobacterium or *Escherichia coli*. Under ammonia or nitrate deprivation, cells of *Anabaena* differentiate into two types, one specialized for nitrogen fixation, and the other for photosynthesis (Golden et al., 1985). To our knowledge, no reports to date have suggested any special mechanism that governs its cell differentiation. It is also shown that *E. coli* cells can differentiate even in a homogeneous environment (Ko et al., 1994) (see Sect. 7.6). Simple as they are, the most

[3] Indeed, the proposed scenario may work, without oscillatory dynamics, as long as some instability is included. Also, in some models, oscillation that existed in initial stage of cells disappears for some cell types determined later.

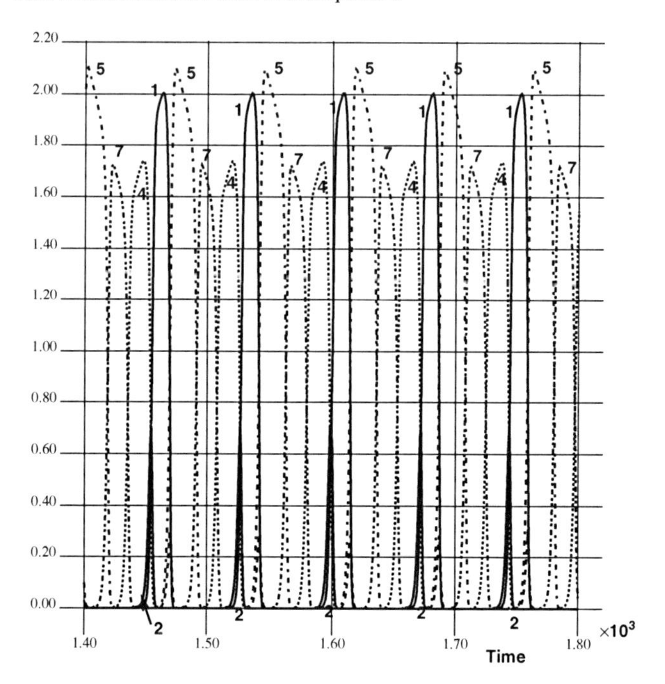


**Fig. 7.8.** An example of oscillation of chemical concentrations in a model (reproduced from Kaneko and Yomo (1997) with permission)

probable fundamental factor to be considered is then the interactions among the cells through the biochemicals in the reaction network of the cells.

Our models, therefore, include a simple diffusion process of biochemicals in and out of each cell. By denoting the concentration of the chemical $m$ at the medium by $X^m$, the diffusion term from the medium to the cell $i$ is given just by $D^m(X^m(t) - x_i^m(t))$, while the chemical in the medium is reduced by $-D^m \sum_i (X^m(t) - x_i^m(t))V_{rel}$, where $V_{rel}$ is the ratio of the volume of each cell to that of the medium.

In some models, besides the simple diffusion process, we have also included active transport from the medium to the cell, which keeps the cellular state out of equilibrium. This gives a transport term $X^m \times F$, with some activity $F$, which generally depends on the concentrations of biochemicals in the cell. For example, the activity $F$ is given by sum of chemicals in the cell (see also the Appendix). Again, this term leads to global cell–cell interaction through the medium.

As is shown below, there will be differences in the concentrations (and composition) of biochemicals between the cells. Existence of a gradient in turn leads the cells to share the chemicals in a simple manner. That is, from the medium, one cell can take up a chemical that is produced by the other cells and diffused into the medium. In short, cells interact with each other through the process of diffusion.

### 7.3.3 Cell Division

To proliferate, a single cell undergoes cell division resulting in its maintenance in the society. In our models, cell divisions occur when the sum of concentrations of some chemicals reaches a given threshold value, following the models in earlier chapters. In one class of models (adopted in this chapter), the concentration of a "final" product determines the next division.[4] It is to be noted that two cells arising from a single cell contain almost identical chemical compositions with some deviation (e.g., 0.1% used in most of the simulations). It will be shown that differentiation occurs among the cells, although a cell in our model divides into nearly identical ones.

Cell division and intracellular dynamics are mutually related. It will be shown below that the intracellular dynamics show oscillation in an ensemble of cells. This oscillation is maintained by cell–cell interaction. If one of the cells divides, the balance that maintained a certain type of oscillation is disturbed, because the new cell introduces new interactions into the system. As a result, the dynamics of the individual cells will continue to change. In fact, in this kind of "open" system, where the number of variables grows as in a cell culture system, several types of different dynamics appear successively, even if each element alone has simple oscillating patterns.

### 7.3.4 Molecular Fluctuation

So far, the rate equations of chemical concentrations are obtained by neglecting molecular fluctuations: this approximation is valid if the number of molecules is large enough. In contrast to the simulations of earlier chapters, we adopted the rate equation for chemical concentrations here. "Continuous" variables $x_i^m(t)$, concentrations of chemicals, are adopted instead of "integer" numbers of molecules. In reality, the number of molecules for each chemical is not necessarily huge, and the number of each signal molecule is typically of the order of 1000 or so, as discussed already. Thus a noise term should be included to take into account (thermal) fluctuations arising from finiteness in the number of molecules. Considering fluctuations of $\sqrt{N}$ for the reaction of $N$ molecules, we have added a noise term proportional to $\sqrt{x_i^m(t)}\eta(t)$, with a random force $\eta(t)$, represented by a Langevin equation. The amplitude of the noise is denoted by $\sigma$.[5]

[4] In the model of the next chapter, the sum of all chemical abundances, giving a cell volume, determines the division. The scenario to be presented holds in this case also.

[5] Of course, another approach is the use of stochastic model, as adopted in earlier chapters, and includes cell–cell interactions. Indeed we have carried out such simulations also, but the results to be presented below are not altered.

### 7.3.5 Model Equation

Specific form of the equation adopted in our computer experiments is not so important. In the Appendix, we list an example of a set of equations, adopted in our computer experiment, mainly for readers who are interested in working in this type of study.

*Remark.* Instead of using catalytic reaction network model, we can adopt a gene expression model, where gene expression takes either high or low value depending on the "input" to the gene. For example, one can adopt a threshold-type response for an element $x^i$ as

$$dx^i/dt = -\gamma x^i + F(x^i_{inp}) \tag{7.1}$$

where $F$ is a monotonous function approaching 0 for small $x^i_{inp}$ and some positive constant for large $x^i_{inp}$, such as the form

$$F(z) = z^n/(1+z^n)\,.$$

The input into a given gene $x^i_{inp}$ is given by activation or inhibition from other genes, and for example given by $x^i_{inp} = \sum_j J_{ij}(x^j - thr_j)$. For some network, the final expression level of genes $x^j$ approaches a stationary pattern corresponding to fixed on/off pattern of genes. However, for many complex networks, there appears oscillatory dynamics in gene expressions. As for intra-cellular dynamics, we can choose this gene expression network model instead of catalytic reaction network model. By including interaction and growth suitably, an intra-intercellular dynamics model as in the present section is derived. Indeed, the results to be presented in the next sections hold for this class of model also.

## 7.4 Results

From several numerical experiments of the model, the logic mentioned in Sect. 7.2 is confirmed. Here we first show numerical results for each step of the differentiation. Indeed, the scenario, originally extracted from simulations without molecular fluctuations, works completely well up to some noise threshold, as will be discussed later.

### 7.4.1 Five Stages of Isologous Diversification

1) *Synchronous oscillations of the chemicals in the cells.* Only up to a certain number of cells (e.g. eight cells) can the dividing cells from a single cell have the same characteristics. Although each cell division is not exactly symmetrical because of the accompanying noise-level of perturbation in the biochemical composition, the phase of oscillation in the concentrations

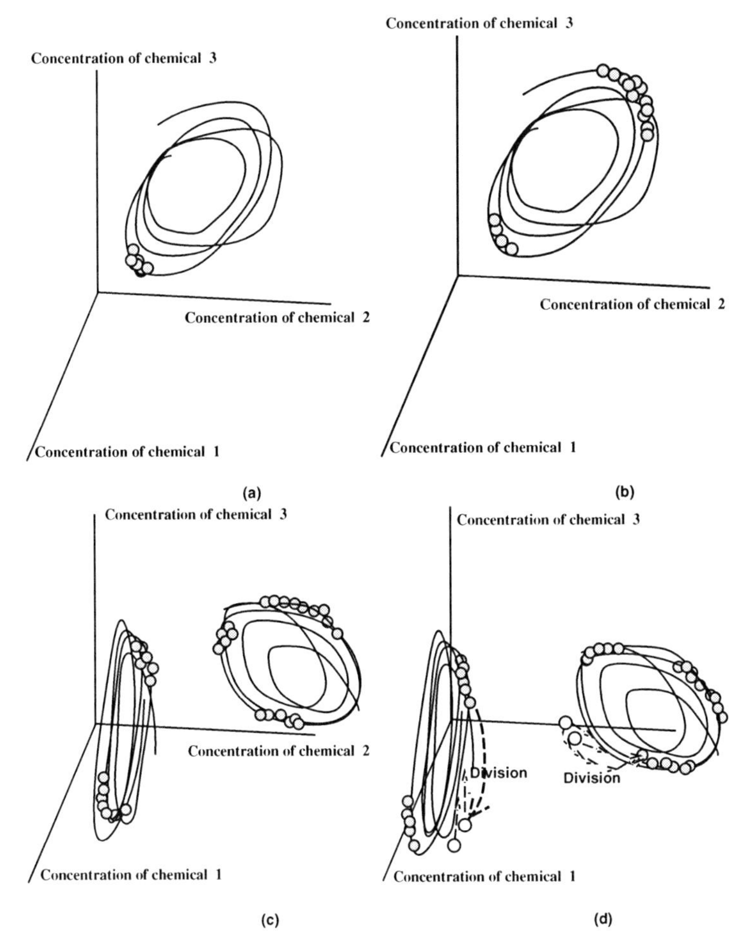


**Fig. 7.9.** Schematic representation of the stages of our differentiation scenario. These results are supported by numerical simulations of several models. See Kaneko and Yomo (1997) and Furusawa and Kaneko (1998a) for direct numerical results

is synchronized (see Fig. 7.9a). Consequently cells divide almost at the same time.

Such synchronous cell division is also observed with the cells in the embryogenesis of mammals up to eight cells. In our model, the synchronous division is kept until small differences due to molecular fluctuations or slightly unequal distribution of metabolites in each cell division are amplified.

2) *Clustering in the phases of oscillations.* As slight differences among cells are amplified, the synchronous oscillation breaks and hence, cells start to show different phases of oscillations. Cells split into few clusters where the cells belonging to each cluster are identical in phase.

However, this diversification in the phases is not to be mistaken as cell differentiation, since taking the average concentration of biochemicals in time reveals that all the cells are almost identical. Hence, in the second stage of isologous diversification, the cells are different only in their phase of oscillation, brought by the difference in the concentration of biochemicals.

The time course of changes in the concentration of biochemicals, since it is governed by nonlinear reaction network, is sensitive to slight differences brought by molecular fluctuations or cell division. Once these small differences are amplified, the newly divided cells will produce different phases of oscillations. For instance, if two cells from a single cell have slightly different compositions, rates of biochemical reaction are affected by the difference in the phase of metabolic oscillations between the cells, since the reactions involved are autocatalytic. Hence the phases of metabolic oscillations start to be different between the cells (see Fig. 7.9b).

Sensitivity of the time course (of chemical concentrations) to tiny perturbation is nothing but orbital instability as explained in Sect. 3.1. Although perturbations caused by cell division would be amplified so that all the cells would have their own different phases of oscillation, our models show that the cells are not completely diversified in their phase but rather tend to resolve into an ordered clustering as schematically shown in Fig. 7.9(b).

This clustering of phases is due to cell–cell interaction. In our model, the state with identical cells (i.e., $x_i^m(t) = x_j^m(t)$ for different cells $i$ and $j$) is unstable at the second stage. With this orbital instability, small differences between cells are amplified. However, by forming clusters with different phases, such instability is smeared out. Because of the cell–cell coupling, the dynamics in each group retains stability. Now small differences in each group are no longer amplified.

Differences in the metabolite concentrations between the clusters are being balanced by the biochemicals secreted from all the cells. The sensitivity to small change is smeared out by this balance. This event in turn leads the cells to stay at a certain phase of oscillation, resulting in the clustering. In other words, it is the transport of biochemicals from one cell to another or cellular interactions (regarded as cell signals) that gives rise to the different clusters in phase.

The mechanism governing the clustering of the phases is nothing but the clustering discussed in Sect. 3.5, where robustness of the state is mathematically clarified. Such robustness is expected as stability is achieved by the clustering. The number of cells in each cluster has to be within a certain range, to have such stability. In other words, developmental process leads the number to stay within the range, independent of initial conditions and perturbations. Hence, the distribution of cell numbers into each cluster is rather robust against perturbations.

3) *Differentiation in metabolite composition.* As the cells with different phases continue to undergo cell division, the average concentrations of the

biochemicals over a cell cycle become different between clusters of cells. That is, the composition of biochemicals as well as the rates of catalytic reactions and transport of biochemicals across membrane become different between the clusters.

The composition of biochemicals as well as the rates of catalytic reactions and transport of the biochemicals become different for each group. The orbits of chemical dynamics plotted in the phase space of biochemical concentrations lie in a distinct region within the phase space, while within each group, the phases of oscillation vary from cell to cell (see Fig. 7.9c).

Hence distinct groups of cells are formed with different chemical characters. Each group is regarded as a different cell type, and the process to form such types is called differentiation. In biological terms, the third stage is no other than the division of labor of several biochemical reactions in the cell, since the use of nutrients is different for each group.

The temporal difference in the phases leads to the difference in chemical concentrations between cells. The difference of each biochemical activates its transport by a simple diffusion process and makes a small difference in its composition among the cells. The small difference is then amplified through the nonlinear nature of reaction network, leading to clusters of the cells with different intracellular dynamics.

The composition of biochemicals of each cell is not yet an inherent property, since the intracellular dynamics governed by the nonlinear reaction network, generally, if not in all cases, may vary in time. Hence, the composition of a newly divided cell can be different from that of the parent cell. Therefore, each newly divided cell has a different rate of biochemical reactions depending on the composition. In short, cells are at the intermediate stage of the differentiation process.

4) *Determination of the differentiated cells.* As the cells continuously undergo further cell divisions, they start to have their own inherent composition that is preserved by the next generation of cells (see Fig. 7.9d). That is, the cells come to a stage where the reaction dynamics and the chemical composition are not much influenced by the environment and other cells. Hence, the biochemical properties of a cell are inherited by its progeny, or in other words, the properties of differentiated cells are stable, fixed, or determined over the generations. This is the fourth stage of isologous diversification.

Now the recursive production is achieved. In Fig. 7.10, chemical averages of cells between successive divisions are plotted as the "return map," that is the relation between the chemical averages between the mother and daughter cells. Up to around the 6th division here, chemical averages differ by divisions. Later, the recursive production is seen as points lying around the diagonal ($y = x$) line.

The balance between the clusters, each having its own compositions in the third stage, is attained through cellular interactions in the same way as in the dynamical clustering of the oscillation phase at the second stage.

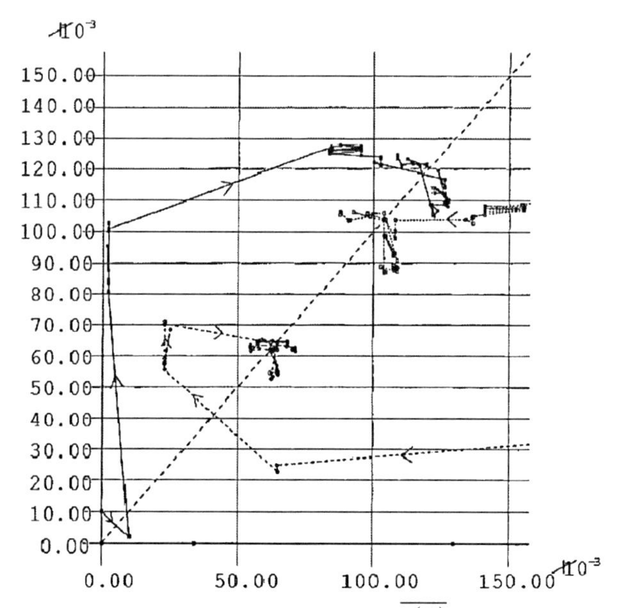


**Fig. 7.10.** Return map of chemical concentrations $\overline{x_i^{(m)}}$ averaged over two successive divisions. A daughter cell's average concentration is plotted versus its mother cell's average before the division to the daughter. Chemicals 2, 5, and 8 are plotted with different marks, while the *dotted lines* are drawn only for convenience. The lower column is the expansion of the upper column. See Kaneko and Yomo (1997) for details

Although the metabolite composition of the cells tends to change over the generations, this tendency is compensated by the intake of biochemicals secreted from the cells belonging to the other clusters.

In the present theory, the inheritance is achieved through the transfer of "initial conditions" at the division. The discreteness in the types of chemical characters, noted previously, is relevant to preservation of the type to daughter cells, since continuous change (like the phase of oscillations) may easily be blurred by the division process and cannot be transmitted to daughters robustly.

As soon as the features of the cells are inheritable, a cell lineage map can be drawn, to show where the cell types branch out from their origin. Generation of cell lineage gives rather useful information to be compared with that obtained in cell biology. From the cell lineage, one can see how the differentiation processes hierarchically, and how inheritance of cellular state is realized.

In Fig. 7.11, we have plotted the cell lineage diagram, where the division process with time is represented by the connected line between mother and daughter cells. The gray scale in the figure shows the cell type determined according to the chemical averages in Fig. 7.9.

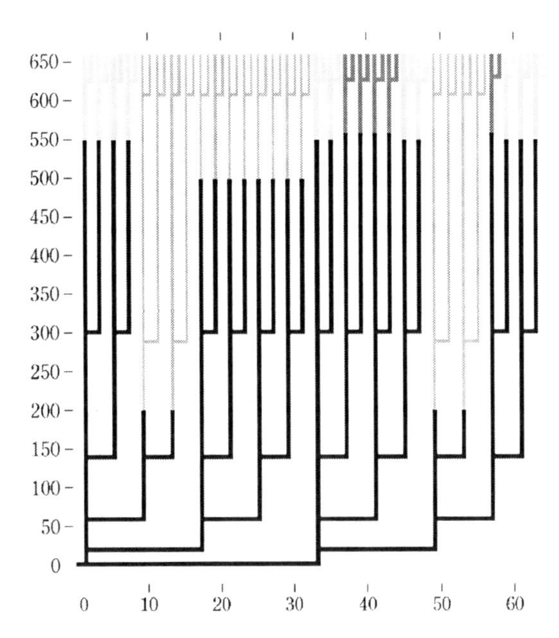


**Fig. 7.11.** Cell lineage diagram obtained from a simulation. The gray scale of the line represents a cell's character defined as the average chemical composition of the cell. The *dark line* corresponds to undifferentiated type (i.e., the particular composition of this cell type may change from division to division). The *medium* and *light lines* correspond to determined cell types with different chemical compositions that do not change after the cell division. Reproduced from Kaneko and Yomo (1999)

Cellular memory is clearly seen in this figure, where medium and light lines are preserved through divisions for $t > 550$. As shown in Fig. 7.11, emergence of certain cell types at different branches is observed in the model. Such convergence of cell types from different branches is also known in cell biology. In fact, such lineage is similar to that in tissues such as the hypodermis, neural tissue, and muscle in the development of the *C. elegans* (Kenyon, 1985). In Fig. 7.11, cells of a certain phenotype (e.g., given by "light lines") arise in isolation individually at different points in a lineage. Therefore, not only the history of each cell lineage accounts for the maintenance of different cell types over the generations. Instead, the global interaction among cells is important. In short, the emergence of certain cell types at different branches is but one of the features of the interaction-driven society.

The obtained cell lineage is not changed even if some fluctuations are included in chemical reaction process and at the cell division process. With the noise term mentioned in Sect. 7.2, the same cell types and the same differentiation process are observed up to a certain noise strength, as will be discussed in the next section.

5) *Hierarchical organization of cell types.* By means of global cellular interaction among the clusters of different cell types, the cellular phenotypes are stable at stage 4. However, as the cells continuously proliferate, cellular interaction within each cluster can result in further differentiation to the cells. Subgroups, undergoing the cycle of stages 1–4, are formed within a cluster.

   Successive differentiation and determination of cells are seen in the cell lineage (Fig. 7.11). After two types of cells are differentiated around $t \approx 550$, one of the types is later differentiated. Once this differentiation occurs, this character is fixed again, and after some time, such characters are determined by the daughter cells. We will discuss the hierarchical differentiation in more detail in the next chapter.

## 7.5 Further Results on Robustness and Dynamics of Differentiation

### 7.5.1 Robustness of Developmental Process

In our model, the robustness of the cell society is shown in different stages of developmental process. When the system is perturbed externally by changing the concentrations of the metabolites, still a similar developmental path is taken by the cells in the society. As a result, similar cell lineages are obtained.

To demonstrate this robustness, we discuss the influence of the molecular fluctuation (given by the Langevin equation), in more detail. In Fig. 7.12, we have plotted the average of two chemicals $(x^2(i), x^3(i))$ at time step 2000, when 64 cells split into two distinct clusters with different biochemical compositions, in a simulation without molecular fluctuations. As shown in the figure, the formation of two distinct clusters is robust while their biochemical compositions are hardly modified, as long as the noise strength $\sigma$ is less than $\sigma_{thr} \approx .008$. The number distribution of cells at the two clusters lies within the range $25 \pm 1{:}39 \mp 1$ again when the noise strength is less than $\sigma_{thr}$.

As discussed already, the final cell types are dynamically selected by cell–cell interaction, overcoming the instability brought about by the intracellular dynamics. If the chemical composition or the number of each cell type deviated from that of the selected cell society, the instability would reappear, which would enforce the system to revert to the original composition and distribution. Hence robustness against molecular fluctuations is a logical consequence of our scenario.

It may also be interesting to study how the differentiation pattern is destroyed when the molecular fluctuation is too large. As shown in Fig. 7.12, the two types start to merge at $\sigma \approx \sigma_{thr}$, and for larger $\sigma$, no distinct types are observed any more. In this case, the biochemical characteristics of each cell are continuously distributed.

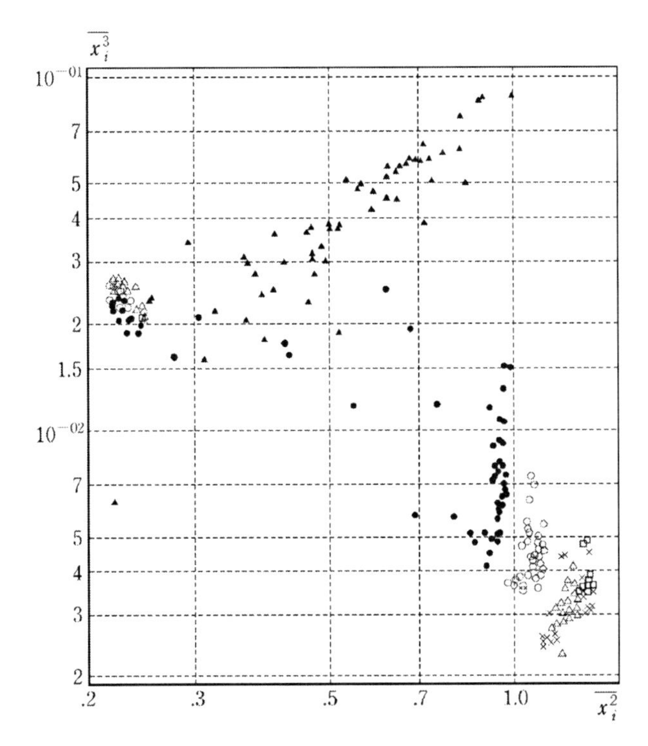


**Fig. 7.12.** The average concentrations of $(x_i^2(t), x_i^3(t))$, at the time step 2000. Plotted are the temporal averages, taken from the latest division to a time step 2000. The noise amplitudes $\sigma$ are 0.0001(□), 0.0003 (×), 0.001 (△), 0.02 (○), and 0.05 (▲). For each symbol, there are the number of points corresponding to each cell existing at the time step 2000 (64 up to $\sigma = 0.005$, 62 for $\sigma = 0.02$, 53 for $\sigma = 0.05$), although for $\sigma \leq 0.001$, some of the points are overlaid and may be invisible since the plotted concentrations of the cells are rather close by each other. Reproduced from Kaneko and Yomo (1999) with permission

If the cells cannot form different types exploiting different chemical "niches," the competition between cells for nutrients is higher. In fact, with the increase of noise amplitude $\sigma > \sigma_{thr}$, the cell replication is suppressed. With the further increase of $\sigma$, (say $\sigma > 0.06$), some cells lose their chemical activity, and the amount of chemicals starts to decrease toward zero with time. In other words, if the noise becomes too large, a developmental process with the increase in the cell number does no longer occur.

From the threshold noise $\sigma_{thr}$, it may be possible to estimate the minimal number of molecules in a cell, required to have a robust development process. For it, note that the concentration variable $x_t^m(i)$ is the number of molecules $N_t^m(i)$ divided by the cell volume $V$ ($x_t^m(i) = N_t^m(i)/V$). Assuming the fluctuation of the order of $\sqrt{N}$ for a system with $N$ molecules, the noise strength $\sigma$ for the concentration equation is estimated as $\sqrt{1/V}$. Then the number of molecules necessary to have robustness is given by $N = Vx = x/\sigma_{thr}^2$. In our simulation, the concentration $x$ for typical biochemicals is of the order of $0.01 \sim 0.1$. Hence, the threshold number of molecules in our model is estimated as $100 \sim 1000$. Of course, we have to admit that this is a rough estimate and that the present model is too simple to claim the number as a value comparable to the number of signal molecules in a real cell. Still, it should be stressed that a huge number of molecules are not required to have robustness, according to our scenario.

### 7.5.2 Transplant Experiment and Cellular Memory

Our cell society is not just robust against molecular fluctuations. It is even robust against "macroscopic" perturbations, for example, against removal of some cells. To study this problem, we first discuss how memory in differentiated cell types is sustained. Recall that the differentiation in our theory originates in the interaction among cells, but later, at the third stage, chemical characters of a cell are memorized through the initial condition after division. The differentiation at the initial stage is reversible, while the latter mechanism leads to determination.

In the natural course of differentiation and in the simulations so far, it is not possible to separate the memory in the inherited initial condition from the interaction with other cells. To see the tolerance of the memory, one effective method is to choose a determined cell and transplant it in a novel environment, surrounded by a variety of cells. Let us discuss the results of these "transplantation" experiments.

Here, we have made several numerical experiments taking such "artificial" initial conditions. Numerically, transplantation experiments are carried out by choosing determined cells (obtained from the normal differentiation process) and putting them into a different set of surrounding cells, to make a cell society that does not appear through the normal course of development.

When a determined cell is transplanted to another cell society, the offspring of the cell keeps the same type, unless the cell-type distribution of the society is strongly biased (e.g., a society consisting only of the same cell type as transplanted). When a cell is transplanted into a biased society, differentiation from a "determined" cell occurs. For example, a homogeneous society only of one determined cell type is unstable, and some cells start to switch to a different type. Hence, the cell memory is preserved mainly in each individual cell, but suitable intercellular interactions are also necessary to keep it.

Thus the cellular interactions play the role not only of the trigger to differentiation, but also of the maintenance of diversity of cells. Now, the relevance of cellular interactions to robustness is again confirmed. The cellular interactions do not only trigger differentiation but they also make a society with diversified cell types robust.

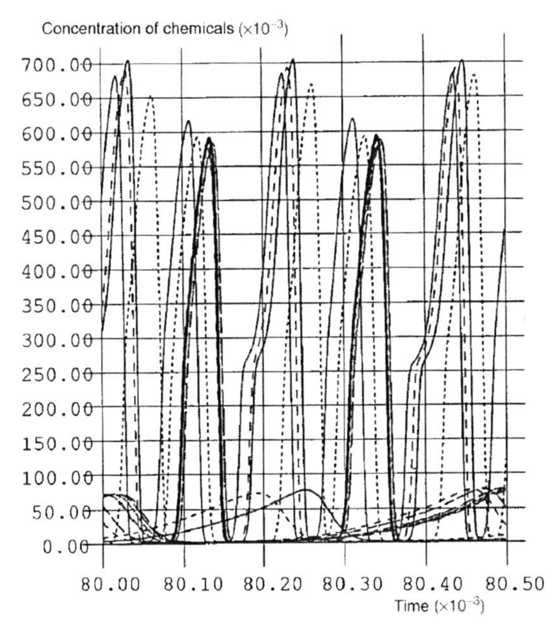


**Fig. 7.13.** Time series of a concentration of chemical component 2 ($x_i^{(2)}(t)$) plotted over $800 < t < 805$, overlaid over all cells $i$. Each line corresponds to the time series of each cell. Oscillations with a larger amplitude correspond to cells with larger activity ($\sum_m \overline{x_i^{(m)}}$). Reproduced from Kaneko and Yomo (1997) with permission

### 7.5.3 Separation of Inherent Time Scales

Another important feature here is the differentiation of the frequency. One group of cells oscillates faster than the other group. Typically cells with low activities oscillate more slowly in time with smaller amplitudes and divide slowly (see Fig. 7.13). Hence inherent time scales differ by cells, which is also seen in the differentiation of speed of division. Indeed, one group of cells divides faster than the other group of cells. It should be noted that the inherent time scales of cells are created spontaneously through cell divisions and differentiations.

### 7.5.4 Relevance of Low-Concentration Chemical to the Initiation of Differentiation

In our simulation, the differentiation starts after some divisions have occurred. Since the division leads to almost equal cells, a minor difference is enhanced to lead to macroscopic differentiation. We have found that small difference of chemicals with very low concentration leads to the amplification of the difference in the concentration of other chemicals. (see Fig. 7.14). At the initial stage of differentiation, difference in chemical concentration of chemical

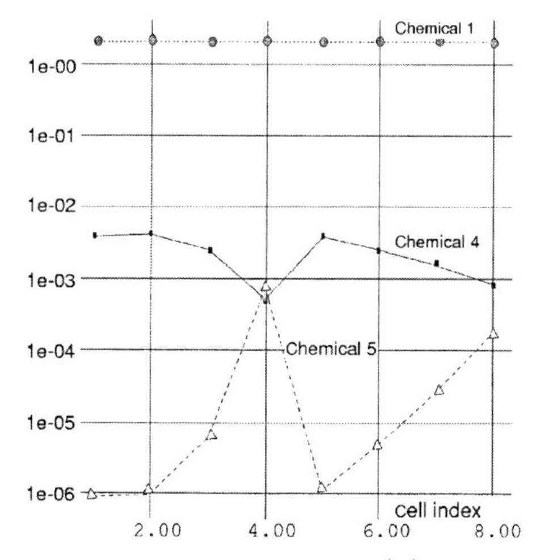


**Fig. 7.14.** Snapshot chemical concentrations of $x_i^{(m)}$, at $t = 63$, just the onset of chemical difference by cells (clustering). Chemicals 4, 5, and 7 are plotted in a logarithmic scale, in the order of cell index, the order that the cell is born. Reproduced from Kaneko and Yomo (1997) with permission

species with high concentration is negligible while the difference in the concentration of the species with very low concentration is remarkable. Such chemical with a low concentration in fact triggers the differentiation and thus plays a much more important role than that with a high concentration. Through catalytic reaction, change in the concentration of such minor chemical species is amplified. The importance of species with such low concentrations reminds us of minority control in Chap. 4.

## 7.6 Experiment

If we take seriously the consequences of the theory so far, we do not necessarily need to study the differentiation process of a multicellular organism. Any cells, under suitable conditions, may start differentiating to form a multicellular society.

### 7.6.1 Differentiation of *E. coli* Through Interaction

Here we discuss an experimental study made about a decade ago by Ko et al. (1994), who observed populations of *E. coli* in liquid culture over long times. The bacteria live in a culture with rather high density of cell populations. This experiment is "constructive," since a rather high-density liquid culture of mutant *E. coli* is studied, to see a differentiation process that is not commonly observed in nature.

They found that *E. coli* differentiate into two groups, with regards to the enzymatic activity of xylanase. They also selected out the bacteria to see what type of colony the bacteria form. The two groups of bacteria are found to show a quite different form of colony. This also supports the differentiation of the characteristics of *E. coli.*

One might still ask whether these differentiations may be brought about by mutations. They first checked the genes with regards to this enzyme activity, and no mutations were observed. Next, they carried out the experiment of the same liquid culture, but by selecting out only one type of *E. coli* from the two groups. Then, again, as the cell number is increased to reach the high density, the cells differentiate again to form the initial two groups with different enzyme activities. If the differentiation were due to just rare mutations, this result would not be explained. Hence the result also suggests that the differentiation is not due to the genetic change, but due to cell–cell interactions. One may conclude that some of *E. coli* (prokaryotes cells) with identical genes can be differentiated. We note that these cells are under liquid culture, and thus they are in an identical environment. No spatial factor is relevant as in the theory discussed in this chapter.

The fraction of the number of each cell type exhibited a complex oscillation. The amplitude of oscillation was higher than statistical expectation, that is, $\sqrt{N}$ for the population size $N$. This result suggests existence of complex dynamics due to cell–cell interaction.

The above experiment was carried out about a decade ago. Now with the use of fluorescent proteins, this type of experiment is much easier. By inserting a gene for fluorescent protein just at the downstream of a gene, and measuring the degree of fluorescence of each cell by flow cytometry (see Chap. 3), one can obtain the distribution of expression of the gene per cell which gives a measure for the concentration of the corresponding protein. If the distribution of fluorescence over many cells has a double peak, one can detect the differentiation. With the cell sorter, it is also possible to select one group of cells with higher (or lower) fluorescence.

The differentiation of bacteria at a colony on a plate has been extensively studied by Shapiro. Indeed, Shapiro and Dworkin (1997) edited a book *Bacteria as Multicellular Organisms*, where several forms of differentiation of bacteria in a colony are discussed in depth.

In cells cultured on a plate, one can also see how differentiation progresses as spatiotemporal pattern. Fractal patterns of cells are studied by Matsushita, Ben-Jacob, Visceck, and so forth (Ben-Jacb et al., 1994; Matsuyama & Matsushita, 1993). Quite recently Elowitz and Shapiro (2003) have observed the change of gene expressions of *E. coli*, by inserting the Green Fluorescent Protein into a specific gene, as mentioned above. An example of development of these bacteria is displayed in Fig. 7.15. As the colony develops, one can see differentiated *E. coli* forming a pattern, indicated by changes of the fluorescence strength.

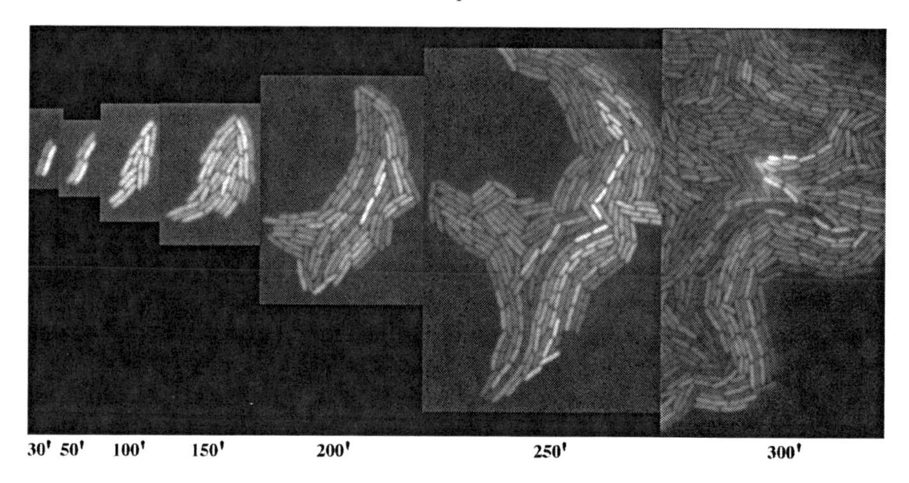


**Fig. 7.15.** Differentiation of *E. coli* in colony. Time-lapse images of *E. coli* cells expressing GFP from a promoter upstream of the serA locus. The *gray-scale* levels indicate the relative amount of GFP present in the cells at each time point. The numbers under each frame indicate the time of incubation on LB agar at 37 C in minutes. The time-lapse sequences were recorded by Michael Elowitz and James Shapiro. The GFP – promoter fusion constructs were prepared by Alon Zaslaver, Mike Surette, and Uri Alon as referenced in Ronen et al. (2002) and James Shapiro (2003). Under the courtesy of Elowitz & Shapiro (2003)

Of course, each gene expression fluctuates, even at the level of a single cell (Elowitz et al., 2002). Still, in the figure one can see some specific pattern that is not random. Hence it is reasonable to assume that the cell–cell interaction has some relevance to differentiation of these gene expressions. Of course, in this experiment one has to seriously take into account the spatial configuration. We will discuss this problem in Chap. 9 again.[6]

The gene expression is a result of complex intracellular reaction dynamics. Some part in the network has a positive feedback process leading to differentiation, while other part has a negative feedback process. Hence it is expected that the differentiation in fluorescence pattern depends on the position in DNA where the gene for the GFP is inserted. Indeed, Elowitz and Shapiro have carried out several experiments changing the position of the insertion of GFP gene. Now at the downstream of some genes, the differentiation is not observed, whereas for some others, differentiation patterns as in Fig. 7.15 are observed.

Importance of cell–cell interaction in bacteria has been extensively studied as "quorum sensing" (Miller & Bassler, 2001). There are also studies to synthesize such interaction by embedding gene network in bacteria (You et al., 2004), while synthesis of oscillatory gene expression has been achieved by Elowitz and

[6] See also Suel et al. (2006), for recent report on transient cellular differentiation in *Bacillus subtilis* (bacterium), induced by an excitable gene regulatory dynamics.

Leibler (2000). Bistable gene expression related with quorum sensing is also reported for some mutant bacteria (Veening et al., 2005). Hence it will be possible to synthesize "isologous diversification," by embedding a gene network to bacteria with suitable gene expression dynamics and cell–cell interaction.

### 7.6.2 Interaction-Dependent Tumor Genesis

Rubin (Yao & Rubin, 1994; Chow et al., 1994; Rubin 1994a, 1994b), in a series of experiments, has shown that formation of a type of tumors strongly depends on the density of cells, not on their number. In general, mutation is believed to be a trigger to the genesis of tumors. If the tumor formation were totally dependent on random event of mutation as independent events by cells, the formation would depend on the number of cells, not on the density. The experiment shows that even if the mutation may be relevant, at least the rate of formation of tumor also depends on the cell interaction. This suggests that some type of tumor formation depends on epigenetic factors, and cell–cell interaction has to be seriously considered there.

Note that the tumor formation most certainly is not a well-controlled, highly programmed event. The interaction dependence on the tumor formation suggests again the relevance of cell–cell interaction to the fate of cell type.

## 7.7 Relevance to Biology

### 7.7.1 Summary of the Result

To discuss the relevance of our isologous diversification to biology, let us briefly summarize our results:

- Cell differentiation occurs through the interplay between intracellular chemical reaction dynamics and the interaction among cells through chemical media.
- Cell differentiation is initiated by the clustering of chemical oscillations, appearing at some cell number, which is explained by general features of coupled nonlinear oscillators.
- Chemicals with tiny amounts in cells are relevant to trigger differentiations.
- Distinct and memorized cell types are formed as discrete sets of chemical states, characterized by chemical compositions, while intermediate states among these types are unstable, which are not sustained. The distinct types are later transferred to their offspring cells, as determination.
- Inherent time scales, given by the oscillation period, depend on the cell. Generally active cells oscillate faster and divide faster. This separation of timescale brings about the separation of growth speed of cells and leads to the disparity between rapidly growing and inactive cells.

- Determined cell types formed at the later stage are preserved by their transplantation to another cell society as long as there are not too many cells of the same type. On the other hand, cells may de-differentiate when surrounded by a biased distribution of cell types.

It should be mentioned that these results naturally appear as a general consequence of our isologous diversification theory without preprogrammed implementation, and are independent of details in the model.

We should also mention that our theory is compatible with the genetic switching mechanism for the differentiation. Here such switching-type expression appears naturally through cellular interactions. Note that these results are consistent with the experimental observations so far, while they include some predictions related to the generality of these features. We discuss a few more points here.

### 7.7.2 Robustness from Unstable Dynamics

Even though the model disregards space as a factor, a society of diversified cell types exists. This indicates that spatial bias, for example, a morphogen gradient, is not necessary for cell differentiation. It is to be noted that this statement does not deny that in the developmental process, the gradients of morphogen are imperative for constructing the body in space. Under a certain threshold of the morphogen concentration shown in Fig. 7.1, cells differentiate. If the concentration difference of the morphogen between two adjacent cells exceeds the threshold, the cells differentiate. However, if a differentiation process would simply rely on gradients created and maintained by a diffusion process, a developmental process with successive differentiations would not succeed because the diffusion of a small number of molecules is stochastic. The threshold mechanism based on the gradient cannot by itself overcome the stochastic effect of diffusion, regarding cell differentiation.

On the other hand, in our isologous diversification, there is no uncertainty in cell differentiation process. Even though the gradient of morphogen around the adjacent cells is perturbed, the concentration difference is amplified and gives rise to cell differentiation. Here, some experimental studies on equivalent group of cells proved that cellular interaction is necessary for cell differentiation (Greenwald & Rubin, 1992), as proposed in the present isologous diversification theory.

In the same way as each of the cell differentiations, the final cell-type distribution is reproducible. If the development process works like a programmed machine, then the signals and switches that trigger each cell differentiation event must work as programmed. Nevertheless, since only a small number of biochemical molecules trigger each cell differentiation, stochastic errors due to molecular fluctuation are inevitable. Hence, the machine-like mechanism for development cannot cope with the errors in each cell differentiation. In our theory, even without a "fixed" mechanism for controlling cell differentiation, the system shows a reproducible distribution in cell-types during the development. Moreover, the whole development is robust against any perturbation such as removing and adding some cells. The robustness of our model cells indicates that the embryo is not necessarily a programmed machine that requires strict control. Rather, it has a certain flexibility to develop a precise cell-type distribution.

### 7.7.3 Relevance of Molecules with Very Low Concentration to Development

The relevance of minority molecule reminds us of a certain protein that is known to trigger a switch of differentiation with only a small number of molecules. The relevance of chemicals with low concentration is also seen in the determined differentiation. As noted in Fig. 7.14, the difference is most remarkable for chemicals with low concentrations. It is interesting to check this proposition from experimental cell biology. The relevance, on the other hand, is a consequence of our theory. Since the theory is based on the amplification of tiny differences by a nonlinear mechanism, difference of "rare" chemicals by cells can be easily amplified to lead to a macroscopic difference at the cell level.

### 7.7.4 Oscillation

In our model, the differentiation results from nonlinear oscillation in chemical reaction dynamics. Although our theory is applicable even without oscillation, as long as there exists instability in the state. Still, it is interesting to discuss the relevance of oscillation. Biochemical oscillations are studied in $Ca^{2+}$, NADH, glycolysis, and so forth, and theoretical models for them are also developed (Goldbeter, 1996; Tyson et al., 1996). Still, relevance of oscillation to cell differentiation was not seriously discussed.

Here, it is interesting to note a report by Dolmetsch et al. (1998). They have shown that the $Ca^{2+}$ oscillation in a cell depends on the cellular state, and the gene expression pattern can be changed by externally varying the frequency of oscillation. According to their experiment, cell-types are tightly correlated with the frequency of $Ca^{2+}$ oscillation. Studying biochemical dynamics in relationship with cell differentiation should be important in the future.

### 7.7.5 Control of Tumor Formation

In simulations with a larger diffusion coupling, a peculiar type of cells appears. They destroy the ordered use of nutrients, which makes the cell society disorganized. This type of cell is a different type of a specialized cell and has a much higher concentration of one chemical than other species. The cells destroy the mutual relationships among cells, which was achieved through the successive stages of the isologous diversification. Taking into account the fact that these cells destruct the chemical order sustained in the cell society, they may be regarded as "tumor"-type cells.

The formation of these "tumor"-type cells may be triggered by the mutation, which, in our model, is represented by the "noise" term in the division process. Still the growth of tumor cells depends on the cellular interactions, for example, on the diffusion constant or the density of differentiated cells. Depending on the interaction term, errors in the division process may or may not lead to the "tumor"-type cells.

Of course, the high concentration of a single chemical species here is due to the simplicity of our biochemical network consisting only of a few chemical species. For a network with more components, the bias must be weakened. However, it is still expected that there appears a cell-type with biased composition of chemicals.

Indeed potentiality to form "tumor"-type cells is a consequence of isologous diversification theory. In the theory the differentiation process is not programmed explicitly as a rule but occurs through the interaction. Thus, when a suitable condition of the interaction is lost, for example, by the increase of density, "selfish" cells destroying the cooperative use of nutrients may be formed.

Then, to reverse these tumor-type cells to a normal state in our model, the diversity in the metabolite composition should be recovered. Indeed, if the "tumor"-type cells in the model are mixed with the undifferentiated cells containing various biochemicals, most of them regain a diverse biochemical composition and reverse to a normal state. Similarly, by introducing cytosol from undifferentiated cells or egg, into some human cancer cells, through the liposome, their malignancy may be cured.

## 7.8 Appendix: An Example of Model Equation

Here we give the explicit model equations used in the simulations in Sect. 5. The dynamics of the $\ell$th biochemical concentration $x_i^{(\ell)}(t)$, in cell $i$ consists of an internal reaction term $React$, an active transport $Transp$, a diffusion term $Diff$, and molecular fluctuations $Fluct$, as

$$dx_i^{(\ell)}(t)/dt = React_i^{(\ell)}(t) + Transp_i^{(\ell)}(t) + Diff_i^{(\ell)}(t) + Fluct_i^{(\ell)}(t) \,. \quad (7.2)$$

In this chapter, the reaction term $React$ consists of autocatalytic reaction network with a set of reaction paths from $j$ to $\ell$, written as

$$
\begin{aligned}
React_i^{(\ell)}(t) = & S(\ell)e_0 x_i^{(0)}(t)x_i^{(\ell)}(t) \\
& + \sum_j Con(\ell, j)e_1(x_i^{(j)}(t))^{\alpha_{\ell,j}} x_i^{(\ell)}(t)/(1 + x_i^{(\ell)}(t)/x_M) \\
& - \sum_{j'} Con(\ell, j')e_1 x_i^{(\ell)}(t)(x_i^{(j')}(t))\alpha_{\ell,j'}/(1 + x_i^{(\ell)}(t)/x_M) \\
& - \gamma P(\ell)x_i^{(\ell)}(t), \qquad (7.3)
\end{aligned}
$$

where $Con(\ell, j) = 1$ if there is a reaction path from the chemical $\ell$ to the chemical $j$, and $Con(\ell, j) = 0$ otherwise. This network is chosen randomly. The value $x_M$ comes from the Michaelis–Menten's kinetics, while the exponent $\alpha_{\ell,j}$ is the degree of catalytic reaction, which is set at 1 for simple catalytic reaction which is adopted in this chapter. $j$ to $\ell$, and $Con(\ell, j) = 0$ otherwise. Here we assume that there is a source chemical 0, from which other chemicals are formed by the term $S(\ell)e_0 x_i^{(0)}(t)x_i^{(\ell)}(t)$. There is a flow to a product required for cell division, given by the term $\gamma P(\ell)x_i^{(\ell)}(t)$. Again, $S(\ell)$ or $P(\ell)$ is 1 if such a path exists and 0 otherwise. (The equation form for $x_i^{(0)}(t)$ is obtained straightforwardly.)

With the flow at the rate of $\gamma P(\ell)x_i^{(\ell)}(t)$, the product initiating cell division is accumulated. Hence, cell $i$ divides if the condition $\int dt\gamma \sum_\ell P(\ell)x_i^{(\ell)}(t) > Threshold$ is satisfied. After the division, the two cells have the same chemical composition as the mother cell $i$, with some tiny fluctuation (e.g., 0.1% for each chemical $x_i^{(\ell)}(t)$).

The transport term expresses active transport from the medium to each cell. The rate of transport in general depends on intracellular chemical concentrations. As a simple model, we choose that the activity is given by $\sum_{k=1} x_i^{(k)}(t)$. Then, the term is given by

$$
Transp_i^{(m)}(t) = p\left(\sum_{\ell=1} x_i^{(\ell)}(t)\right) X^{(m)}(t)\,, \qquad (7.4)
$$

where $X^{(m)}(t)$ denotes concentration of chemical $m$ in the medium. Apart from the active transport, chemicals diffuse through the membrane as

$$
Diff_i^{(m)}(t) = D(X^{(m)}(t) - x_i^{(m)})\,. \qquad (7.5)
$$

Since the active transport and diffusion processes are just the transportation of chemicals to and from the medium, equations for the chemicals $X^{(m)}(t)$ in the medium are equal to the sum of $Transp_i^{(m)}(t)$ and $Diff_i^{(m)}(t)$ over cells $i$ (divided by the volume ratio of the medium to a cell). (For the source chemical $X_i^0(t)$, we assume a flux from the outside, so that the consumed nutrients are supplied.)

So far we have adopted the model given in the earlier paper (Kaneko & Yomo, 1997) in details. In this Chap. 7, we add a molecular fluctuation term, given by the Langevin equation

$$Fluct_i^{(m)}(t) = \eta_i^{(m)}(t)\sqrt{x_i^{(m)}(t)} \tag{7.6}$$

with Gaussian white noise satisfying $\langle \eta_i^{(m)}(t)\eta_i^{(m)}(t') \rangle = \sigma^2 \delta(t - t')$.

Note that the model choice in this section is just a specific example. In other models such as the model given in Furusawa and Kaneko (1998), the same scenario for the developmental process is confirmed.

# 8

# Irreversible Differentiation from Stem Cell and Robust Development

## 8.1 Question to Be Addressed: Regulation for Differentiation of Stem Cell

**Question: In the development of multicellular organisms, there exists a cell, called stem cell, that either replicates (proliferates) or differentiates to several other cell types. There exists a rule for such differentiation from multipotent stem cells. How are such differentiation rules generated? How can the rule and several cell types coexist stably under fluctuations? Furthermore, such stem cell systems form a hierarchy. In the development of multicellular organisms, there is a successive determination from totipotent embryonic stem (ES) cell, to multipotent stem cells, and to several committed cell types (that are determined to self-replicate). There exists a clear time's arrow in the direction of the loss of potency to form a variety of cell types, at least in the normal development course. How is such irreversibility generated and how is it characterized quantitatively? On the other hand, the developmental process is robust against fluctuations and perturbations. Is this robustness related with developmental irreversibility?**

In a multicellular organism, cells differentiate into several types, as discussed in the last chapter. In the developmental process, cells at the initial stage of embryo have the potentiality to produce all other cell types. This capability to produce all types of cells is named totipotency, while a cell with totipotency is called an ES cell. As differentiation from this ES cell progresses, a differentiated cell loses such potentiality. Some cells can produce only one fixed type of cell in the course of development, while other types of cell still partially keep the potentiality to differentiate. The latter type of cell either produces the same type of cell or differentiates to a few other types of cell, from which some other cells are differentiated. This type of cell is said to have multipotency, and such cells are called stem cells. Stem cells are defined as cells that have a potential to divide throughout the life of an organism and

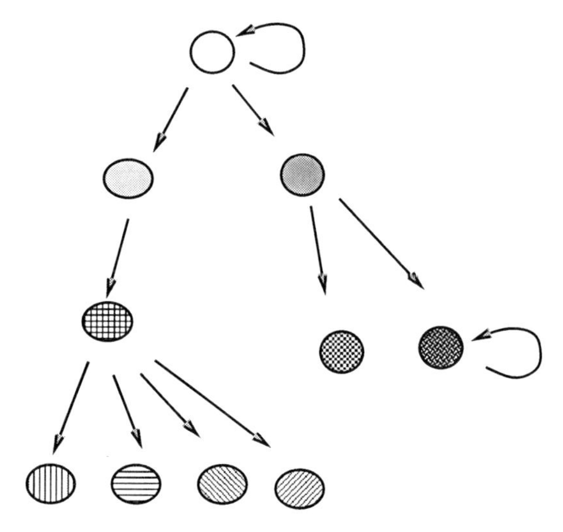

**Fig. 8.1.** Schematic representation of hierarchical differentiation from a stem cell

to produce either stem cells or other differentiated cells. They have both the potentiality of self-replication and differentiation.

For example, the neuro-stem cell has the potentiality to produce all cells in neural systems including neuron, glya, and so forth. The blood stem cells have the potentiality to produce all cells in the blood, such as leucocyte including neutrophil, acidophil and basophil, red blood cell (eythrocyte) blood platelet (thrombocyte), and so forth. In the normal development, the stem cells have this potentiality to produce a limited range of type of cells, in contrast to the totipotency of ES cells. In this way, as shown in Fig. 8.1. The cell differentiation process is organized in a tree-like graph, where the node at a higher level has the potentiality to produce the cells at the lower branches derived from it.

Through the developmental process, the potency, that is, the ability to create different types of cells decreases. Initially, the embryonic stem cell has totipotency and has the potentiality to create all types of cells in the organism. Then a stem cell can create a limited variety of cells, having multipotency. In the normal developmental process, this hierarchical loss of potency terminates the level of "committed" cells, which are *determined* only to replicate itself while preserving its own type. The degree of determination increases in the normal course of development. How can one understand such irreversibility? Hence, in a multicellular organism with a developmental process, there is a definite time's arrow. Indeed, we know that irreversibility is common in biological systems. As an empirical fact, we know that the direction from the alive to the dead is irreversible.

Such universal time's arrow may remind us of the second law of thermodynamics. However, it is clearly impossible to simply attribute this irreversibility to the second law of thermodynamics. One can hardly imagine that the thermodynamic entropy, even if it were possible to define it for a biological system, suddenly increases at death or successively increases at each step of the cell differentiation process. Furthermore, it should be generally very difficult to define a thermodynamic entropy to a nonequilibrium system such as a cell. The irreversibility we discuss here holds at a different level than that of usual thermodynamics.

On the other hand, it is also very difficult to imagine that only some specific molecules or genes play a decisive role responsible for irreversibility or stability (such as "stability gene" or "irreversibility gene"). If there were such gene or molecule, one might imagine the possibility of some long sought-after "miracle medicine" for eternal youth! Probably, the stability and irreversibility are too universal features to be attributed solely to the characteristics of a few molecules. Such a general feature is the property of a system. In this sense, some analogy with thermodynamics might be relevant, even though the concept "thermodynamic entropy" cannot be applied here: Thermodynamic irreversibility is not restricted to a system composed of some specific molecules. Analogously, we need to find a logic that universally holds in all cellular systems.

Note, however, that this loss of potency holds in the case of a normal development process. For example, consider a branching cell differentiation process as displayed in Fig. 8.1. In this case, each cell can replicate or differentiate into cells at a lower level to the original one by an arrow. This process is observed during the normal developmental process. If a stem cell is extracted from the tissue and is transplanted to other tissues, the stem cell may regain the potency to produce cells of other branches.

Here it should be noted that the self-replicating, committed (determined) cell can recover the potentiality to differentiate into other cell types, once its nucleus is transplanted by a suitable operation. For example, the transplanted nucleus of somatic cells can produce all other types of cells. This was first known in plants (Steward et al., 1958). Then Gurdon succeeded in somatic clone experiments by using *Xenopus* (frog) (Gurdon et al., 1975), which demonstrated that committed somatic cells can produce all other types of cells. This was followed by Campbell, Wilmut for mammals (Campbell et al., 1996).

When discussing this irreversibility, one should make sure that the condition(s) under which "normal development" occurs must be fulfilled. Conversely, to reverse the differentiation, some operation violating the condition for "normal development" is necessary. What are these conditions condition, then? How is the condition for irreversibility in the normal development characterized? Once such condition has been stated explicitly, it becomes possible to discuss which operation is necessary to reverse the differentiation.

Let us come back to the discussion of normal development. In the developmental process of the multicellular organism, stem cells play an important role to make the body, since they can replicate as well as produce all other cell types of the down-flow. Stem cells are essential to the development of multicellular organisms. They sustain tissues including cells that have a limited life span and lose the ability of self-renewal (e.g., blood and epidermis), by supplying differentiated cells continuously. For example, in the development of plants, the self-replication and differentiations of stem cells at the apical meristem is responsible for the buildup of the whole organism.

In some tissues, stem cells disappear after the development has progressed, while in some other tissues, they remain to exist, which are responsible to keep the tissues. Asexual reproduction of most primitive multicellular organisms, such as Hydra (David & MacWilliams, 1978), is based on their stem cells. Since stem cells can produce all other cell types, it is important to know whether stem cells remain present or disappear at later stages of development. The ability to regenerate some tissues after damage may be related with the existence of stem cells.

Note that the stem cells have to possess two quite different properties, that is, self-replication[1] and differentiation. For the self-replication, the stem cell has to keep its state through its growth up to division. For the differentiation, on the other hand, the stem cell has to switch to other states so that its original state should be somehow destabilized. These two distinct features for self-replication and differentiation have to be satisfied in a single stem cell: After cell division, one of the cells keeps its state, and the other is differentiated. Now the problem how stem cells determine their fate (to self-renew or to differentiate) is still unsolved.

Roughly speaking, there are two major hypotheses about the determination of stem cell differentiation. The first one is the cell-intrinsic hypothesis, according to which the fate of a stem cell is determined by a program driven autonomously, inside the cell. The other is the inductive hypothesis, in which the differentiation is caused by inductive signals from the outside of the system, which provide asymmetry among the cells. Different cells appear from different signals. A most important model formalizing the cell-intrinsic hypothesis is the stochastic differentiation model, proposed to describe the hematopoietic stem cell system by Till and McCulloch (1964). They observed variations in the number of stem cells in spleen colonies that originate from a single stem cell and explained this heterogeneity in terms of the hypothesis that each decision of a stem cell to undergo either self-replication or differentiation is stochastic, with a given fixed probability. Nakahata et al. extended this stochastic determination and made some experimental observations that support it (Nakahata et al., 1982). They prepared two daughter cells derived from a single multipotent stem cell and cultivated them independently under identical conditions. Repeating this experiment a number of times, they found that the cell number distributions of two such colonies generated by paired daughter cells often turned out to be different from each other. These experiments demonstrate the stochastic character of stem cell differentiation.

[1] Or, sometimes called as "proliferation."

In experiments on several stem cell systems, spontaneous differentiation has been observed, and in terms of the observations, it appears that it occurs in a stochastic manner. Although external signal molecules, such as growth factors, are required to maintain these systems, it has been found that the differentiation of stem cells does not require any signals from the outside, providing inhomogeneity among the state of stem cells.

Then, does the differentiation occur just randomly by some probability? If so, it would be difficult to understand the stability of the stem cell system. Indeed, an important feature of a stem cell system lies in the robustness of a multicellular organism at a cell population level, as, for example, shown in regeneration. When a tissue is damaged, active stem cells reappear through activation of quiescent stem cells or de-differentiation of neighboring cells. The self-replication and differentiation of these cells, by which the damaged tissue can be reconstructed, are regulated depending on the environment. For example, in the hematopoietic system, when the number of committed cells (e.g., red blood cells) is made artificially lower, the replication of and/or differentiation to this cell type have to be enhanced, so as to recover the original distribution of blood cells.

To achieve such stability, the choice of differentiation and proliferation should depend on the surrounding cells. Depending on which types of cells are dominant in a population, the choice can change. Here, regulation of the probability is required as long as a developmental process satisfies some robustness against external perturbation. For example, let us consider the differentiation from a stem cell to some other cell types. Assume that some of the committed cell types is continuously eliminated. If there were no regulation to the rate of differentiation from the stem cell to the damaged cell-type, the distribution of cell types would deviate fatally.

On the other hand, consider the situation in which only a single stem cell exists in an ensemble of cells. If, after the division, the two daughter cells happen to differentiate into different types, there is no more stem cells, and the ability to produce several other cell types will be lost in this cell ensemble. In contrast, transplantation of even a single blood stem cell seems to be sufficient to regenerate the whole blood system. When a single hematopoietic stem cell is transplanted into a mouse whose hematopoietic system has been destroyed through irradiation, this cell can reconstitute the hematopoietic system with high probability (Ogawa, 1993). This result implies that the probability of differentiation is externally controlled at least in the case where there are only a small number of stem cells.[2]

Now, it is expected that the choice of the differentiation or self-replication depends on the cell–cell interactions. But how is this possible if the choice of proliferation and differentiation is stochastic? In other words, it is yet unknown whether the probabilities involved in differentiation are controlled

[2] The regulation of this probability has also been reported in the spontaneous differentiation of hydra stem cells (David & MacWilliams, 1978).

extrinsically. The original form of the stochastic model assumes that the probability of differentiation is fixed, and that a cell population is regulated by controlling the proliferation and survival of cells. The experiments mentioned above suggest that this probability can be controlled by extrinsic factors, and that the stochastic behavior cannot be separated from the complex cell–cell interaction, and is regulated depending on the population distribution of cell types.

Summing up, the following features are necessary for a stem cell system.

1. The diversity of cell types through differentiations from stem cells can emerge without any information from the outside of the system. These differentiations occur stochastically.
2. The self-replication and differentiation of cells in the system are regulated depending on the complex cell–cell interactions. This regulation makes it possible to maintain the robustness of the system.

In other words, the stem cell system has both the features of stochasticity and regulation. Since molecular fluctuations are inevitable to a cell system, it might be natural to assign the origin of the probabilistic behavior to molecular fluctuations. For example, reactions among specific chemical substances, such as a DNA molecule and a regulator protein occur stochastically due to random collision of the molecules. However, if stochastic molecular events are the only origin of stochasticity, it would be very difficult to imagine a mechanism to regulate the probability itself. Now, control to change the rate for some specific reaction in such network should be very difficult.

Then, how are the two features – stochasticity and regulation – understood? In the present chapter, we will show that these features are a general consequence of intracellular chemical dynamics and cell–cell interaction, independently of the details of specific chemical substances. In particular, stochastic differentiation and the regulation of probability of differentiations are naturally derived as a general property of interacting dynamical systems.

## 8.2 Logic: Chaotic Stem Cell

To answer the question in the last section, we have extended our isologous diversification theory, to consider plasticity in the dynamics more seriously.

When a cell includes a large variety of chemicals, the chemical dynamics are not generally stable, as also mentioned in Sect. 7.2. Furthermore, for a cell to grow efficiently, some autocatalytic chemical reactions with positive feedback process must be present to amplify some chemical concentrations. Then instability in dynamics, as well as flexibility is also present at the level of the cell state. Although the description of the cell state may involve a large number of degrees of freedom, considering a large variety of molecule species, its temporal evolution is not generally random. What happens then when such a cell, whose state is unstable, increases its number?

For these reasons we consider a cell with high plasticity, for example, derived from chaotic reaction dynamics. Then, as the number of such cells increases, the intra-cellular biochemical state starts to be unstable, because of the cell–cell interaction. Through this instability, chemical composition starts to be different, and cells with different compositions are formed, while cells with the original chaotic dynamics remain to exist with some fraction.

With cell–cell interactions, it is expected that cells start to form distinct states that mutually stabilize each other. (Recall the results of the previous chapter.) With this process, different cell types are formed, which mutually stabilize their state. The initial state has a higher instability in dynamics and fluctuations in the chemical compositions. It has flexibility. The new state that appears later will be more stable and loses some degree of such flexibility. As mentioned in Sect. 3.3, irreversibility in the transition is expected: Consider two states with and without flexibility. The flexible state can switch to another state possibly less flexible than itself, whereas the less flexible one can only remain as is. Hence irreversibility is expected. Generation of differentiation rules, discrete stable cell-types, as well as stable tissue within a certain range of cell-type distribution is a result of the instability in biochemical states. The initial cell type with flexibility is regarded as a stem cell, from which hierarchical differentiation progresses to a finally committed cell that can proliferate itself.

Now, two fundamental features of stem cell systems – stochastic differentiation of stem cells and the robustness of a system due to regulation of this differentiation – are found to be general properties of a system of interacting cells exhibiting chaotic intracellular reaction dynamics.

Here, the initial cell type S either self-replicates (proliferates) or differentiates, with some probability due to spontaneous fluctuations generated by the flexibility in the dynamics. This "probability" depends on the internal cellular state that also depends on the surroundings of cells, since the stabilization to each cell state occurs through cell–cell interaction. For example, as the fraction of the number of cell type S is increased, the original instability that existed for the cell state in the condition consisting only of the type S cells appears again, and the probability for differentiation from the type S to other types increases. Hence the probability of the differentiation is regulated so that it increases (decreases) as the fraction of type S cells is decreased (increased). In other words, there should generally exist a spontaneous regulation of differentiation probability so that the number distribution of each cell type is stabilized against perturbations. Note, as will be discussed later, this stabilization is similar to the Le Chatelier principle in thermodynamics.

Here, it should be noted that unstable (chaotic) dynamics provides a plausible mechanism for such control of probability, since it leads to stochastic behavior as to the fate of an orbit, starting from a set of deterministic equations that can model the biochemical reaction networks of a cell (see Chap. 3). As long as the complete description of the internal cellular state is not precisely known, the differentiation can look like stochastic. Since "stochastic" behavior is generated by a deterministic mechanism, it naturally opens up the possibility of self-generated regulation of differentiation. The developmental process in these dynamical systems must be stable against molecular fluctuations. First, intracellular dynamics of each cell type are stable against such perturbations. This stability is not solely determined by intracellular chemical reaction dynamics. With the increase of the cell number, the cellular state of chemical compositions is destabilized, but then through cell–cell interaction, the differentiated states are stabilized. Given cell-to-cell interactions, the cell state is stable against perturbations on the level of each intracellular dynamics. Even if the cell type number distribution is perturbed slightly, each cellular dynamics keeps its type. Hence discrete, stable types are formed through the interplay between intracellular dynamics and interaction, starting from unstable dynamics. The recursive production is attained through the selection of initial conditions of the intracellular dynamics of each cell so that it is rather robust against the change of interaction terms as well.

As mentioned, this also allows for the stability at a macroscopic level, that is, with regards to the number distribution of cell types. When macroscopic perturbation is applied, to change the number distribution, the probability of differentiation from stem cells is altered leading to recovery of the original cell type distribution. Hence macroscopic robustness arises (Fig. 8.2).

### 8.2.1 Direction of Determination: Logic

Then, is the differentiation from the type S cell irreversible? As the process mentioned above progresses, the differentiated cells tend to adopt biased

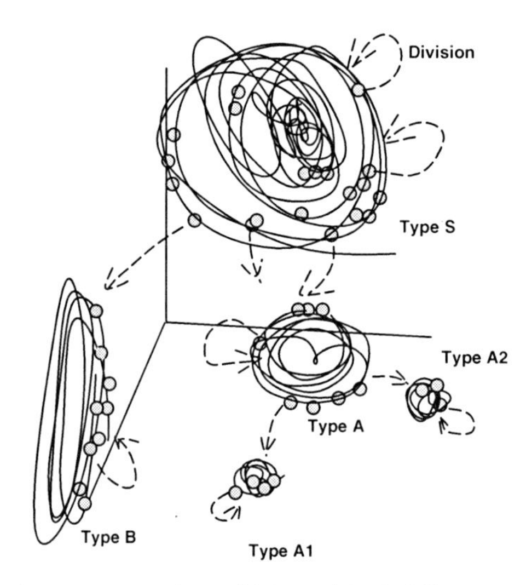


**Fig. 8.2.** Schematic representation of hierarchical differentiation from stem cell, represented in the state space. By using the model described in the next section, we obtain such differentiation process

chemical concentrations: a few chemicals have higher chemical concentrations. For example, consider differentiation from the type S cell to differentiated types A and B. The type A cell possesses a higher concentration of some chemical(s) than other types, and the type B cell possesses a higher concentration of different chemical(s). With this, cells are differentiated in roles. One group of cells (say type A) synthesizes some molecules, which are used also for another differentiated type of cells (say type B), which synthesizes different molecules that are also used for the other type. Then, the diversity of chemicals within each differentiated cell is lower than that of the initial stem cell. In general, recovering the diversity is harder than losing it, which brings about a source of irreversibility. As the remaining degrees of chemical species are decreased, the reaction dynamics become simpler (are lower dimensional), so that the variation or fluctuation of chemical concentrations in the cells will decrease. Hence the decrease of "plasticity, or changeability of cells" is naturally expected. This is nothing but the loss of plasticity discussed in Chap. 3. As discussed there, switch from a state with higher plasticity to that with a lower plasticity is irreversible. In this sense, the irreversibility in cell differentiation is discussed in terms of the loss of diversity in intracellular chemicals and plasticity.

## 8.3 Model

Here we basically follow the modeling strategy of Chap. 7, while intracellular dynamics are slightly revised to allow for chaotic dynamics, as given in Fig. 8.3a. We choose a catalytic network among the $k$ chemicals. Each reaction from some chemical $i$ to some chemical $j$ is assumed to be catalyzed by a third chemical $\ell$, which is determined randomly.

We again denote that the rate of increase of $x_i^{(m)}(t)$ (and decrease of $x_i^{(j)}(t)$) through a reaction from chemical $j$ to chemical $m$ catalyzed by $\ell$ as $ex_i^{(j)}(t)(x_i^{(\ell)}(t))^\alpha$, where $e$ is the coefficient for the chemical reaction and $\alpha$ is the degree of catalyzation. For simplicity, we use identical values of $e$ and $\alpha$ for all paths. In this chapter, instead of $\alpha = 1$ (as in the last chapter), we adopt $\alpha = 2$, which implies a quadratic effect of enzymes. This specific choice of a quadratic effect is not essential in our model of cell differentiation, but this higher nonlinearity introduces a higher degree of instability in the steady state of chemical concentrations, and often leads to chaotic dynamics. That is the necessary part. (See also Appendix of Chap. 7 for details of the equation, as well as (Furusawa & Kaneko, 1998a, 2000b, 2001)).

Furthermore, we take into account the change in the volume of a cell, which varies as a result of the transportation of chemicals between the cell and the environment. For simplicity, we assume that the total concentration of chemicals in a cell is constant, $\sum_m x_i^{(m)} = const$. It follows that the volume of a cell is proportional to the sum of the quantities of all chemicals in the cell. The volume change is calculated from the transport, as discussed below.

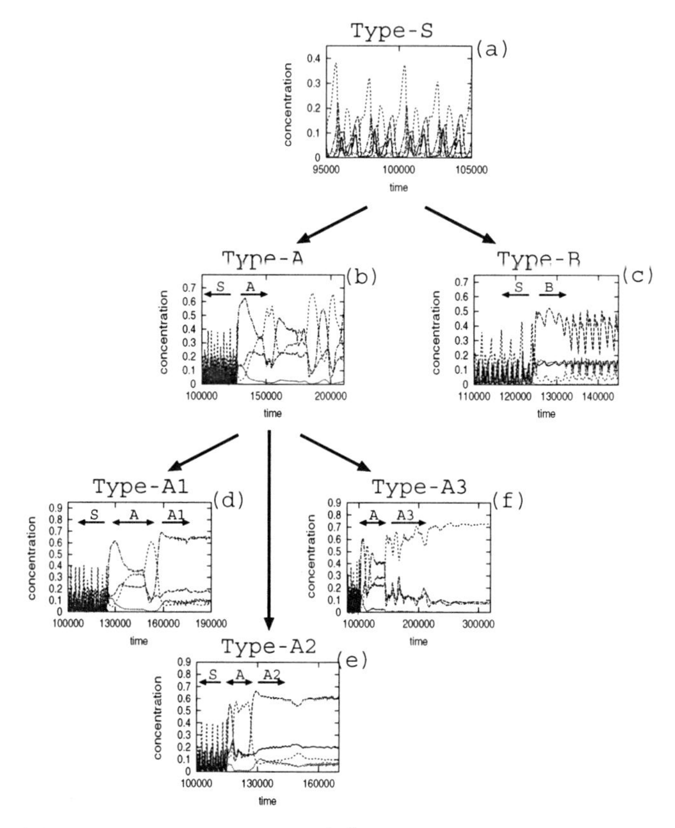


**Fig. 8.3.** (**a**): Overlaid time series of $x^{(m)}(t)$ for the type-0 cell. The vertical axis represents the concentration of chemicals and the horizontal axis represents time. In this figure, we have plotted the time series of only 6 of the 32 internal chemicals, for clarity. (**b**)–(**f**): Time series of $x^{(m)}(t)$ in a cell, representing the course of differentiation to type A, B, and type A1, A2, A3 cells, respectively. Reproduced from Furusawa and Kaneko, (2001) with permission

## 8.4 Results

### 8.4.1 Hierarchical Differentiation

First, we shall give numerical results obtained by using a fixed reaction network. It is a typical example, when initial intracellular concentration dynamics is chaotic. As an initial condition, we take a single cell, with randomly chosen chemical concentrations of $x_i^{(\ell)}$ satisfying $\sum_\ell x_i^{(\ell)} = 1$. In Fig. 8.3, we have plotted a time series of concentration of the chemicals in a cell, when only a single cell is in the medium. This attractor of the internal chemical dynamics is a limit cycle, whose period is longer than the plotted range in Fig. 8.3. We call this state "attractor-S" or "type-S" in this chapter. Here the abbreviation

$S$ is used in considering the correspondence to a stem cell, to be clarified below. This is the only attractor that is detected from randomly chosen initial conditions.

As already discussed in Chap. 7, the cell volume increases as external chemicals flow into the cell through diffusion. Thus the cell is divided into two, with almost identical chemical concentrations. Chemicals of the two daughter cells oscillate coherently, with the same dynamical behavior as the mother cell (i.e., attractor-S). Successive cell divisions occur simultaneously, and the cell number increases as $1-2-4-8\ldots$, up to some threshold number. After a few divisions, internal dynamics of each cell still belongs to the same attractor (i.e., attractor-S), but the oscillations are no longer synchronized. The microscopic differences introduced at each cell division are amplified to a macroscopic level through cell–cell interaction that destroys the phase coherence.

When the number of cells exceeds some threshold value, some cells begin to display different types of dynamics. In Fig. 8.3, the time series of the chemical concentrations in these new types of cells is plotted. We call here this new type of cell as "type A". The orbits of the chemical concentrations of cells in the phase space are displayed in Fig. 8.4. As the cell number further increases, another type of cell, type B appears here, again differentiated from the type S cell, as shown in Figs. 8.3 and 8.4. These figures show that each of the these dynamics occupies a distinct region in the state space, and that each

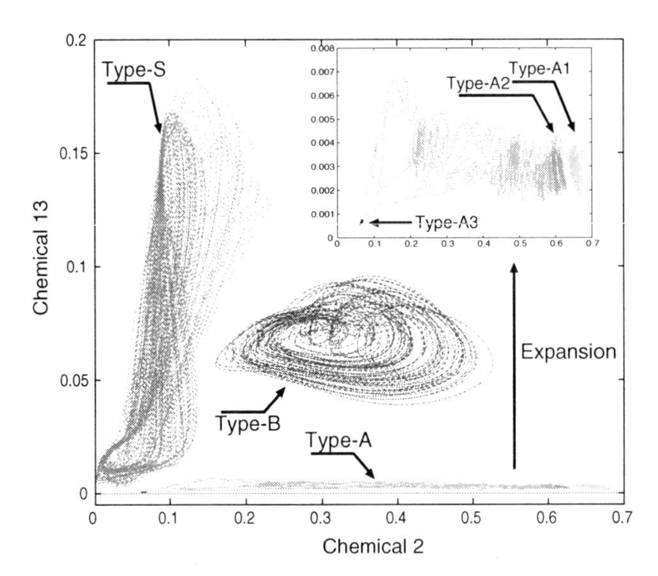


**Fig. 8.4.** Differentiation process from a stem cell, represented in the state space. From the previous figure, we chose two components, and their chemical concentrations $(X^1, x\ 2)$ are plotted in the two-dimensional plane. The orbits for cell types S, A, B, and A1, A2, A3 (in the inset as magnified) are plotted. As shown, these types exist in a "discrete" manner, and no intermediate state exists stably

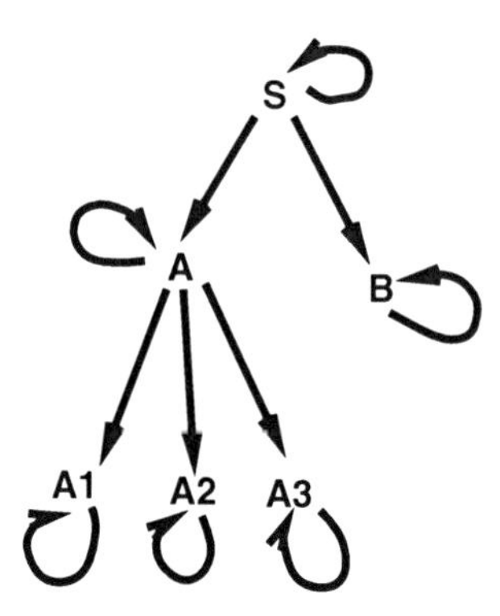


**Fig. 8.5.** Rule of differentiation generated in the model

can be clearly distinguished as a distinct state. The type S cells have the potentiality to differentiate to either "A" or "B," while some of the type "S" cells preserve their type, that is, proliferate. These transitions of the cellular state are interpreted as differentiation. Note that these discrete states are not attractors of the intracellular dynamics of a single cell (Fig. 8.5).

Then, with the further increase of cell numbers, cell type "A" further differentiates into either of three groups represented as "A1," "A2," or "A3." The time series of these three types are shown in Fig. 8.3, while representation at the phase space of each type is also plotted in a projected phase space in Fig. 8.4. The orbit of type A cell itinerates over the three regions corresponding to "A1," "A2," and "A3."

In the simulations starting from a single (or a few cells) without external operation, cells of the types "B" and "A1," "A2," and "A3" reproduce only themselves, without further differentiation. The type S cell either reproduces the identical type or forms different cell types, denoted for example as type A and type B. The type A cell reproduces or further differentiates to A1, A2, or A3 cells but none of the offspring ever becomes a type S cell again.[3] The self-replication (proliferation) and differentiation that occur in this developmental process is summarized as $S \to S, A, B$, $A \to A, A1, A2, A3$, and $B \to B$, $A1 \to A1$, $A2 \to A2$, and $A3 \to A3$. The rule that governs differentiations into cell types is generated from the reaction dynamics.

Note that this differentiation is not induced directly by the tiny differences introduced at the division. The switch from one cell type to another does not occur simultaneously with the division, but occurs later through the interaction among the cells. This phenomenon is caused by an instability in the total system consisting of all cells and of the medium. In the present example, six distinct cell types are observed. The number of cell types appearing in the cell

[3] Among these five cell types, only the cell type S and type A3 cells are attractors of intracellular dynamics, while other types appear and replicate only when surrounded by different types of cells.

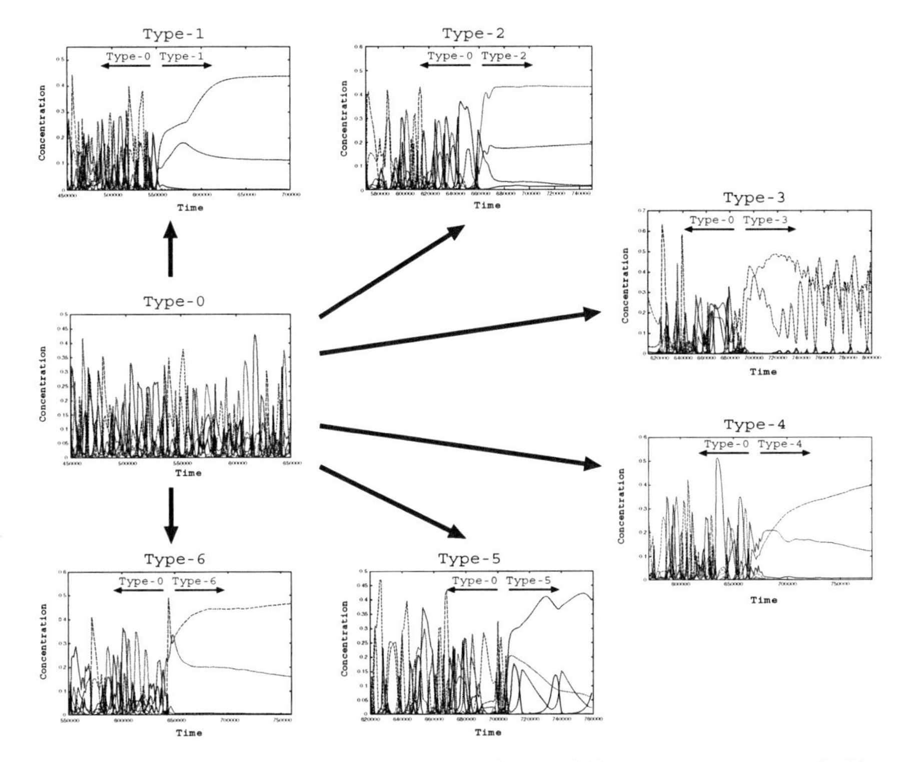


**Fig. 8.6.** Another example of differentiation from a different reaction network. Here six types of cells are generated from the stem-type cell $S$. Plotted are the timeseries of chemical concentrations of 6 species among 20. Reproduced from Furusawa and Kaneko (2001)

society depends on the nature of the catalytic network we choose. The hierarchical differentiation represented by tree-like rules, however, is often observed when the internal dynamics has sufficient complexity with oscillation, as will be the case when dynamics is chaotic. Another example of differentiation is given in Fig. 8.6.

### 8.4.2 Dual Coding of Cellular State: Discrete Types and Continuous Modulation of Each Type

One might think that our definition of types is rather ambiguous and is not clearly defined. However, one can clearly distinguish them by plotting and comparing the time series of various chemical concentrations and check how these orbits are separated. Some differences remain within the same cell type. However, such difference is clearly much smaller than that between different cell types. There are no reproducing cells and chemical concentrations intermediate between those of displayed cell types never correspond to reproducing cells. This demonstrates that the differentiated cell types are well-defined as "digitally" distinct states.

Note that the cell state of each type is slightly modulated depending on the number distribution of cell types around it. For example, as the numbers $N_s, N_A$, $N_B$ for type S, A, B cells are changed, the chemical state (i.e., the average composition or location of the orbit in the phase space) is modified slightly. This change, however, is much smaller than the change between the types. Hence the cell system has both discrete changes of types and continuous change of the state of each type depending on the number distribution.

### 8.4.3 Regulation of Stochastic Differentiation

Now a differentiation rule has been generated, which can be represented on a tree structure. When there are multiple choices of differentiation process (as in "*S*," → "*S*," or "*S*" → "*A*," and "*S*" → "*B*" (see Fig. 8.5), the question remains as to how each arrow for differentiation is selected. Here, when the cell types that exist at a given time are specified, whereas all values of $x_i^{(k)}(t)$ for all chemicals over all cells are not known precisely, the choice of each arrow from the type S cell is probabilistic. For a stem cell also, whether to replicate or to differentiate is regarded as probabilistic.[4]

Since such stochasticity is not due to external fluctuation but is a result of the internal dynamics, the probability of differentiation can be regulated by the intracellular state. To consider this problem, we focus on the differentiation from type S to types A and B. Then after a cell division, one can assign the probabilities $p_S$, $p_A$, and $p_B$ for the type S cell to self-replicate ($p_S$) or to differentiate to $A$ or $B$, respectively. Now we study how these rates of self-replication or differentiation from the type S cell ($p_S$, $p_A$, or $p_B$) depends on the surrounding distributions, that is, the number distribution of cell types $(n_S, n_A, n_B, ..)$.

When the number of type A (resp. B) cells is decreased artificially, we find that the probability $p_A$ (resp. $p_B$) of differentiation to $A$ (resp. $B$) is increased. The frequency of differentiation from the cell type S to other cell types increases almost linearly with $n_S$ when the fraction of type S cells is larger than approximately 40%. By decreasing the number of type S cell, on the other hand, the self-replication rate $p_S$ increases.

To sum up, this stochastic branching is accompanied by a regulative mechanism, to the direction that suppresses external perturbations. When some type of cells is removed by some external operator during the developmental

[4] Of course, if one determined completely the chemical composition and the division-induced asymmetry, then the fate of the cell would be determined, as long as there were no fluctuations. Here, however, such possibility is practically irrelevant, since as one knows from chaos theory, such deterministic dynamics leads to stochastic behavior, as long as one cannot know the initial condition of the internal cell state in every detail.

process, the rate of differentiation changes so that the number of the cells of the removed type is increased and that the final cell distribution is recovered. Through this emergent regulation, the cell distribution is attracted toward the proportions approximately equal to $(n_S, n_A, n_B) = (40, 30, 30)$.

How is this interaction-dependent rule formed? Note that depending on the distribution of the other cell types, the orbit of internal cell states is slightly modified.[5] Such modulation is possible, because of the "dual" coding mentioned in the last subsection. The state of a cell is mainly characterized by its discrete type. However, a continuous modulation between cell types also exists and carries global information about cell type distribution. For example, when the number of "A" type cells is reduced, the orbit of an "S" type cell is shifted toward the orbits of "A," with which the rate of switch to "A" is enhanced. The information of the cell type distribution is represented by the internal dynamics of "S" type cells, and it is essential to the regulation of differentiation rate (Furusawa & Kaneko, 1998a).

### 8.4.4 Differentiation of Colony

The result of the last subsection implies the emergence of a higher level dynamics, which controls the rate of cell division and differentiation according to the number of each cell type. In other words, the stochastic dynamics on the number of each cell type $n_S$,$n_A$,$n_B$ that depends on $\{n_k\}$ $(k = S, A, B, \ldots)$ possesses a stable fixed point at about $(n_S, n_A, n_B) \approx (40, 30, 30)$. Recall that the differentiation rate of cells (each arrow in Fig. 8.5), and accordingly the higher level dynamics of $n_k$ depend on the distribution of cells $n_k$. The result of the last subsection implies that the state reached through development is an attractor on this higher level dynamics on $n_k$. Indeed, the "attractor" of this higher level dynamics is stable against perturbations whereby some cells of each type are removed.

Then, is such a "macroscopic" attractor always unique? Or, can the dynamics on the number of cells of each type also possess several stable attractors with autonomous control of the rate of differentiation? In a biological term, this corresponds to the existence of different tissues. If the answer is positive, then different types of cell colonies consisting of different stable distributions of cell types can be developed from a single stem cell.

To study this problem, we have performed the simulations, by simulations starting from a variety of single-cell initial conditions, with different internal chemical concentrations, and measured the distribution of cell types $n_S, n_A, n_B$ when the total cell number reaches a given threshold value (set at 300 here). In Fig. 8.7, the number of initial configurations (among 100 runs) leading to a cell type distribution with a given range of $n_B$ is plotted as a

[5] This is a continuous modulation of the orbit depending on the distribution (or environment), and is compatible with discreteness in cell types without intermediary state.

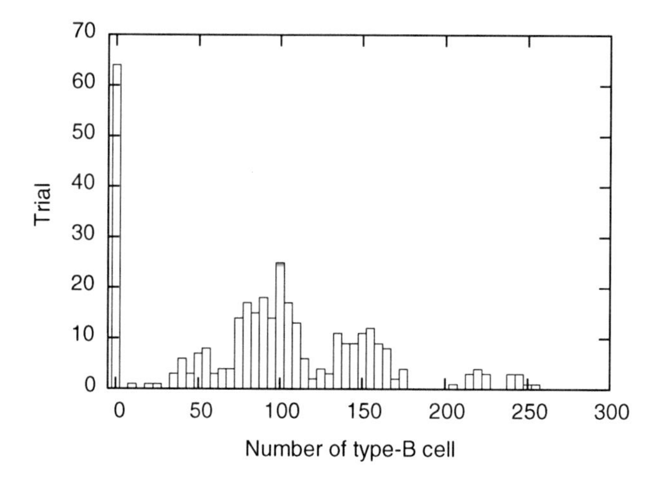


**Fig. 8.7.** Histogram of the number $n_B$ of type B cells. Starting from a single cell with randomly chosen chemical concentrations, the simulation is carried out until the total cell number reaches 300, when the number $n_B$ of type B cells is measured. Repeating the runs 347 times, we have counted the number of such initial conditions that $n_B$ falls onto a given bin (with the size 5). The histogram of $n_B$ is obtained from the count. There are four peaks at $n_B = 0$, 100, 150, 220, each of which corresponds to a stable distribution of the cell colony. Reproduced from Furusawa and Kaneko (1998a) with permission

histogram, by computing the number of type B cells $n_B$ among the total cell number 300. Four peaks are clearly visible at $n_B = 0$ , ~100, ~150, and ~220, which correspond to possible distinct sets of cell distributions. As mentioned, the possible set of cell types (from "S" to "A3") and the temporal ordering of their appearance (e.g., $S \to S$, $S \to A$, $S \to B$) are independent of the initial conditions. However, at a later stage, several types of cell groups emerge depending on the initial conditions. To each peak corresponds a cell colony consisting of different cell types, for example, that consisting only of type S, A, and B cells, of only type S, B, and A2 cells. Here, four types of cell colonies, that is, four stable distributions of cell types, appear through developmental processes with different initial conditions (see Furusawa & Kaneko [1998a] for details). The choice of each colony state is history-dependent, that is, depends on how the development process unfolds.

Recall that the differentiation rate of cells (each arrow in Fig. 8.6) and, accordingly, the higher level dynamics of $n_k$ depend on the distribution of cells $n_k$. The above result implies that there are several attractors on this higher level dynamics on $n_k$. To see this population dynamics of cell numbers in the colony, we have studied the flow chart of the change of $(n_S, n_A, n_B)$, where the direction of change of $n_S$ and $n_A$ is represented by an arrow, starting from the initial distribution given by $(n_S, n_A, n_B)$ of the corresponding site.[6] The chart shows that cell colonies on the cell type distribution $\{n_S, \ldots n_{A3}\}$ have at least 5 stable states around $(n_S, n_B)$ = (0,0), (38,32), (30,50), (18,58), and (0,78) respectively. Each state has a basin of attraction, and the corresponding cell type distribution is stable against external perturbations, as is supported by the higher level dynamics on $\{n_S, \ldots, n_{A3}\}$.

The fixed point at $(n_S, n_B) = (0, 0)$ corresponds to a colony consisting only of type A3 cells, while the fixed points denoted by "II," "III," and "IV" correspond to colonies consisting only of S,A,B, of S,B,A1, and of S,B,A2, respectively. Indeed, these cell type distributions correspond to each of the peaks of Fig. 8.7, respectively.

Still we note that one of the "attractors" at the distribution level observed here is not realized in the simulation starting from a single cell: The state "V" consisting of types B and A2 cannot be obtained from the developmental process from a single cell. This discrepancy is caused by the conjunction of cell number change with the population dynamics of cell types. Through the change of the number of cells, the population dynamics shifts from one flow chart for a given cell number (e.g., shown in Fig. 8.8) to another for a different number of cells. The organized cell colony from a single cell has such developmental constraints.

Now the coexistence of several stable cell colonies is clear. Depending on the initial cell condition, different cell colonies are obtained. The result here means that several types of tissues can appear through the interactions among cells.

### 8.4.5 Stability

As has been discussed already, stability of development process is miraculous when one considers the number of involved molecules, cells, and processes. In spite of molecular fluctuations, and other external perturbations, the "error" in the development is quite small. In a developmental system, there are at least three kinds of stability, which are mutually interrelated.

- Microscopic: The developmental process is robust against molecular fluctuations.
- Macroscopic: The process is robust against macroscopic perturbations such as the removal of some cells or somatic mutations.
- Path: Not only the final states but also the developmental path resulting in such states is stable against microscopic and macroscopic perturbations.

[6] To draw the figure, we kill the division rule for the moment, and the total cell number $(n_S + n_A + n_B)$ is fixed to 100, and the division rule is removed for the moment. This change of rule is only for the convenience, to eliminate the change in the dynamics due to the increase in the cell number, and is not important. From the two-dimensional plane, the number of cells of types A, A1,A2, and A3 are given by $100 - n_S - n_B$.

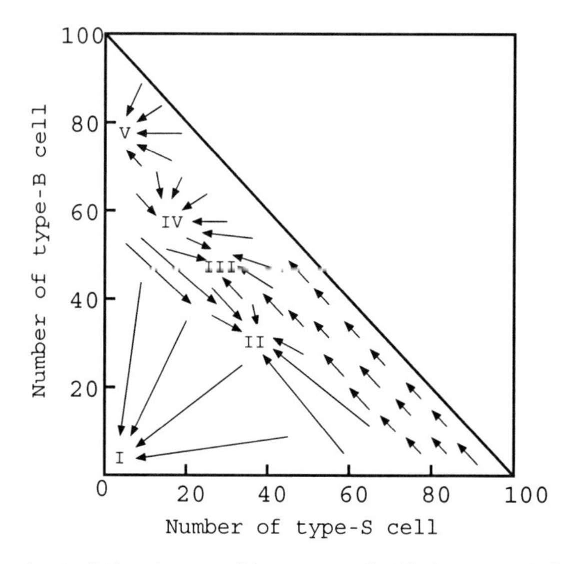


**Fig. 8.8.** Flow chart of the change of $(n_S, n_A, n_B)$. We have carried out the simulations starting from the initial condition at each $(n_S, 100 - n_S - n_B, n_B)$ by fixing the total cell number to 100 (by removing the cell division process). Change of the number of cell types is measured from simulations, from which the direction of changes of $(n_S, n_B)$ is shown as an arrow in the $(n_S, n_B)$ space. As is seen, there are five fixed points, each of which corresponds to the stable population distribution of cell types. Reproduced from Furusawa and Kaneko (1998a) with permission

Our theory, being based on unstable state dynamics, might look inappropriate to generate such stability properties. Still, the result of our model is stable both microscopically and macroscopically. To check the microscopic stability of our model, we have included molecular fluctuations as a noise term. Again, the cell types and number fractions obtained remain the same.

Furthermore, macroscopic stability is also present. For example, after removal of some types of cells, the number distributions are recovered, thanks to a change of differentiation rate as mentioned in Sect. 8.4.2. In general, stability at the ensemble level is a rather general consequence when units with unstable dynamics interact with each other (Kaneko, 1992; Kaneko & Ikegami, 1992). In the present case, macroscopic stability is sustained by the change of the rate of differentiation, that is achieved by the change of internal dynamics through the interaction. But why is the regulation oriented to keep the stability, instead of the other direction? In the example discussed here, the differentiation from S to $A$ is enhanced when the type A cell is removed. Assume that the regulation worked in the other way (i.e., when removing $A$, the rate $S \to A$ decreases). Then, at the initial stage when type A cells are produced for the first time, their number would in fact decrease to zero, since this number is lower than that in the final state at that time. Hence, the type A cell would in fact never appear. In other words, only the cell types that have a regulation mechanism to stabilize their coexistence with other types can appear through the developmental process.

It should be stressed that our dynamical differentiation process is always accompanied by this kind of regulation process, without any sophisticated programs implemented in advance. This autonomous robustness provides a novel viewpoint to the stability of the cell society in multicellular organisms.

The regulation of differentiation probability we have discovered has the same structure as the Le Chatelier priniciple (or Le Chatelier-Braun principle in a generalized form) in thermodynamics, in the sense that a response occurs in the direction to resist external perturbation. Since the basis of thermodynamics lies in the stability of a system, it may be natural that a similar response appears for the stability of developmental system. We will come back to this problem in Chap. 12.

### 8.4.6 Irreversibility

Often, stability and irreversibility are two sides of the same coin. The stability properties of a thermodynamical system are modeled by introducing a thermodynamic potential (e.g., the free energy) that takes a minimal value when the system reaches equilibrium. When perturbed, the system relaxes to equilibrium by evolving toward the state with a minimal value of the potential. This evolution toward the minimum, as well as the approach to the equilibrium state is not reversible, as is beautifully described by the second law of thermodynamics. This is the expression of the irreversibility in thermodynamics.

In a similar way, the stability of the development process is likely to be related to some kind of irreversibility. In fact, there is a clear "time's arrow" in the cell system, as already discussed. Initially cells have totipotency, that is, the potentiality to differentiate to all other types of cells. Later this potentiality is successively limited, and in a committed cell type, only the same type of cell can be reproduced (Kaneko & Furusawa, 2000; Furusawa & Kaneko, 2001; Kaneko, 2003a). The evolution of the degree of determination is an irreversible process.

- (U)"Undifferentiated cell class": cells that change their chemical composition from generation to generation.
- (S) "Stem cell class": cells that either replicate faithfully or switch to different types with some rate.
- (C) "Committed (determined) cell class": chemical compositions are preserved after the division process[7] (Fig. 8.9).

[7] *Germ-line cell class*: As a special case of a determined cell, the following cell type can appear: In the course of development, it replicates as a committed cell type.When isolated, however, it comes back to the undifferentiated cell state, from which all the other cell types are formed. We come back to this problem in Chap. 9.

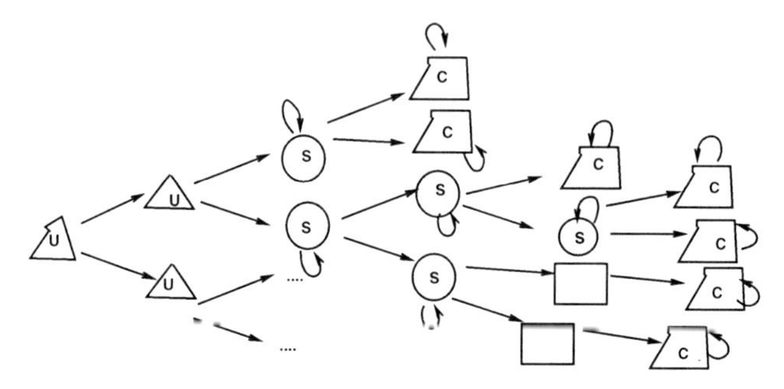


**Fig. 8.9.** Schematic representation of the hierarchical cell differentiation process. The degree of determination increases as (U) → (S) → (C), where the letters U, S, and C stand for undifferentiated cells with a varying state, stem cells, and committed (determined) cells

The degree of determination increases as (U) → (S) → (C). In the class (U), the cell is in a transient state as a dynamical system and changes its character (e.g., the average chemical composition) in time (and accordingly by division). The class (S) is a stable state and is regarded as a kind of "attractor" *once interaction is fixed.* Still, the type is only weakly robust against a change of interaction, and can switch to a different state, as mentioned. The determined type is both an attractor for the internal dynamics and a stable state under cell–cell interaction: it remains stable over a wide range of interactions and environments. Unless the interaction term is changed to a large extent, it neither becomes unstable, nor switches to (U).

The temporal flow ((U) → (S) → (C)) that accompanies the development is characterized quantitatively as follows (see also Fig. 8.10 for a schematic representation).

- (I) the diversity of chemicals decreases
- (II) the plasticity (variation) of the states decreases
- (III) the stability of the states increases

First, we study the diversity of chemicals in the cellular dynamics. In our model, a variety of chemicals participate in the oscillatory dynamics of the stem-type cell. On the other hand, in differentiated cell types, the concentrations of some chemicals fall approximately to zero. As a simple measure to study the decrease of chemical diversity quantitatively, we define this diversity for the $i$-th cell by

$$S_i = -\sum_{j=1}^{k} p_i(j) \log p_i(j), \tag{8.1}$$

with

$$p_i(j) = \left\langle \frac{x_i^{(j)}}{\sum_{m=1}^{k} x_i^{(m)}} \right\rangle, \tag{8.2}$$

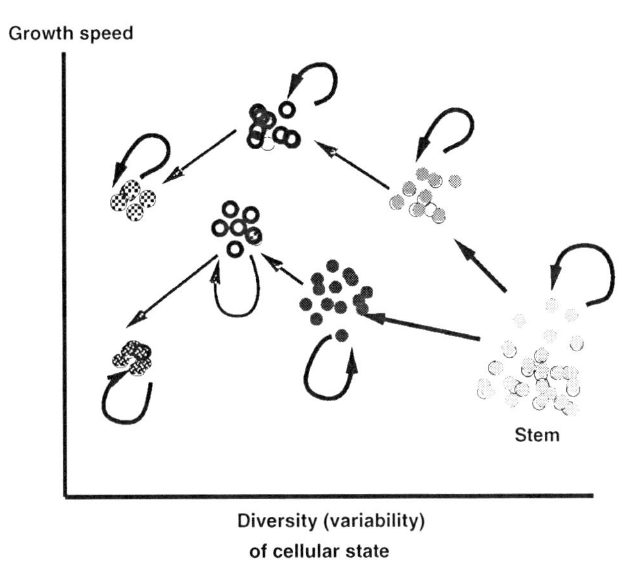


**Fig. 8.10.** Schematic representation of irreversibility in differentiation with regards to the loss of diversity in chemicals and the growth speed of a cell

where $\langle \ldots \rangle$ represents the temporal average taken over successive cell divisions.

In Fig. 8.11, we display some examples of the change in the chemical diversity accompanying the developmental process. In the figure, the chemical diversity of a cell, averaged over its lifetime, is plotted at each cell division event. As shown, with the course of hierarchical differentiation, the diversity of chemicals decreases successively through the developmental process.

Next, the variations in the cellular state decreases with the loss of multipotency. The totipotent cell in our model shows large temporal variations of chemical concentrations. These variations decrease as the cells lose the multipotency. For example, one can see clearly in the phase–space plot of chemical concentrations of each cell that the variations decrease with the differentiation from $S$ to $A$ and then to $A1$, $A2$, and $A3$.[8]

Step (III) could be characterized by determining the minimal perturbation necessary to switch a cell state (see Kaneko [2002b] for a definition of such "attractor strength"). (U)-type cells can switch state without any perturbation of the interaction term, while (S)-type cells require a tiny change. Indeed, a tiny change in the interaction made during a single cell division event suffices

[8] The decrease in complexity of the dynamics along with the loss of multipotency is also confirmed by measuring quantities characterizing variety of dynamics or instability in the dynamics. For example, the Kolmogorov–Sinai (KS) entropy is computed, characterizing the variety of orbits and the degree of chaos. We have found that this local KS entropy of differentiated cells is always smaller than that of stem-type cells before differentiation.

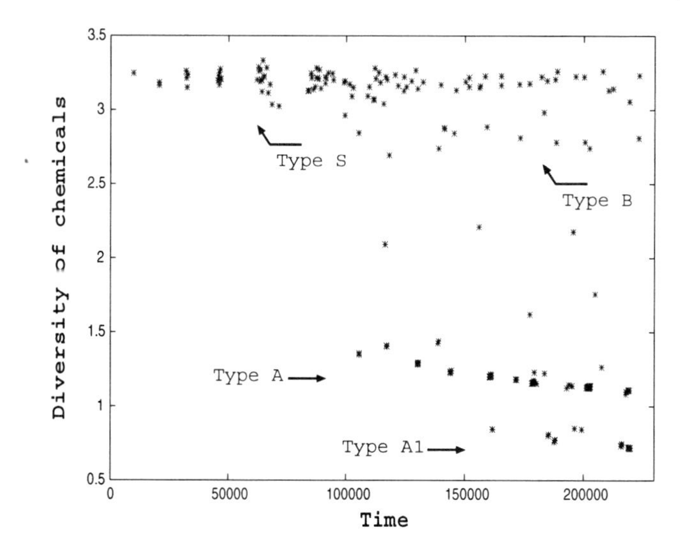


**Fig. 8.11.** Change of chemical diversity in a cell, through the differentiation from type S to A, B, and A1. During the first stage of development (*time* < 10000 in Fig. 8.3(a)), cells have not yet differentiated, and the dynamics of cells are those with a high diversity of chemicals. At a later time, differentiated cells begin to appear from the initial cell type. Reproduced from Furusawa and Kaneko (2001)

to switch an (S)-type cell to some other cell type.[9] On the other hand, (C) cells require a sufficiently large change in the interaction (or environment) to make a switch to another cell state. Indeed, the degree of determination of (C) cells can be measured as the minimum perturbation strength required for a switch to a different state.

## Change in the Growth Speed of Cells with the Developmental Process

According to our results here, the growth speed of stem-type cells is found to be smaller than that of differentiated cell that appear up to some stage. In Fig. 8.12, we display an example of the relationship between the growth speed of a type of cell and the chemical diversity within the cell. Through several simulations, we have found that the growth speed increases up to some stage of differentiation. In other words, there is a negative correlation between the growth speed and the chemical diversity in a cell, up to some

[9] By any small change in the interaction term, the stem cell state changes its stability, as it reproduces or differentiates. In this sense the state is marginally stable. The minimum distance between the attractor and basin boundary is zero (see also Chap. 3). In dynamical systems, such state is studied called a "Milnor attractor" (Milnor, 1985), while its prevalence in coupled dynamical systems has been discussed in Kaneko (1997a, 1998a).

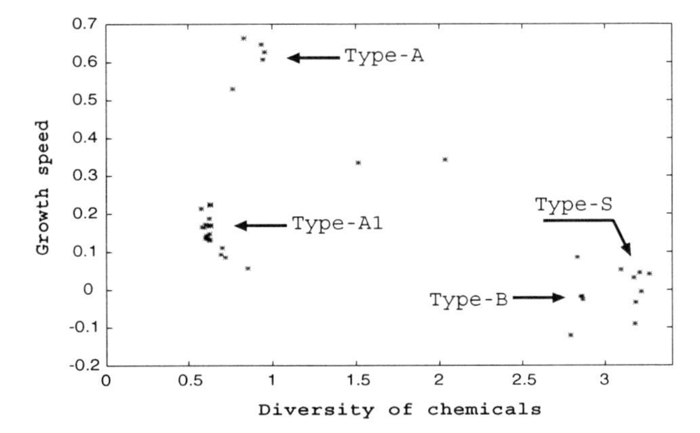


**Fig. 8.12.** Relationship between diversity of chemicals and growth speed of a cell, obtained from the model, through the course of differentiation from type S to A, B, and A1. Reproduced from Furusawa and Kaneko (2001)

stage. Then the speed decreases again for the final committed cell. Consider the successive differentiation process represented as $S \to I_1 \to I_2 \to \cdots \to C$, where $I_1, I_2, \ldots$ are intermediate types between the stem cell S and the final committed cell type $C$.[10] Then, the growth speed has a single humped structure in the above order of cell types, that is, has a maximum at some intermediate cell type $I_j$.[11] The increase of growth speed from the stem cell to the next level of differentiated cell, and the decrease of the growth speed at the final terminal cell type, is generally observed in our model simulations (Yoshida et al., 2005).

The increase in the growth speed from the stem cell to the next differentiated cell can be explained in terms of a change in the requirement for nutrients. A stem-type cell with high chemical diversity absorbs a variety of nutrients from the medium. On the other hand, a differentiated cell type with simple internal dynamics and lower chemical diversity is specialized in the use of only a few kinds of nutrients. In such a cell with simple internal dynamics, a small number of reaction paths is used dominantly and they often form a simple autocatalytic reaction network. Since each reaction rate is governed by products of the involved chemical concentration, the internal reaction dynamics of such simple cell progresses faster, when the chemical concentration is biased. Then, required nutrients are transformed into nonpenetrating chemicals much faster than those in stem-type cell with complex cellular dynamics. When the stem-type cells and differentiated types of cells coexist, differentiated cells can efficiently obtain the nutrients they require, while stem-type

[10] In the example in Sect. 8.4, $A = I_1$ and $A_m = C$ $(m = 1, 2, 3)$. In general, there are several intermediate types in our model and in natural cells.

[11] It should be noted that the growth speed of stem-type cells decreases only when the differentiated cell types begin to coexist with them. That is, the growth speed of an isolated stem-type cell is not always smaller than that of an isolated differentiated cell.

cells, which are not specialized in taking specific nutrients, cannot do it so efficiently.

As the differentiated cell has a higher growth speed, the number density of stem cells generally decreases as the total cell number increases. If the total number of cells does not increase indefinitely, because of probabilistic deaths of each cell type, then the stem cell may disappear. Indeed, there are cases where the stem cells either continue to constitute a small fraction of the total number of cells and maintain a slow division speed or disappear in the steady state. In the former case, the spontaneous regulation of differentiation from the stem cells insures the stability of the cell number distribution with respect to changes in the death rates of each cell type. In this case, the decrease in the growth speed of the final committed cell is relevant to sustaining the stem cell.

### 8.4.7 Universality of Differentiation

So far we have demonstrated the possibility of a hierarchical differentiation from "stem"-type cells by choosing some parameter values and specific reaction networks. But is this behavior universal? To confirm the universality of our differentiation scenario, we have carried out numerical experiments, using several different sets of parameter values by choosing thousands of different reaction networks. The results found then are changed little when we change the model parameters. They are essentially unchanged as long as the magnitudes of the internal reaction term and the cell–cell interaction term are of comparable order, and provided that the set of equations describing the internal reactions are sufficiently complex to exhibit nonlinear oscillatory dynamics that in some case can be chaotic.

Next, we would like to discuss whether most of randomly chosen networks lead to hierarchical differentiation from a unique stem-type cell. This turns out not to be the case. For most networks chosen randomly, the reaction dynamics are stable, and show neither temporal oscillation of chemical concentrations nor cell differentiation. Of course, in the present cells, concentrations of some chemicals, for example, the $Ca^{2+}$ ion, oscillate. Still, we are not making a model to "imitate" a natural cell but rather try to answer general questions. As emphasized in Chap. 2, we do not wish to select models uniquely on the basis of the reality of existing cells.

To answer the question about the choice of networks, we have carried out numerical experiments by taking a huge number of networks and examined the growth speed of the cell and an ensemble of cells (Furusawa & Kaneko, 2000a). In Fig. 8.13, we plot the growth speed of a single cell versus the growth speed of an ensemble of cells.[12] In the figure, the horizontal axis denotes the

[12] The data used here are based on the model with spatial pattern that will be discussed in the next chapter. Still, the result is identical even if one adopts the model with all-to-all cell interaction as in this chapter

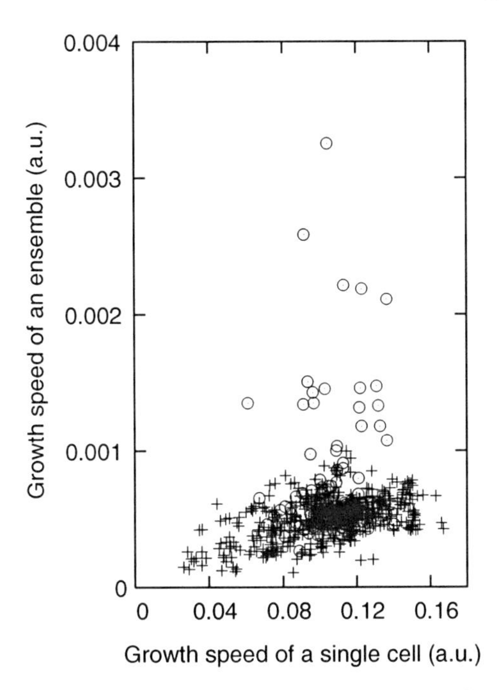


**Fig. 8.13.** Relationship between the growth speed of a single cell and that of an ensemble. The ordinate shows the growth speed of an ensemble, measured as the inverse of the time for the cell number to double from 100 to 200, while the abscissa represents the inverse of the time for a single cell to divide. Each point is obtained by using a different chemical reaction network. The points + are the results of the case without differentiation, where the concentration dynamics are mostly fixed points. The points with ○ are results from the case with differentiation, where the chemical concentration of initial cell type shows chaotic dynamics. Adapted from Furusawa and Kaneko (2000a) with permission

inverse of the time for a cell to divide, and the vertical axis is the inverse of the time for cell aggregates to reach 200 cells. In other words, we plot the speed of the growth of a cell ensemble versus the growth speed of a single cell. Each point is a result of a simulation carried out for a different network, chosen randomly (we used 800 samples). For each example, the parameters are identical.

The points plotted by the circles are the results of the simulations including initially chaotic dynamics and differentiation. Here, the growth speed of a single cell is not so large, but the growth speed as an ensemble is higher than in the case without differentiation. In each cell, a variety of chemicals coexist, supporting complex reaction dynamics and cell differentiation.

For the other points represented by +, there are much simpler intracellular reaction processes, with only a few autocatalytic reactions used dominantly, that can produce the rapid replication of a single cell. In this case, the growth speed of a single cell is often large, while the growth speed of an ensemble always remains small. In some sense, simple cells with rapid growth are "selfish": Although such simple cells with low diversity of chemical species can exhibit large growth speeds as single cells, they cannot grow cooperatively, and their growth speeds as an ensemble are suppressed because of strong competition for resources. On the other hand, those cells with initially chaotic dynamics and differentiation, even though they have a lower growth speed as a single cell, the growth is not much hindered in cell aggregates. With differentiation, cells adopt different roles and do not compete for nutrients or cooperate for synthesis of chemicals, which is the reason why the growth is sustained in an ensemble.

To sum up, our study has provided evidence that an ensemble of cells with a variety of dynamics and stable states (cell types) has a larger growth speed than an ensemble of simple cells with a homogeneous pattern, because of the greater capability of the former to transport and differentiation in the use of nutritive chemicals. Note that no elaborate mechanism is required for the appearance of such heterogeneous cell ensembles. Some fraction of the randomly chosen biochemical networks we considered exhibit dynamics sufficiently complex to allow for spontaneous cell differentiation. Once such networks appear by chance, they are selected as a result of higher growth speed as an ensemble. The result here suggests that the complexity of multicellular organisms with differentiated cell types is a necessary course in evolution, once a multicellular unit emerges from cell aggregates.

*Remark*: Instead of using catalytic reaction network, one can adopt gene network as also mentioned in Chap. 7. Even though each single gene expression dynamics possesses only bistable fixed points, when coupled, the gene network dynamics can exhibit chaotic dynamics. Indeed, the fraction of the network that exhibit chaotic dynamics is larger when more genes interact globally with each other (Ishihara & Kaneko, 2005). Hence gene networks that possess unstable dynamics of the present stem-cell type is rather large.[13]

## 8.5 Experiment

### 8.5.1 Constructive Experiment from Embryonic Stem Cell

What type of "constructive experiment" is possible corresponding to the theory in the present chapter? As an experiment with a small amount of construction, one can use a stem cell in the present multicellular organisms and then study the differentiation process under externally imposed, well-controlled

[13] To be precise, we do not necessarily need the chaotic dynamics. As long as there exists some instability in the dynamics leading to differentiation such cells can remain, and the argument in the present section holds (Takagi & Kaneko, 2005).

conditions. For example, the embryonic stem cells of mammals have the potentiality to produce all types of cells of the organism. Then, it is desirable first to devise an environment where the culture of ES cells is well controlled (without the use of "serum" whose components are not completely controlled), and in a second stage, to change the condition (e.g., by varying the concentration of some nutrient in the medium or the number density of cells) to see what impact is made on the differentiation process. This process can be studied with the use of flow cytometry, by using some fluorescent proteins. With the recent advances in the fluorescent protein studies and the measurement by flow cytometry, this direction of research should be promising. The difficulty, at the moment, lies in the precise control of the culture system in the above sense, to carry out a precise, reproducible experiment. Still, there are several progresses toward this direction also. Another related possibility is the use of blood stem cells where construction of a well-controlled proliferation system is in progress (Zandstra & Nagy, 2001).

### 8.5.2 Differentiation from Callus of Plant

In plants when a piece of tissue is cultivated in a suitable medium, an aggregate of undifferentiated cells, called callus, is formed. In other words, dedifferentiation from committed (differentiated) cells easily occurs compared with animal cells (Steward et al., 1958). By adding some hormone (some chemicals) such as auxin or cyokinin into these callus cells, it is possible to force these cells to differentiate into root or bud cells, or even to produce the whole body of plant. In this case, how this differentiation depends on the concentration of the hormone has also been studied. Hence, by using these hormones as control parameters for differentiation, and measuring some indices of cells, it should be possible to study the differentiation process in the state space of chemical concentrations. With this experiment, one can study the degree of plasticity in relationship with the reversibility of the differentiation. Such study is essential to answer a very naive question why or how plant cells have a higher plasticity than animal cells.

### 8.5.3 Construction of "Stem-type Cell" from Bacteria

Of course, as a higher level of constructive experiment, it is desirable to use cells of uni-cellular organisms and make "stem cells" from them. For example, take a bacteria (*E. coli*) cell, and embed a gene network (into a plasmid) within the cell. Such introduction of genes is now experimentally feasible. Then by embedding a network that can show plastic dynamics, we can examine the scenario presented here. Although such experiment has not yet been completed, a related experiment with embedded network has recently been carried out by Kashiwagi et al. (2005). Here, we have introduced the feedback network as shown in Fig. 8.14. In this network, two genetic subsystems A and B that mutually suppress each other are embedded. When one gene

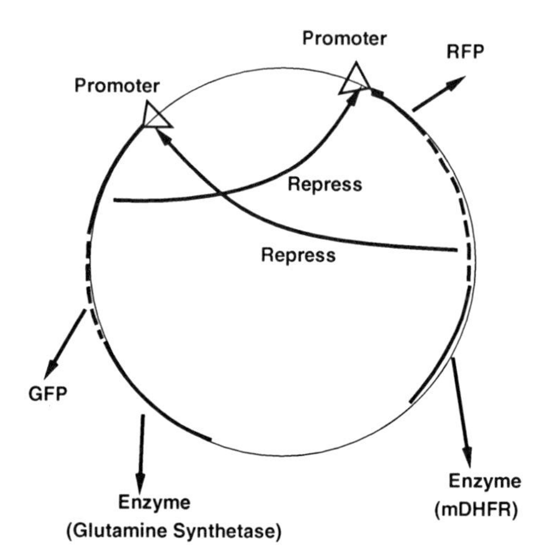


**Fig. 8.14.** An example of embedded gene network with mutual inhibition, into *E. coli* (see Kashiwagi et al. [2005] for details)

is expressed, it suppresses the expression of the other, and vice versa. To be specific,

(1) when the system A is on, enzyme A (glutamine synthetase), protein A, and Green fluorescent protein A are synthesized. Protein A works on the promoter of system B to suppress the expression of system B.
(2) When system B is on, enzyme B (tetrathydrofolate synthetase), protein B, and Red fluorescent protein are synthesized. Protein B works on the promoter of system A to suppress the expression of system A.

This gene expression process can be modeled by [Hasting et al. 2000]

$$
\begin{aligned}
dx_1/dt &= syn(s)/(1 + x_2^2) - deg(s)x_1 \\
dx_2/dt &= syn(s)/(1 + x_1^2) - deg(s)x_2
\end{aligned}
\tag{8.3}
$$

where $x_1$ and $x_2$ are the expression of each gene, that is, the concentration of the corresponding mRNA, and the mutual inhibition is expressed by the term $1/(1 + x_i^2)$, that shows the synthesis of each mRNA is suppressed by the other gene, while the last term represents the decomposition. Here the rates of synthesis ($syn$) and decomposition ($deg$) depends on the cellular activity s. As the synthesis process is complex, it is easily saturated with regards to the growth of the activity, while the degradation process increases more straightforwardly with the activity. For example, let us take $syn(s) = as/(1+s)$, and $deg(s) = s$. Then, as the activity $s$ decreases, eq. (8.3) shows bifurcation. When the activity $s$ is small, there are two fixed point solutions $x_1 > x_2 \sim 0$ and $x_2 > x_1 \sim 0$. When the activity $s$ is larger, there is only a solution $x_1 = x_2$, which is not close to zero. As the resource (nutrient)

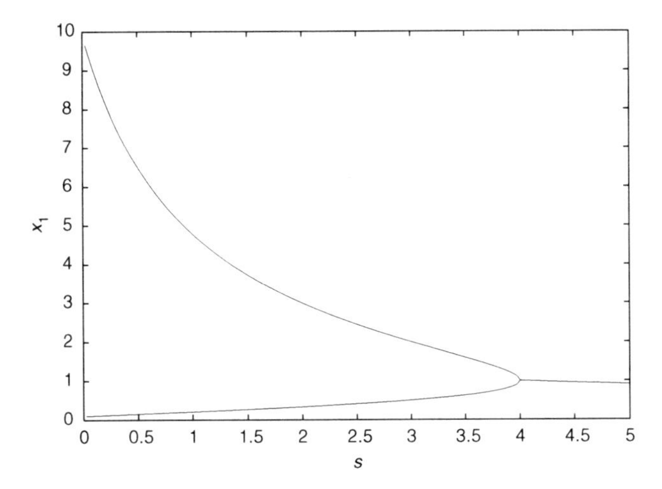


**Fig. 8.15.** The stationary solution $(x_1, x_2)$ of the model (8.3), plotted as a function of $s$, for $a = 10$. Here the solution bifurcates into two solutions $(x_L, x_S)$ and $(x_S, x_L)$ with $x_L > x_S$ from the solution with $x_1 = x_2$, at $s = 4$ (see Kashiwagi et al. [2005] for details)

for the growth is decreased, and the activity of the cell is decreased, there is a bifurcation, into two multiple states occurs, as shown in Fig. 8.15:

(i) For large S (for $syn(s)/deg(s) \leq 2$) $x_1 \sim x_2$ with small values. In other words, both the genetic systems A and B are weakly expressed.
(ii) For small S (for $syn(s)/deg(s) \geq 2$) there are two stable states with $x_1 \gg x_2 \sim 0$ and $x_2 \gg x_1 \sim 0$. In other words either the genetic system A or B is expressed.

By using the fluorescent proteins, this change is clearly detected. Indeed, when the resource (nutrient) is supplied sufficiently, the bacteria show weak fluorescence, while for poor resource condition, either of the genetic subsystems *A* or *B* is on, as measured by green or red fluorescence. Now in the environment without glutamine, glutamine synthetase by system A is essential for survival, while in the environment lacking tetrahydrofolate the expression of system B is essential. Surprsingly it is found that in the medium lacking glutamine, the state in which the gene network A is expressed appears in spite of the symmetry of (8.3).[14]

In other words, when the medium is rich in both glutamate and tetrahydrofolate, the cell state with weak expression of both A and B appears. When one of these resources is insufficient, the cell state differentiates into either the A-expressed or the B-expressed state, thus compensating the lack of resources.

[14] A plausible and general mechanism for this adaptive attractor selection is proposed in Kashiwagi et al. (2005).

Finally, when the medium lacks both glutamine and tetrahydrofolate, it is expected that the differentiated cells with A-expressed and B-expressed types coexist and support each other for their survival. Now it will be possible to construct an experimental system showing differentiation from A&B weakly expressed state into a specialized states with the expression only of A or B. This will lead to a minimal construction of a differentiation system from a stem type cell to differentiated cell types.

## 8.6 Relevance to Biology

The stem cell system is basic in multicellular organisms to produce diversity in cell types. We expect that the properties we have found through Sect. 8.3 hold also in the real stem cells of the present organisms, since the results here are rather universal. Some of these universal features we have found are consistent with some experimental data, while some have not yet been examined experimentally. Hence we make some predictions here concerning the general features that have to be commonly satisfied in real stem-cell systems and discuss the possibility of experimental verification.

(i) **Interaction-based differentiation**. In our theory, the differentiation of a stem cell is induced by interactions between cells in the system. Thus, the behavior of each cell in a stem cell system depends strongly on the existence of surrounding cells as well as on the states of these cells. For example, we expect that if a stem cell is maintained in an isolated environment by continuously removing divided cells, it will not be able to differentiate and will replicate only itself.

   Recall that in our model, differentiation of a stem cell is not necessarily concurrent with its cell division. Rather, this differentiation occurs when the instability of the cellular dynamics exceeds some threshold, and this is caused by the increase of cell number through cell division.

(ii) **Regulation of the differentiation probability leading to the robustness of the cell society**. As mentioned, the probabilities governing differentiation of a stem cell are not fixed in our theory, but are autonomously regulated by the distribution of surrounding cell types. An important point here is that this regulation of differentiation occurs in such a manner that the effect of an external perturbation is compensated, in analogy to the Le Chatelier principle in thermodynamics. This regulation is that which creates the robustness of the system. By controlling the population density of each cell type, one can examine experimentally if the rate of differentiation of stem cells is regulated so as to compensate for external change. For example, if the death rate of one type of cell is increased, or cells of this type are continually removed, the differentiation rate from the stem cell into that type of cell should be increased.

(iii) **Irreversible loss of multipotency characterized by some characteristic indices of intracellular process**

(iiia) **Decrease in diversity of chemicals**. In our theory, stem cells possess a higher diversity of chemicals, while differentiated cell types always possess more biased chemical compositions. In terms of cell biology, the decrease in diversity of chemicals and diversity of use of reaction paths in cellular dynamics can be interpreted as the decrease of the number of expressed genes. Considering the generality of our conjecture, it is relevant to check how the diversity of gene expressions changes in the course of development from a stem cell. The confirmation should be possible with the use of microarray data, and by checking how the expressions are biased with the course of differentiation. Indeed, such decrease in the number of expressed genes has been observed in real systems, for example, multipotent progenitor cells in hematopoietic systems (Hu et al., 1997). Quantitative analysis in diversity change will be possible by adopting the method in Sect. 8.5.1, and then examining the gene expressions of cells taken from a given range in some indices using the flow cytometry.

(iiib) **Loss of variation of cellular state.** In the theory variation in chemical concentrations in time and by cells decreases as the differentiation from stem cell progresses. Such variation can be measured by flow cytometry, as the distribution in some chemical concentration, for example, with the use of fluorescent proteins.

The direct observation of the intracellular dynamics, on the other hand, may be difficult. It was reported that the $Ca^{2+}$ oscillation in a cell depends on the cellular state, and that can change the pattern of gene expressions (Dolmetsch et al., 1998). It may also be interesting to study the oscillatory reaction dynamics, such as $Ca^{2+}$ oscillation, in a cell and check how the irregularity in the oscillation differs between a stem cell and a differentiated cell. Studying the dynamic change of gene expression patterns is also important.

(iv) **Recovery of the loss of plasticity.** While during the normal course of development, the loss of multipotency is irreversible, it is possible to recover the multipotency of a differentiated cell through perturbation, by changing the diversity of chemicals or by putting the cell into a suitable environmental condition. For example, by forcing the expression of a variety of genes in some committed cells, the original multipotency may be regained.

(v) **Smaller growth speed of stem cells**. According to our result, the growth speed of stem cells is generally smaller than that of daughter differentiated cells, when both coexist. In general, there is negative correlation between the chemical diversity (or the diversity in gene expressions) and the growth speed, up to some stage of differentiation, which can be experimentally examined.

Of course, in a real biological system, the activation or suppression of cell growth and division involves several factors, including signal molecules called "growth factors", in addition to the condition for nutrients. Still, we believe that this smaller growth speed of stem cells is a general characteristic of the diversification process in stem cell systems.

In real multicellular systems, stem cells, such as neural or hematopoietic stem cells, often enter a quiescent state (i.e., a $G_0$ state) without division. These stem cells resume division actively when some external condition is changed, for example, an injury that decreases the population of differentiated cells. On the other hand, differentiated cells, such as progenitor cells in the hematopoietic system and interstitial crypt, generally have a faster growth speed. Here, as an experimentally verifiable proposition, we conjecture that the growth of stem cells generally is hindered if they are surrounded by a sufficient number of differentiated cells (that are produced by that stem cell), while the decrease of the number of surrounding cells leads to a resumption of the division and differentiation.

(vi) **Positive correlation between growth speed of cell ensemble and cellular diversity.** As shown, an ensemble of cells with a variety of cell types that is sustained by differentiations from stem-type cells generally has a larger growth speed than an ensemble of homogeneous cells, because of the greater capability of the former to transport and share nutritive chemicals. This *positive* correlation between the growth speed and the diversity of cell types (at the cell ensemble level) can be experimentally verified in any stem-cell system. For example, if a stem cell loses the ability to differentiate through mutation, tissue consisting only of the homogeneous cells has smaller growth speed as a whole than tissue of the wild type with various cell types.

Since even primitive organisms such as *Anabena* (Yoon & Golden, 1998) and *Volvox* (Kirk & Harper, 1986; Tam & Kirk, 1991) exhibit differentiation in cell types and some spatial pattern, the relationship between complexity and growth can be verified. In fact, for a mutant of *Volvox* that possesses only homogeneous cells, the growth becomes slower in comparison to the wild type. Further study of this correlation will also be important to understand the evolutionary process of multicellular organisms.

To sum up, it is important to study that the stem cells have higher plasticity, and the differentiation is a result of cell–cell interaction. Stem cells can change flexibly according to the environment, while differentiated cells have no such flexibility. This flexibility of stem cells results from the complexity of their internal dynamics with a variety of chemical species.

Indeed, there are several recent attempts that seriously discuss the biological relevance of our theory from the experimental side. With the recent advances in microarray analysis, fluorescent techniques, flow cytometry, and controlled cultivation of (embryonic) stem cells, we hope that the predictions here will be confirmed experimentally in the near future.

In relationship with the topic (iv), recall that in somatic clone experiments mentioned in Sect. 8.1, the "irreversible" differentiation to committed cells is "reversed" to recover the totipotency, by external operation. Following the theory presented here, this means that the cell plasticity (measured by diversity and stability) is regained through the operation. It is important to reveal the condition to reverse the time's arrow in cell differentiation. This problem will be discussed in Chap. 12.

Finally we briefly discuss about the phenomenon of metamorphosis that is often seen in the development of insects and shellfish (*crustacea*) and others. The process of metamorphosis is quite mysterious. In complete metamorphosis, a large number of cells are destroyed at some developmental stage, and novel tissues consisting of new cell types are formed from remaining stem cells. Whether such metamorphosis occurs or not should depend on how many stem cells remain. At some stage of differentiation, many cells show apoptosis, and a novel course of differentiation from stem cells starts, leading to new types of cells and a new attractor at a cell ensemble level.

Indeed, Takagi and Kaneko (2005) have revised the model in this chapter, to include more stable states in intracellular dynamics, and to add the cell death process depending on the cell state.[15] In this model again, cells differentiate as the cell number increases, and a stable state consisting of various committed cell types is formed, as discussed in this chapter. Then, after further temporal evolution, some type of cells start to die when the total number of cells reaches some value. With these cell deaths, the plasticity of cells (as measured by diversity in chemicals and attractor stability) can increase, and a new differentiation process progresses to reach a new type of cell ensemble, as postulated in the metamorphosis. In the future, it will be important to study the metamorphosis as a dynamical change in the plasticity, both theoretically and experimentally.

[15] In the model, cells die when the total chemical abundances in a cell decrease below some threshold.

# 9

# Pattern Formation and Origin of Positional Information

## 9.1 Question to Be Addressed: Origin of Positional Information

**Question: Biological pattern formation (morphogenesis) is understood as a change of genetic expression, depending spatially on the gradient of chemicals. To encompass the macroscopic pattern formation, it is often argued that the positional information is formed according to the gradient of a chemical, leading to the change in the concentration of the signal molecules, and finally leading to the change in genetic expressions. However, how is such positional information generated? To form positional information, spatial differentiation of cells (at least at some part) is necessary. If the gradient of chemicals is required for cell differentiation, how is the bootstrap between this gradient and differentiation resolved?**

In the morphogenesis, cells differentiate as the developmental process progresses and these differentiated cells form an ordered spatial pattern. Recent experimental studies have clarified how the change of gene expressions takes place in the process of differentiation, and in the spatial ordering process. However, the description of genetic changes only is not sufficient to fully explain the whole morphogenesis, since how some genes are on or off depends on the chemical states of a cell, which are also influenced by interaction with surrounding cells.

Note that the pattern formation process during normal development is rather stable. Considering possible fluctuations of molecules and perturbations on cell behaviors, they are quite robust in the normal developmental process. Almost identical final patterns are formed and the temporal course for the pattern formation is also repeated. Furthermore, some experiments to perturb externally the process by removal or addition of cells suggest that the cells somehow "know" their position within the whole system of an ensemble of cells so that the damage is repaired. Following these observations, Wolpert

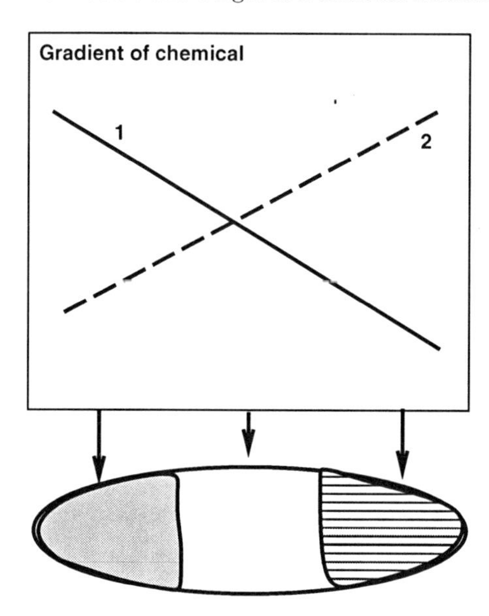


**Fig. 9.1.** Schematic representation of positional information

proposed the concept of positional information, which states that the cell knows its position by determining the spatial dependence of concentrations of diffusible chemicals (see Fig. 9.1).

However, some questions remain unanswered by this positional information idea. The first problem is with regards to the origin of positional information. If some chemical gradient is externally given, then the position can be specified. However, the question of the generation of the chemical gradient is not addressed within the positional information theory. To form a gradient in chemicals, cells at one end and cells at the other end have to behave in a different manner. However, to trigger such differentiation, the existence of gradient is required in the positional information theory. In other words, how polarity is generated spontaneously is not solved. In this sense, it is desirable to set up a theory in which the formation of positional information and the cell differentiation reinforce each other.

Of course, in some cases, the gradient is given as a condition from its mother, and exists in a fertilized egg, as the concentration of some proteins (such as bicoid), by which the pattern formation processes at the first stage of development is believed to be controlled.

However, the axis for the pattern is not completely determined by the mother cell. Also, in mammals, the homogeneity seems to exist up to four divisions. For example, even though the initial concentration gradient of bicoid in *Drosophila* egg is changed experimentally, final pattern of the entire body is robust with respect to this perturbation up to certain range. This means that the positional information is not always embedded in the initial condition of developmental process, but rather it is generated during developmental process.[1]

The second problem is on that of robustness. As a standard picture of developmental biology, cell differentiation progresses depending on the concentration of signal molecules. Since this concentration depends on the space, the cell state changes according to its position. In this way, a cell is believed to "read" the positional information. However, as we have already discussed in Chap. 1, there should be large fluctuations in the chemical concentrations, which may cause a serious problem.

Recall that these signal molecules work often with very low concentration. The quantity of chemical "concentration" is given by the number of molecules per tiny volume around the cell. Here, the chemical concentration of relevance is the number of molecules per unit volume in the tiny region around the cell. Hence the number of signal molecules associated with cell differentiation is often quite small, while in diffusion processes, each molecule moves randomly. Thus there can be large fluctuations in the concentration of molecules affecting a single cell in general. In fact, as discussed in Sect. 7.1, Houchmandzadeh et al. (2002) have demonstrated such large variation of bicoid concentration in the *Drosophila* eggs with the same developmental stage. Such large fluctuations no longer appear at later steps in the developmental process, and the pattern formation is robust. Accordingly the threshold mechanism, only by itself, is vulnerable under the fluctuation and cannot resolve the error arising from the fluctuation.

Prior to positional information theory, Turing proposed a theory in which spatial inhomogeneity of diffusible chemical concentrations emerge spontaneously (Turing, 1952). In fact, the term morphogen was introduced by Turing. In this pioneering study Turing introduced a system of coupled elements interacting diffusively with each other. He has shown how a spatial pattern is formed with regards to the chemical concentration, starting from spatially homogeneous initial condition.

The first one activates the reaction ("activator"), while the second one inhibits the reaction ("inhibitor"). When the diffusion constant of the inhibitor is larger than that of the activator, under a certain condition, a homogeneous state with a constant concentration of the chemical is destabilized. Because of the instability of the homogeneous state, a pattern with some wavelength is formed that stabilizes the system.

Consider two types of chemicals $X$ and $Y$, where $X$ activates the synthesis both of $X$ and $Y$, and $Y$ inhibits the synthesis both of $X$ and $Y$. Such relationship can be considered by suitably taking a reaction system. In general, the diffusion coefficients can differ between $X$ and $Y$. Consider the case

[1] Even among insects, the basic segmentation process is not necessarily governed by genetic control, depending on the gradient. Importance of such mechanism is true for the so-called long-germ egg, but not necessarily so for short-germ ones, which are quite common in insects.

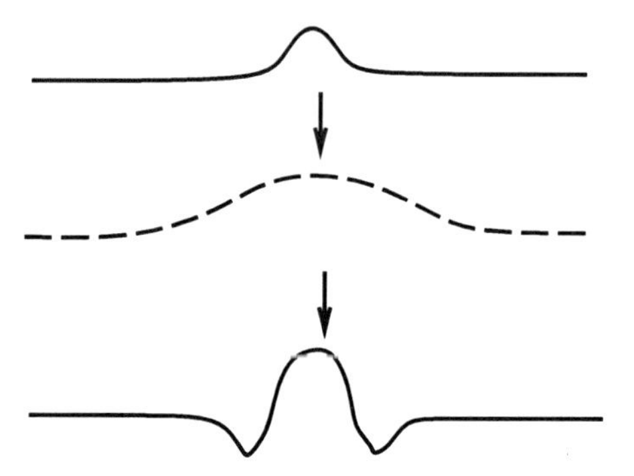

**Fig. 9.2.** Schematic representation of Turing pattern. Fluctuation in the activator concentration (given by thick line) leads to the change in that of inhibitor (given by dotted line), which leads to the wave pattern

that $Y$, the inhibitor, diffuses much faster than $X$. In this case, a spatially homogeneous state is unstable, under some (but rather general) choice of parameters, and through this reaction and diffusion, and a spatially periodic pattern of concentrations is formed. Although this instability can be shown mathematically by linear stability analysis, here we simply describe an intuitive explanation (see Fig. 9.2 for schematic representation). First consider a stationary state with a homogeneous concentration of $X$ and $Y$, and then assume that there is small perturbation to increase the concentration of $X$ at some localized region. Then, since the concentration of $X$ is larger, both the synthesis of $X$ and $Y$ are enhanced. Then, since $Y$ diffuses faster, at the neighboring region, the concentration of $Y$ first starts to increase without the increase of $X$. Since $Y$ suppresses the synthesis of both the chemicals, this results in the decrease both the concentrations of $X$ and $Y$ therein. Then, this decrease of the concentration $Y$ will be propagated to the further neighboring domain, where the increase in both the concentrations of $X$ and $Y$ is enhanced. With this process, spatially periodic pattern of concentrations with high, low, high,... are formed. This is a rather simplified sketch, but by carrying out stability analysis, one can show that the homogeneous state is unstable against some perturbation for a given range of wavelength, and a periodic pattern with such wavelength is formed, if some conditions for the values of reaction and diffusion coefficients are satisfied.

Indeed Turing discussed generally the pattern formation in a diffusion–reaction system. There is a case that the instability leads to temporal rhythm. With more than two chemicals, there is also a case with spatiotemporal pattern dynamics.[2] Turing proposed to call such diffusive chemical relevant to the morphogenesis as "morphogen." Although his theory has not yet been fully appreciated by developmental biologists, the theory, on the other hand,

[2] Although in his pioneering paper, chaos was not discussed, in a general case there appears spatiotemporal chaos in a class of reaction diffusion equations.

is fully appreciated by physicists and chemists interested in nonequilibrium systems.

Turing's theory partially answers the first question for the positional information. However, in spite of his pioneering work on morphogenesis, some basic problems remain unsolved, that is, on the origin and the robustness of the positional information. Furthermore, Turing's theory was not seriously considered by biologists, since the genetic control mechanism discovered since then is thought to answer the basic questions on development.

First, although a periodic stripe pattern with alternation of chemical concentrations is generated as a Turing pattern, how the difference in concentrations leads to cell differentiation to "discrete" cell types is not solved yet. Indeed, the change in chemical concentrations is rather continuous, and it is not clear how discrete types represented also by switching on or off genetic expressions emerge from it. As differentiation to discrete types is not fully discussed, the irreversible loss of potentiality is not studied either, in Turing's theory.

Second, in the standard Turing model, the developmental process with the increase of cell numbers is not considered. In his original model, the system size, that is, the number of cells, is fixed in time. Hence, neither the robustness nor irreversibility of the developmental process is discussed.

Accordingly, the irreversible loss of potentiality, as has been discussed in Chap. 8, cannot be discussed. In the development, the initial few cells have potentiality to differentiate to all other cell types. In the normal developmental process, some cells lose the potentiality. This irreversible loss of potentiality is influenced by interaction with surrounding cells, and is related with the pattern formed by cells. Indeed the potentiality generally depends tissues. For some tissues, when some cells are externally removed from the pattern, regeneration can occur. For some other tissues, cells are determined in their type, and such regeneration is not possible. How such irreversible loss of potentiality for differentiation is related with the morphogenesis is not answered yet.

The third problem concerns the discovery of gene regulatory mechanisms made after the publication of the Turing's paper. For example, in *Drosophila*, each segmentation corresponds to the expression of some gene. From the gene regulatory dynamics, gene expressions are on or off according to the concentration of some signal molecules. Hence Turing's mechanism has not been considered seriously. Still, these observations do not necessarily lead to reject Turing's idea itself. If Turing's idea is generalized, one can say that the interplay between intracellular reaction dynamics and the diffusion of chemicals leads to morphogenesis. This generalized viewpoint is consistent with the present molecular biology view based on the interplay between intracellular gene regulatory dynamics and diffusive signal molecules. What we need to add to the original Turing's mechanism is sufficient complexity in the intracellular reaction dynamics. This is another missing link between Turing's pattern and gene-regulatory network dynamics.

In spite of extensive studies in Turing patterns (Meinhardt & Gierer, 2000), the above three problems were not completely resolved yet. In Chaps. 7 and 8, we proposed a theory to overcome such drawbacks and to explain the spontaneous differentiation of cells to discrete types with the increase in the number of cells. In the theory so far, we have disregarded spatial factor for simplicity. By including spatially local process, it is possible to study how dynamic differentiation process and spatial pattern formation are mutually reinforced, and form stable pattern and stable cell types. This is the topic we wish to discuss in this chapter.

## 9.2 Logic

To discuss morphogenesis, we need consider also the spatial structure of these differentiated cell types. We thus take into account the spatial locations of cells as well as the diffusion of chemicals in space. Then, it is rather natural that the cell types differentiated by the mechanism discussed in the previous chapters are organized spatially to form a pattern, that is robust against perturbation. With this pattern formation, gradients of chemicals are formed that consolidate the differentiation of cell types. Here, cell differentiation by intracellular dynamics lead to a gradient of chemicals, and form positional information, while positional information strengthens differentiation of internal states. With this reinforcement, differentiation by inter–intra-cellular dynamics is transferred to a spatial pattern, while by it the differentiation is further stabilized. Cell differentiation is consolidated with the spatial pattern formation. Interplay among internal dynamics, inter-cellular diffusion of chemicals, cell division process, and spatial pattern formation leads to such consolidation.

## 9.3 Model

In the models of Chaps. 7 and 8, cells interact globally with each other through a well-stirred medium. Accordingly, the spatial position of cells does not matter. We extend the model to consider a spatially local interaction. The position of each cells must be specified, and chemicals diffuse in space, so that nearby cells interact more strongly as a result. Chemicals that are penetrable from the membrane diffuse through the space with a given diffusion constant, while nutrients are supplied with such diffusion rate from the outside.

Here as a form of intracellular dynamics we use the model of Sect. 8.3. Cells have a given size within this space, and when a cell divides, its offspring cells are located in the neighborhood of the original cell. In the numerical computations, cells are assumed to repel each other up to a given length that is the diameter of the cell, so that adjacent cells are separated by this distance (the direction of putting cells is chosen randomly). See Furuswa and Kaneko (1998b, 2000a, 2001, 2002) for detailed account of the model.

## 9.4 Results

In this case also, as the number of cells increases, the cells start to differentiate as already seen in the previous chapters. The different cell types correspond to different chemical compositions. Together with this process of differentiation, these different types of cells start to form some spatial pattern. Here, first intracellular reaction dynamics and cell-cell interaction lead to cell differentiation, which later are fixed into a spatial pattern.

Let us discuss three examples. The first one is the case in which cells are located on a one-dimensional chain (see Fig. 9.3). Here, stem-cell type S reproduces itself up to a certain number, and then as the number goes beyond some value, there starts differentiation from this type S cell to different types

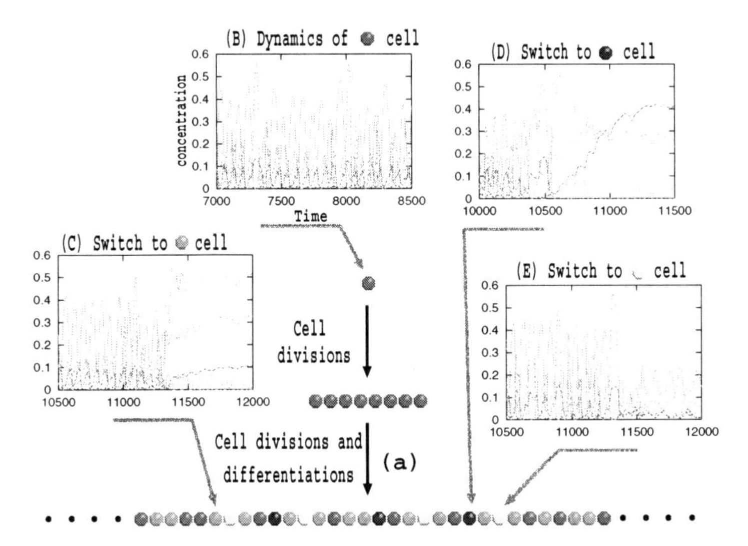


**Fig. 9.3.** An example of the developmental process. (**a**) The developmental process of the spatial pattern with differentiated cells, starting from a single cell. Up to several divisions, a single type of cell reproduces, maintaining its characteristic type of dynamics. This type is represented as "black" cells. The time series of the chemical concentrations of this type of cell are plotted in (**b**), where the time series of only 6 chemical concentrations among the 20 are overlaid, to avoid indistinct figure. (The other chemicals exhibit similar dynamics for each.) With further increase of cell number, some of the type S cells start to exhibit different types of chemical dynamics, plotted as different levels of gray scales. The transitions from the type S cells to three distinct cell types are represented by the plots in (**c**), (**d**), and (**e**). The time series of concentrations of the six chemicals are again plotted. In this example, only these four types of cells appear in the developmental process, while no intermediate types exist. Here, the type S cells are regarded as the stem cells that have the potential both to reproduce themselves and to differentiate into other cell types, while the differentiated cells have lost such potential and reproduce only the same type. A part of the spatial pattern formed by these four types of cells is shown in (a). Adopted from Furusawa and Kaneko (2000a) with permission

A, B, and C. These types S, A, B, and C form a complex spatial pattern. Here the initial type S cells have a larger chemical diversity in composition, while differentiated types A, B, C have less diversity. In this example, the pattern is rather complex, while in some other cases the pattern is more regular with regards to the configuration of cell types. Generally, there is a minimum distance between the closest type S cells, that is, they are located only farther than some distance. Note that one-dimensional pattern with differentiated cells as seen here is observed in *Anabena*, where the cell called heterocyst, which fixes the nitrogen, is located with some distance at minimum (Golden et al., 1995; Yoon & Golden, 1998).

The next two examples involve a developmental process in a two-dimensional space. In the example of Fig. 9.4, from the initial stem type S, there appears first a type-X, at the inside of the circle. The concentric circle pattern with type-X cells inside is formed first. As these cells increase, there appears a new type of cell Y, between the X and the S cells. These cell types

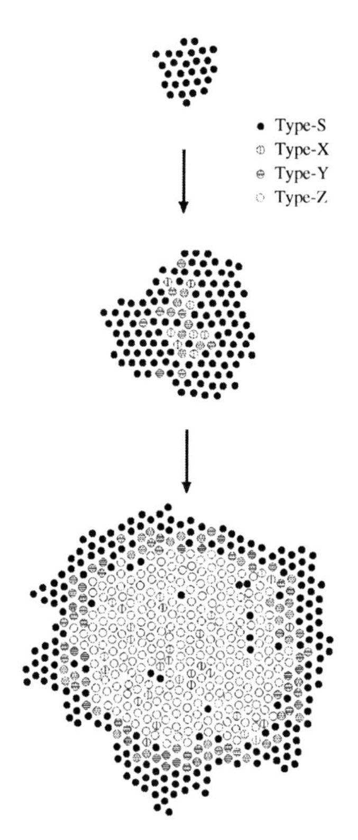


**Fig. 9.4.** Development of a cell cluster toward a "ring pattern" in a two-dimensional medium. Each mark corresponds to a particular cell type, with different cell types distinguished by significantly different internal dynamics. Adopted from Furusawa and Kaneko (1998b) with permission, under the courtesy of Chikara Furusawa

again take distinct chemical compositions. The type S cells have greater chemical diversity and variation. Here successive differentiation ($S \rightarrow S, X, Y$), ($X \rightarrow X, Z$), ($Y \rightarrow Y$), ($Z \rightarrow Z$) can be observed.

The third example that we wish to discuss at length here is that of stripe formation. In this case, a single cell placed at the center of the medium exhibits oscillatory reaction dynamics, and has high chemical diversity. There, type A cells first differentiate from the original S cells, at one side, as shown in Fig. 9.5. Along a specific axis S and A are differentiated. At this stage the differentiation rule is given by $S \rightarrow S, A$ and $A \rightarrow A$. As the number further increases, the state of type S cells at the other side of A is destabilized, to differentiate to a new type B. Hence the stripe pattern with A,S,B is formed, Here the differentiation rules are $S \rightarrow S, A, B$, and $A \rightarrow A$, and $B \rightarrow B$. As a result of these differentiations, a spatial pattern of cells consisting of three stripes, each containing cells of just one kind, is formed, as shown in Fig. 9.5. Later another type $C$ is differentiated near the border of $S$ and $A$.

In a system exhibiting the striped pattern, point symmetry is broken, in contrast to the ring pattern (Fig. 9.4). This is because of the development that takes place at the beginning of the differentiation process from type S to type A cells when there are still only a few cells (Fig. 9.5(b)). It is at this stage that point symmetry is lost, and this asymmetric small cluster of cells expands through further cell divisions. Because of the difference regarding both rates and types of nutrients absorbed and released by type S and type A cells, the asymmetric distribution of cells brings about asymmetric concentration gradients of nutrients in the medium, as shown later.

### 9.4.1 Generation of Positional Information

Now we discuss how the positional information is generated with this differentiation process. Before going into specific discussion, we recall again the requisite for "information." As discussed in Chap. 4, the information carrier has to satisfy (1) controllability and (2) preservation. If a given chemical concentration gradient works as positional information, we need to discuss (i) how it is generated from dynamics, (ii) how it controls the differentiation of cells, and (iii) how such gradient is preserved under molecular fluctuations as well as external disturbances.

Hereafter we focus on the third example with a stripe pattern formation. In the upper part of Fig. 9.6, we have plotted the concentration of nutrients (resources) along the longer axis of this cell aggregate. A chemical gradient arises, although the boundary condition to supply this chemical is constant in time and homogeneous in space. Here the type of cell that occupies a given region is related with the concentration gradient. For example, the nutrient 0 has a higher concentration where we find type S and type B cells, whereas the region where type A cells are present corresponds to a higher concentration of chemical 1.

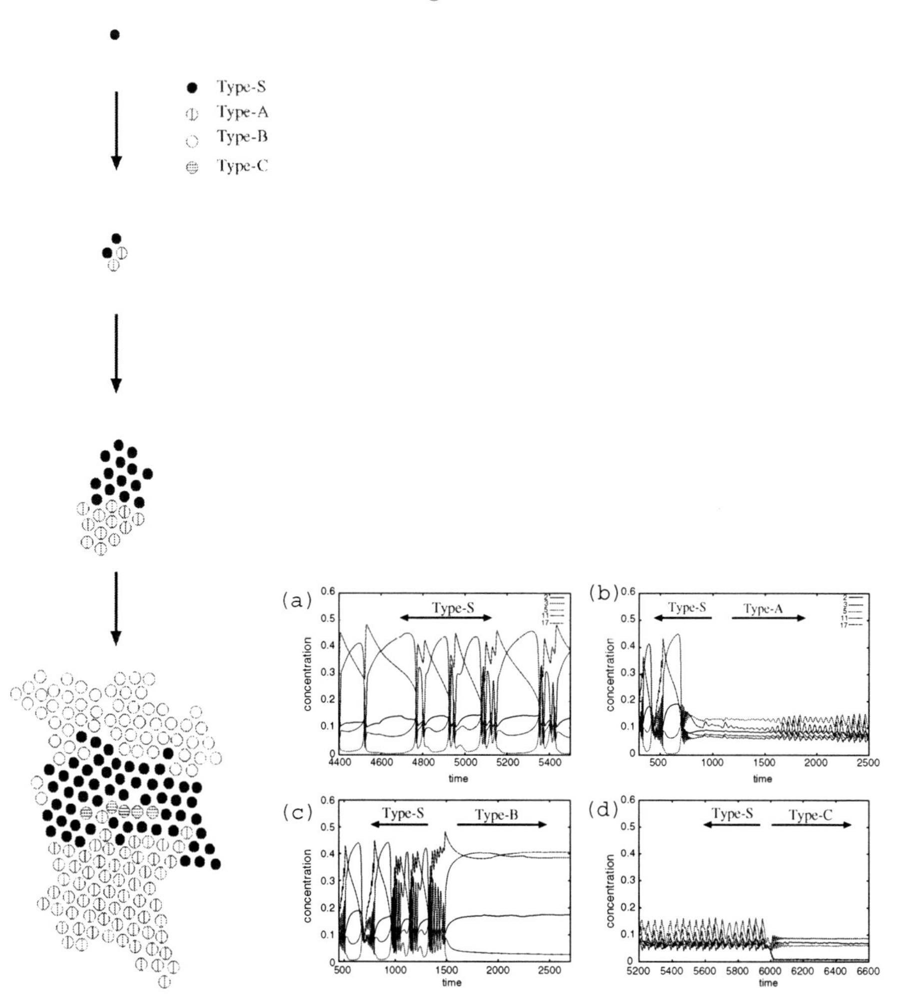


**Fig. 9.5.** Development of a cell cluster toward "striped pattern" in a two-dimensional medium. Each mark corresponds to a particular cell type, distinguished by its particular type of internal dynamics. From the stem-type cell S (*black circle*), type A is differentiated, and forms a cluster (in the lower part of the figure). With further increase of the number of cells, another type of cell B is differentiated which forms a cluster at the upper side. With these processes, a stripe pattern is formed. Plotted also are the change of the chemical concentrations. Adopted from Furusawa and Kaneko (2003b) with the courtesy of Chikara Furusawa

Conversely, if we place a type S cell in a region with a higher concentration of chemical 1 (or 2), differentiation to type A (or B) follows. In this sense, the gradients of these chemicals serve to control the differentiation of a cell into each type. Accordingly, the chemical concentrations are "read" by each cell, and act as its positional "information."

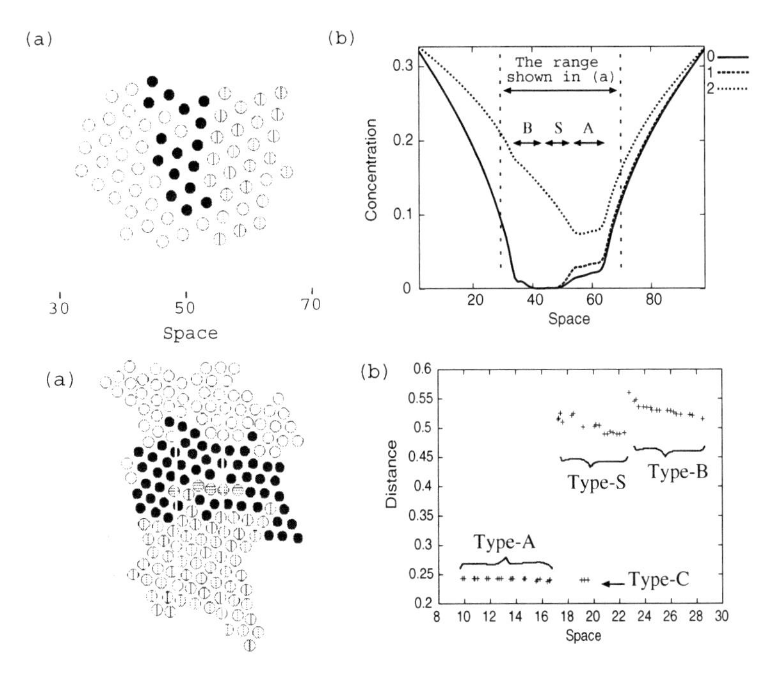


**Fig. 9.6.** (*Upper*) Generated chemical gradient, corresponding to the stripe pattern. Inside the dotted region the cells are present. (*Lower*) Variation of intracellular states with respect to position. To measure the change undergone by the intracellular dynamics when the position of a cell is changed, we determined the average position of the orbit of each cell in the $k$-dimensional phase space, which was calculated as the averages of the k concentrations $c^{(m)}(t)$ over a certain period for that cell. Based on Furusawa and Kaneko (2003b)

Depending on the concentration of the particular chemical environment for each cell, the cell state will change slightly, even if it belongs to the same type. Although differentiation into types is discrete, there arises a slight modulation of each cell's state depending on its position. In the lower part of Fig. 9.6, we have plotted $\sum_j (x_j)^2$ of each cell along the axis, where $x_j$ indicates the intracellular concentration of the $j$th chemical. One can first see clear discrete differentiation into types as the states of the cells of the same type are slightly modulated by their individual positions. This modulation (analogue difference) is well discernible for type S and B cells. It is interesting to note that, besides the digital information on type, there is analogue information provided by cell position, which brings about a continuous modulation of the cell state. Note that, for the type A cells, this modulation is much smaller. We will return to this point later.

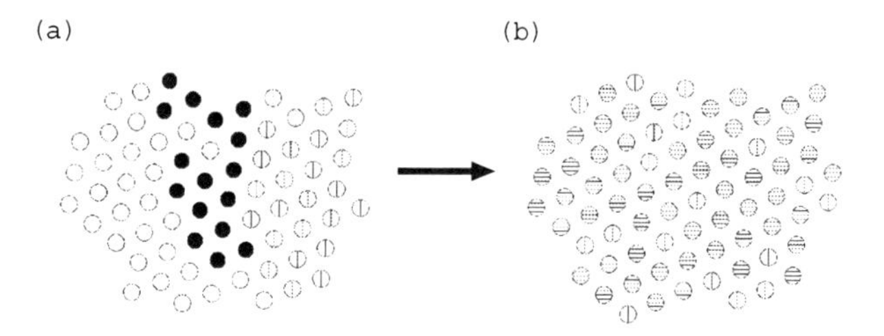


**Fig. 9.7.** Spatial patterns of cells when multiple cells, instead of a single cell, are placed in the generated positional information (i.e., chemical gradient). In this simulation, the reaction network and parameter values of Fig. 9.5 were used. The initial chemical concentrations in the medium were set to be the same as in Fig. 9.5. Note that positional information only is not sufficient to reproduce the original pattern. Adapted from Furusawa and Kaneko (2003b), under the courtesy of Chikara Furusawa

### 9.4.2 Complementary Relationship between Internal Cell State and Positional Information

Once a pattern is formed together with the generation of positional information by a concentration gradient, both the pattern and information are stable against some perturbations, such as the removal of some cells, or perturbations of chemical concentrations into some cells. After perturbations, both the pattern and chemical concentration gradient return to the original states because the cell state and gradient are mutually stabilized. Thus, the property required for preserving information is realized.

However, this does not necessarily mean that only the positional information by itself is sufficient to produce the pattern. For example, if we randomize the cell state, by changing each intracellular concentration, while maintaining the environmental chemical gradient, then the subsequently generated cell pattern is quite disordered, as in Fig. 9.7. The original regular stripe pattern is totally lost. Hence positional information alone is insufficient to regenerate the pattern. If the intracellular state is too different, the cell cannot read the information. Such a complementary interaction between the internal cell state and positional information is important.

### 9.4.3 Regeneration Process

Now we discuss the problem of regeneration in relationship with irreversible differentiation. In the present model case of stripe pattern formation, the type A cell proliferates both the same type of cell, and the type B. What happens if some cells are removed to destroy the field of original positional information? Does the cell ensemble have the potential for regeneration?

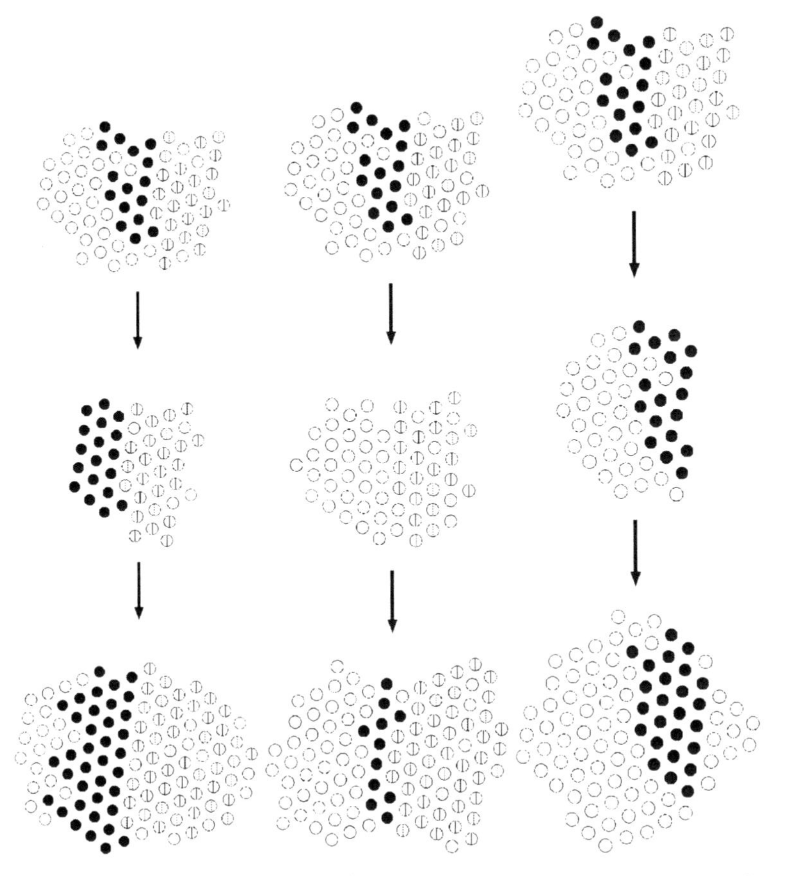

**Fig. 9.8.** Regeneration process (*left*) when the type B cells are removed, the type B cells are differentiated from the type S cells to reproduce the original stripe pattern. (*Middle*) When the type S cells are removed, the type S cells are de-differentiated from the type B cells to reproduce the original stripe pattern. (*Right*) When the type A cells are removed, the type B cells are differentiated from the type S cells to produce a novel BSB pattern. Adapted from Furusawa and Kaneko (2003b) with the courtesy of Chikara Furusawa

(I) **Recovery from the removal of the entire type B region**: In one case, we removed all the type B cells (Fig. 9.8(left)), after the stripe pattern had developed. After this operation, the rate of transitions from type S to type B cells was enhanced at the side of type S region farthest from the type A region; as a result, the stripe pattern with three layers gradually re-appeared.

(II) **Recovery following the removal of the entire type S region – de-differentiation of type B cells into type S, induced by the interaction with type A cells**: In this case, we removed all the type S cells located in the middle of the stripe pattern, and combined the remaining cell clusters consisting of only type A and type B cells, as shown in Fig. 9.8(middle). After this alteration, type B cells located at the boundary between the type A and type B regions de-differentiated back into type S cells, and the stripe pattern with three layers was thereby recovered. It is important to note that de-differentiation from a type B cell to a type S cell never occurs during the "normal" course of development (i.e., without perturbation).

As shown, recovery is possible after major damage to the system. Via this damage, positional information by concentration gradient is also largely destroyed. However, positional information is regenerated through the recovery process. This is possible because the positional information is not given externally, but rather is generated through intracellular chemical reaction dynamics and cell–cell interactions.

(III) **Formation of a new pattern resulting from the removal of the type A region**: Here, all the type A cells were removed from a cluster with a striped pattern (Fig. 9.8(right)). In this case, regeneration of the type A region was not observed. Instead, type S cells at the periphery of the cluster differentiated into type B cells. As a result, a sandwich-like B–S–B structure formed and, with further development, a ring structure with inner type S cells and outer type B cells formed. In this case, the final cell society consisted only of type S and type B cells.

In this example, once a type B cell is formed, the differentiation from S to A is inhibited. In the "normal" process, the type A cell is already differentiated before the type B; the A–S–B pattern is generated, and inhibition is ineffective. In the present case, without the existence of type A cells, the existing type B cells inhibit the differentiation into type A from type S. In this way, the ordering (or history) of the developmental process is important.

Why did type B cells, rather than type A cells, differentiate when type A and B cells were attached? Here one should note that the variation of the chemical concentrations of type A cells is much smaller than the other types, as displayed in Fig. 9.6. As discussed in Chap. 3, the state with smaller fluctuations or variations has a smaller response against external change, and has a lower plasticity. The type B cells have larger fluctuations or variations against the external changes, and has a lower plasticity. The type B cells have larger fluctuations and variations against external changes than the type A cells. Hence, the type B cells have a higher plasticity, and are more readily changed. Thus, they have a higher potential for de-differentiation when external conditions are varied.

### 9.4.4 Importance of the Ordering in Development

As shown in the above examples, the history of how an ensemble of cells develops along with increasing cell numbers is essential to the selection of a pattern. The cells that are generated later are put in a field of chemical concentration that has already been produced. The selection of cell state is then determined accordingly. This point is not well discussed, at least within the original theory of Turing patterns. To reexamine this issue, we put 100 cells initially in the configuration that was generated through the original simulation. Here the initial condition was set so that it had a homogeneous chemical concentration. The pattern thus formed and plotted had no regularity, as in Fig. 9.7. In terms of dynamical systems, the selection of initial and boundary conditions through development is essential to form an ordered pattern.

## 9.5 Experiment

Here we discuss a constructive experiment to form a tissue from undifferentiated cells artificially. To be specific, we explain about recent controlled tissue generation by Asashima's group (Ariizumi & Asashima, 2001; Uochi & Asashima, 1996). They removed some cells from a region of a frog's (Xenopus) egg, which is called the animal cap. These cells were isolated from each other, and then put into a solution of activin (a protein) for a while. This ensemble of undifferentiated cells was cultured to examine development. Here the initial cells were almost homogeneous. Surprisingly, merely by changing the concentration of activin, various frog tissues, including notochord, heart, and muscle, were generated. Furthermore, using a solution of retinoic acid and Conacanavalin A, they succeeded in forming over a dozen tissues, including nephrons and sensory organs such as eyes and ears. These results will be important for tissue engineering. Indeed, they have confirmed that such generated tissues are functional, by transplanting them into an adult body. The eye thus synthesized and transported form a neuronal connection with brain, and the frog can see by this new eye. The technological importance of these results goes without saying but here we discuss the importance of the result when trying to better understand the logic of development.

The following two points should be noted in relationship to our theoretical results. The first feature suggests the existence of attractors or attracting states as a dynamical system at a tissue level. The second feature suggests that differentiation is determined through an interplay between intracellular dynamics and cell–cell interaction, as adopted in our study.

(i) **Jump-over – generation of tissues, jumping over normal temporal course of development.** Activin concentration is used as a control parameter in their experiment. Indeed, the molecule activin is important in normal developmental as Asashima discovered. However, the high concentration used in their experiment is not actually achieved in normal development. In addition, during normal development, several other molecules are involved as controllers. Hence the experimental process they adopted progressed in a rather different situation from that in normal development. The most surprising outcome is that despite this difference

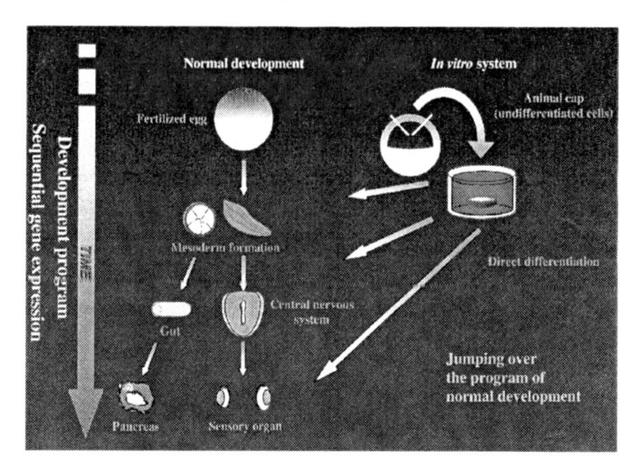


**Fig. 9.9.** Jump-over phenomena discovered by the development experiment artificially controlled by activin. Based on the experiment shown in (Ariizumi and Asashima (2001). Drawings under the courtesy of Makoto Asashima

from the normal course, the tissue eventually generated was the same as those emerging from normal development. This was verified by the morphology of tissues, cell types, and gene expression patterns. Normal function was confirmed as already mentioned.

However, the "process" to achieve these tissues is quite different from the normal case. During normal development, starting from undifferentiated cells, first mesoderm, including notochord and muscles, forms, from which neural systems are generated, and later some sensory organs are formed. In these "constructive experiments," the goal tissue was generated by jumping over processes that would normally form earlier tissues (see Fig. 9.9).

Furthermore by changing the concentration of Concanavalin A and retinoic acid, again sensory organs such as eye or ear are generated, by jumping over intermediate tissues. This is a result of "constructive biology" that even though the normal developmental process is not completely followed final tissue is formed.

The present results contest the viewpoint according to which developmental process should constitute a succession of finely tuned stepwise processes. Rather, it is more natural to adopt the viewpoint that the final tissue is a kind of "attractor" a dynamical system with a huge number of variables, and there can be several paths to reach it other than via normal development. Of course, such paths to attractors are intermingled and complex. Obviously, the basis for such attractors needs some control. The results of Asashima's group suggest that the change of activin concentration they used was not a finely tuned control, but rather served to perturb the initial state by destabilizing the cellular state so that it moved toward a state of attraction to some other tissues.

Of course, some attractors may have smaller basin volumes and are reached only through restricted initial conditions. In the latter case, successive operation to reach the final attractor would be necessary. In this case, suitably controlled time-lag between successive operations is necessary to reach the attractor. Indeed, by changing the timing of the operation of activin and retinoic acid operations, different tissues were generated: When the activin and retinoic acid are simultaneously added to undifferentiated cells, nephros (pronephros) are generated. On the other hand, when the retinoic acid was added only after 5 h of the activin operation, pancrea was generated. With this time difference, novel tissue genesis is possible. Theoretically, one can say that the memory of this operation is at least maintained in cells for a few hours.

(ii) **Community effect – the formation of tissue is highly dependent on the number of cells.** In the experiments of Asashima's group, a number of undifferentiated cells were taken from the animal cap of *Xenopus* embryos, and with these the construction of several types of tissue (e.g., heart, notochord, and muscle) was caused by controlling the concentration of activin. In these experiments, how the construction of tissues proceeded depended on the numbers of cells used. Indeed, normal tissues were generated only for a precise range of initial cell numbers. For pronephros formation by the solution of activin (with 10 ng/mL) and retinoic acid, the normal tissue was formed only if the number of cells was around 300–500. When there were fewer than 100 cells, the cells died, whereas when more than 500 cells were used, inhomogeneous tissues were generated. Construction of heart tissue by activin was also possible only for a limited range of cell numbers (see Fig. 9.10). This experimental result cannot be understood solely by signaling to one of the cells, and cell–cell interaction is essential, as has been discussed in the previous section.

## 9.6 Relevance to Biology

### 9.6.1 Summary

We now summarize the logic of our theory and the results of our simulations, and discuss their relevance to biological morphogenesis.

In our theory, cells possessing internal nonlinear reaction dynamics differentiate into several distinct cell types when the cell number exceeds a threshold value. The transition from one cell type to another is regulated by the position of the cell in question. This regulation leads to an ordered spatial pattern consisting of differentiated cells, as shown in Figs. 9.4 and 9.5. Each cell "reads" information from the external field and reacts to this information by modulating its intracellular dynamics in accordance. This modulation controls the rates of differentiation into various cell types, while such cell differentiations, in turn, alter the state of the environment. With this circular relationship

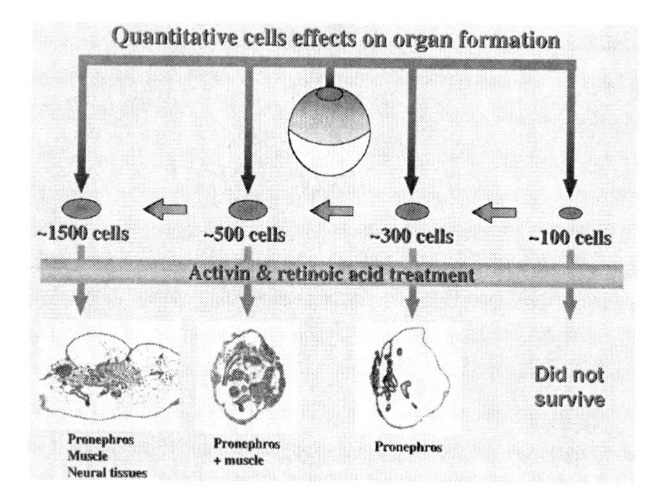


**Fig. 9.10.** The cell number dependence on the formation of nephros by activin and retinoic acid. Drawings under the courtesy of Makoto Asashima, based on the experiment reported in Ariizumi and Asashima (2001)

between intra- and inter-cellular dynamics, the gradients of chemicals in the medium act as positional information controlling the fate of each cell. This interaction between intra- and inter-cellular dynamics is responsible for the robustness of the developmental process.

### 9.6.2 Community Effect

Gurdon showed that cell commitment (i.e., determination to a certain cell type) depends on the total cell number, and termed this cell-number effect as the "community effect" (Gurdon et al., 1993). When the number of cells is small, differentiation is not determined, and the cell state changes in time; whereas when the cell number is larger, the cell state is mutually stabilized and is maintained. The experiments by Asashima's group also show clearly how tissue formation is critically dependent on cell numbers.

Because our theory is based on cell–cell interactions, a "community effect" is to be expected. Indeed, we have studied this problem by taking the stripe case. For example, when a single type B cell (such as that in Fig. 9.5) is placed in a medium containing no other cells, this cell de-differentiates back into a type S cell. However, as shown in Fig. 9.11, the type B cell state is stable when such a cell is surrounded by a sufficient number of other type B cells. To confirm this point, we have carried out simulations with the same model, but placing a large number (more than 10) of type B cells in an otherwise unoccupied medium. In this case, a cell colony consisting of only type B cells grows.

As discussed in the previous section, there can be multiple stable cell colony states. A cluster consisting only of type B cells corresponds to one

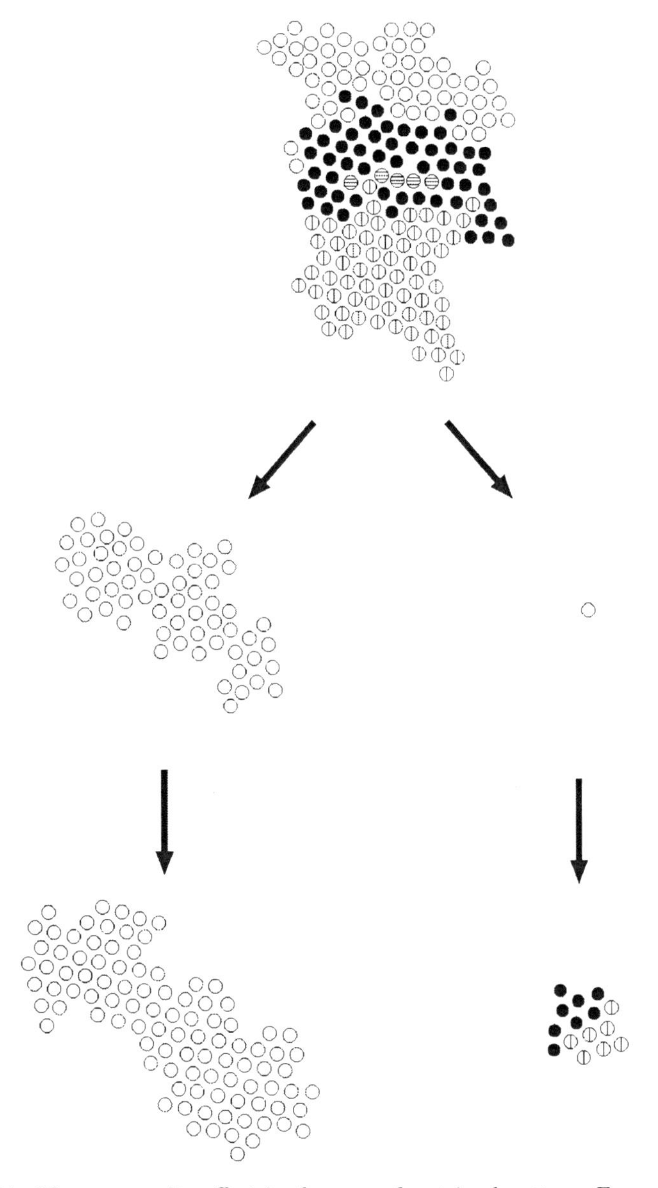

**Fig. 9.11.** The community effect in the case of a striped pattern. From an ensemble of cells exhibiting a striped pattern, a group of type B cells or a single type B cell were transplanted into a new, otherwise unoccupied medium, and they developed under the same rules and parameter values with the previous case. The type B cells transplanted as a group remain type B cells, while the type B cell transplanted as a single cell transforms into a type S cell and then, eventually develops into a striped pattern. Adapted from Furusawa and Kaneko (2003b), with the courtesy of Chikara Furusawa

such stable state, but this state never appears in the ordinary developmental process starting from a single cell. In this case, only when the number of type B cells initially placed in the medium is more than around 15 do these cells remain in their type B states. Otherwise, all the cells de-differentiate into type S cells. This clearly indicates that the states of the cells existing in a cluster can be mutually stabilized by their interactions.

### 9.6.3 Induction and Plasticity

The term "induction" is often adopted to describe development. This means that when some group of cells start to interact with another group, cells of one group change their state to form a different tissue. In fact, both cell groups interact mutually, but only one group of cells is changed, or "induced" by the other group. This means that "changeability" of the cell state is higher for the former group than for the latter group. This is nothing more than the problem of the degree of plasticity in cells we discussed already. Then, it is reasonable to propose the following hypothesis (Furusawa & Kaneko 2006b):

**When two groups of cells with different plasticity meet, the group of cells with higher plasticity is changed more. In other words, the group with lower plasticity brings about the induction on the other group of cells with a higher plasticity.**

Now we can check the validity of this hypothesis from our simulation. In the model we studied here, when the type A and B cells are put next to each other, the type B cells de-differentiate and change to type S cells, whereas when type S and B cells are adjacent, there is differentiation from type S cells. The "changeability" of cells is in the order of S>B>A. Now recall the data of variation of chemical concentrations among cell types. These variations, and fluctuations of each cell type, again decrease in the order of S>B>A. Hence, the hypothesis that the variation or fluctuation gives the degree of plasticity is confirmed. When there are two types of cells, generally those with higher plasticity are changed (induced), and this plasticity can be measured quantitatively as variation or fluctuation of cellular states.

### 9.6.4 Transformation of State Differentiation into Spatial Pattern

It should be noted again that the pattern formation seen here is not predetermined from spatial information, but rather through intracellular dynamics and interaction. Spatial patterns and intracellular states mutually stabilize robust pattern formation consisting of several cell types. Here differentiation by intracellular dynamics is consolidated to a pattern.

Although we have here adopted oscillatory catalytic reaction dynamics for intracellular dynamics, the present mechanism does not necessarily require such dynamics. As long as there is some instability in intracellular dynamics, this mechanism is effective. Indeed, Takagi and Kaneko (2005) have shown that the present mechanism works starting from intracellular dynamics with multiple fixed points. Another choice for intracellular dynamics, which straightforwardly corresponds to gene regulatory dynamics, is the use of threshold dynamics so that each gene expression takes either high or low values, corresponding to "on" or "off" states of genes, respectively (Mjolsness, et al., 1991; Salazar-Ciudad et al., 2000). Even with these dynamics, there can be instability in the intracellular state (Ishihara & Kaneko, 2005). Using such gene regulatory dynamics, and applying the present scheme, it is shown that the present scenario for morphogenesis works. In this sense, the third question addressed when we discussed Turing's mechanism – how dynamical systems mechanism and gene regulatory mechanisms are consistent – is answered.

We should mention here that Newman put forward the idea that Turing patterns will be consolidated to gene expression (Newman and Comper, 1990; Newman, 1994, 2003). The present theory is consistent with this idea.

### 9.6.5 Destabilization of Intracellular State and Regain of Plasticity: An Interpretation of Gastrulation

In the development of vertebrates, there is an important epoch: gastrulation. As the number of cells increases beyond some threshold value, the cells start to move, triggered by mechanical instability such as buckling, so that some cells start to move inside the other group. Although the origin of gastrulation itself could be mechanical, an important consequence of this is that cells that were far apart and that did not interact with each other start to make contact directly, and this novel cell–cell interaction leads to a drastic change in cellular states.

Now, the stability of the cell state is not determined solely by intracellular factors, but also through interaction with neighboring cells. Then, when cells start to contact some other cells that were originally far away, the cell state can be destabilized, which may lead to novel differentiation.

Note that the plasticity in a "closed" system generally decreases (or does not change) over time. All of our simulation results of cell differentiation processes support this proposition. As long as the change of chemical states of cell follows their intracellular chemical reaction dynamics and cell–cell interaction in a given spatial configuration, plasticity will decrease with time. In the case of gastrulation, however, it seems that the plasticity is regained, following the destabilization of the cell state, as it is triggered by novel contacts of cells that were not initially adjacent. Here, mechanical instability of cell configurations leads to cell movement, and to novel arrangements of cells. Hence, a novel type of dynamics that was not included in the original chemical system sets in here. In this sense the system is "open" to a new dimension of interaction and allows for the recovery of plasticity.

Now we reconsider the experiments of Asashima's group from this viewpoint. We propose that, after cells are put in the solution of activin, the cell state must be destabilized to recover some degree of plasticity. Through this destabilization, a new path emerges to states that were hitherto inaccessible. As the concentration of activin is increased, a new path to a tissue that was hard to reach is opened. Now, in the process of gastrulation, contact with new cell aggregates will lead to a similar destabilization. Then, the longer the contact, the greater the instability and a developmental path to a tissue that was harder to reach should be opened. If the argument here is correct, the increase of contact time between cells that were initially separate in the course of gastrulation and the increase of the concentration of activin in the artificial development experiment must have similar effects. In fact, the ordering of notochord, pronephros, muscle, in the order of concentration of activin in the experiments by Asashima et al. corresponds exactly to the order of tissues generated during gastrulation. (Here the cells that first make contact with new cells in gastrulation have a longer time of contact and then develop to form muscle.) Although the discussion here is still premature, it will be important in the future to understand the relationship between the plasticity of cellular state and the ordering of induction as well as the tissue formation by external control.

### 9.6.6 Origin of Individuality

Through Chaps. 7-9, we have shown that the robust morphogenesis with cell differentiation is a natural course in a coupled cell system with intracellular reaction dynamics. However, one important unresolved question here is that of the recursive production of a multicellular organism itself. Any biological unit, albeit whether a cell or an individual multicellular organism, has a specific boundary with some size. How such a unit is formed remains an open question.

Also, for a multicellular organism to reproduce over successive generations, cell aggregates themselves have to replicate recursively (i.e., recursiveness at a cell ensemble level). How is this recursive production at an ensemble level achieved? Here, the next generation of a multicellular organism usually starts not from all cell types of the mother (parents) but only from a specific cell type, called a germ cell. Indeed in the present multicellular organism, the next generation starts from a specific cell such as an egg cell (and sperm) or a seed. Such cells are segregated at the rather early stage of development, as germ cells, which are passed to the offspring. On the other hand, most part of cells, that is, somatic cells, cannot produce the offspring. The germ cells are determined to some specific state, but when they are separated, they have totipotency to produce all cells of the next generation.

For some multicellular organisms, it is possible to leave the offspring from cells other than the germ cells. In plants, a new generation sometimes can be produced without passing through the seed. In planaria, when a part of the body is cut out, a new organism is formed from it, and hence the reproduction without passing through the germ cell is possible. However, the cut part of cells has to include stem cells to produce the next generation. Also, as long as nutrition condition is satisfied, they reproduce sexually so that offspring are generated from the germ cells. For some primitive organisms like *Volvox*, the germ cells do not exit, while the cells are differentiated here, and only some specific cells can produce the next generation.

To sum up, only specific cell types can produce the next generation in multicellular organisms in general. Now, to maximize the number of the offspring cells, it is a waste of resources that only the specific (germ) cells produce the offspring. Usually the number of somatic cells is much larger. However, most cells in a multicellular organism cannot produce the offspring for the next generation.

Why does the development of multicellular organisms "waste" most somatic cells? Why does the process for the next generation have to pass through such a narrow bottleneck?

To answer this question, we extend the study made in this chapter, by further considering how cells are united into one aggregate. Note that without cell adhesion the cells will leave from the aggregate when random force acts on them, as in the Brownian motion. To form one united aggregate, cells adhere with each other. In cell biology, each cell adheres to its neighbor cells through some adhesion molecules on the membrane surface. The nature of membrane proteins depends on the internal state of the cell, and it is natural to assume that whether adhesion occurs or not depends on the pair of neighboring cell types.

As the cell aggregate grows, however, the nutrition cannot come efficiently to the inside cells, as the increase of the surface area of cell aggregates that are contacted with the external environment providing resources is slower than the increase of the volume of the cell aggregate. Then the growth of this cell aggregate will become slower.

Here, if a few cells are detached from the aggregate and move out of the aggregate, then they can get external resources (nutrients) more and restart the growth process. This detachment is possible, if the cell adhesion depends on the cell types that are contacted, and between some cell types the adhesion does not work. Now, by adopting the model of this chapter, and by introducing the cell-dependent adhesion force, the following course of the development of the cell aggregate is naturally expected.

As cells grow, they start to differentiate, and form a pattern, as shown in this chapter. Up to this stage, cells adhere with each other (otherwise, they cannot form a multicellular organism). Then some specific cell type appears as a result of differentiation, which loses the cell adhesion to the neighboring cells, so that such cells are released from the cell aggregate. The number of thus released cells is just a few. After being released, these cells are de-differentiated (as in the type C cell in Fig. 9.11), and become a stem cell to create a new cell aggregate. Recursive production of a cell aggregate is thus possible.

Note that the de-differentiation occurs only if a small number of cells is released as given in Fig. 9.12. If more cells are released together, they keep their cell type so that a new cell aggregate is formed only of the same cell type,

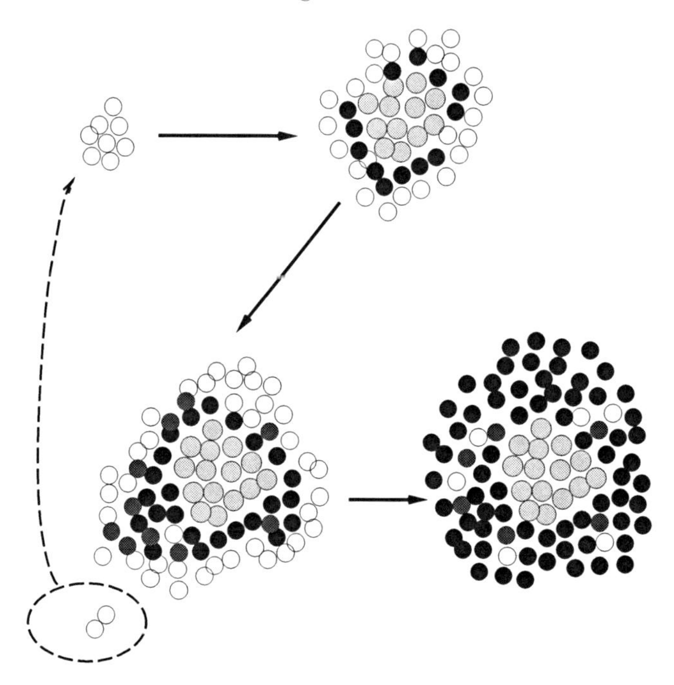

**Fig. 9.12.** Schematic diagram of the cycle of multicellular organism: (A) a number of cells start to increase from the initial one or two cells. (B) They further increase their number. (C) Cells start to differentiate with the further increase in the number. (D) A few cells of a specific differentiated type are detached from the aggregate (E). After several releases of cells, the remained cell aggregate ceases growing. Adapted from Kaneko (2003a)

and easily loses the ability of the cell to grow. Hence, it is relevant that one or only a few cells are released to make the recursive growth of multicellular organism. Note that these released cells are of a specific cell type, and when detached from the aggregate they recover the totipotency to form all the cell types of the next generation. Indeed, we have shown that this scenario works well numerically. (See the Appendix for results of model simulation.)

As shown, the next generation of a multicellular organism is generated only by passing through a few specific cells, which can be regarded as a basis of germ-line segregation. Now we briefly discuss the biological significance of this result.

If the organism of the next generation grows from an initial aggregate consisting of many cells, it will depend on the initial distribution of cell types. If this distribution includes only specific types of determined (committed) cells, then the next generation consists of only cells of the same types. Furthermore, we have also observed cases where an irregular pattern is formed when the initial condition consists of many cells (see Fig. 9.7). On the other hand, if the next generation starts only from one or a few cells, the initial condition for the next generation is well determined, and the recursive production of cell aggregates can be realized (see also Yoshida et al. (2005)).

Indeed, for some organisms that adopt two types of reproduction, those both from germ cells and from somatic cells, there is a tendency that errors accumulate to the point of damaging the recursive production, unless reproduction from germ cells is adopted from time to time. In fact, it is often argued that sexual reproduction is relevant to remove fatal mutational errors (Muller, 1964).

The proposition according to which recursive production of multicellular organisms is possible because the next generation starts only from one or a few cells may be rephrased simply as control by minority cells. This reminds us of the minority control discussed in Chap. 4. Instead of the control by minority molecules, here minority cells control the recursive production. Probably, there may be a common law by minority control in a system that reproduces itself. In Chap. 4, we pointed out that minority control is relevant to the evolvability. Is this true in the present case?

In fact, the argument in Chap. 4 will be valid here. When some mutational change occurs in the germ cell, this definitely affects the next generation. On the one hand, some mechanism to preserve the character of the germ cell will be developed so that the fatal mutation does not occur often, but on the other hand, there may appear mutation to produce novel function, which affects the offspring. If the transfer to the next generation comes through many somatic cells, then the average change by mutation will be decreased as in the discussion of the central limit theorem in Sect. 4.5. Hence the minority control by one or a few germ cells is relevant to evolvability at the level of a multicellular organism.

The significance of germ-line segregation to evolution was pointed out by Weisman (1893) and has been discussed since then (Buss, 1987). Indeed, this is one important evidence that Lamarckism does not occur in evolution – (according to Lamarckism if some biological state is acquired during one's lifetime, it is transferred to the offspring). Because change in somatic cells is not transferred to the next generation, change in them by use of them is not transferred to the offspring (even if you use your brain hard, the neurons are not transferred: see also Chap. 10). Since the mutation in germ cells is independent of which type of somatic cells are adapted in a given environment, the Lamarckian mechanism is prohibited.

## 9.7 Appendix: Model for Recursive Growth of Multicellular Organism

Cell differentiation itself is given as a general consequence of our theory as shown in Chaps. 7–9. Following the argument in the last section, recursive production of a multicellular organism may be possible if some mechanism is included into the model in the present chapter, so that some cells can be separated from an aggregate of cells. For it, it is necessary to introduce either cell-cell force or motility or cell-death.

As a minimal model for adhesion, we assume that cells within a given threshold distance have a "connection," where a "spring" is put between them so that they adhere within the natural length of the spring if the combination of the two cell types satisfies a given condition. For example, cells with the same cell type will be connected by the same spring (with the same strength and natural length) if a distance condition is satisfied, while pairs with any combination of two different cell types do not adhere. Here we change the condition of adhesion between the cells, to see continuous growth in our cell society.

In addition to the adhesion force, a random fluctuation force is applied to all cells that is expected from molecular Brownian motion. We seek a configuration that is stable against perturbations including these fluctuations. When a cell divides, two daughter cells are placed at randomly chosen positions close to the mother cell, and each daughter cell makes new connections with the neighbor cells.

Here we adopt the model that leads to the ring pattern with three layers of differentiated cell types as given in Fig. 9.4. When all the cell types adhere with each other in this model, the growth of cell aggregate stops at a certain stage, and new cell clusters are not formed. Thus, such a cellular system cannot be sustained for long. If a change in the adhesion properties allows for the continuous growth and formation of a new generation of cell clusters, such cellular systems will come to dominate.

To study this problem, we introduce dependence of the adhesion force on cell types. Since the force of adhesion should depend on the membrane proteins on the cell surface, it is natural to include dependence of adhesion on the relative internal states of two adjacent cells. As a simple example, we assume that no connection is allowed between a type Y cell and a type Z cell in Fig. 9.4, while the connections for all other combinations are preserved.

We have made several simulations with these adhesion rules and found that cell clusters divide into multiple parts during development. The first stage of the developmental process is unchanged from the previous example. A cluster of type S cells grows through cell divisions, and type X and type Y cells appear at the inside of this cluster by differentiations until the inner core is formed as a result of further differentiations. When the growth of the inner core that consists mainly of type Z cells reaches the edge of the cell cluster, however, a small cluster of cells, or a solitary cell, is released from the periphery of the mother cluster, as shown in Fig. 9.13(c). This figure depicts the process that gives rise to the fourth generation from the third generation of our multicellular organism. As will be shown, the formation of the second generation proceeds in the same way. The peripheral layer of type S and type Y cells is cut off by the growth of the inner core, and the type Y and type Z cells at the contact surface of these layers do not adhere any more by our model assumption.

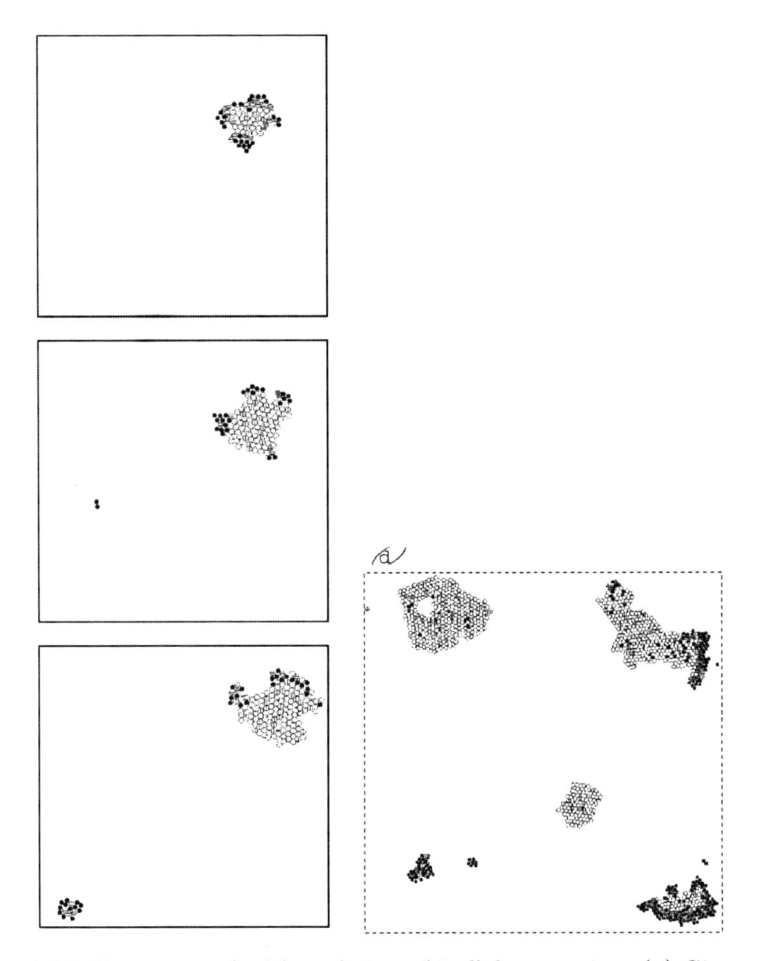

**Fig. 9.13.** Emergence of a life cycle in multicellular organism. (**a**) Gives a part of snapshot pattern of our model, while each of (**b**) and (**c**) is an expansion of the corresponding area in (a). The cluster in (b) has already fallen into a halting state, where the oscillation and the growth of almost all cells have stopped. The cluster in (c) is the second-next generation from the cluster in (b). It is just releasing its peripheral cells (the two type S cells at the upper part of (c)), which will lead to the next generation. Adapted from Furusawa and Kaneko (2002)

The released small clusters move away by the diffusion through the random force. They encounter a new environment with rich chemical substances and start to divide actively. In the new clusters, development proceeds as in their mother-cluster: The cells at the inside of a type S cluster differentiate to type X and type Y cells, while the type Z core is formed through further differentiations, until their peripheral cells are released again (Fig. 9.13c)). Hence a life cycle of multicellular replicating units is observed, which emerges without explicit implementation. Thus, we observe the emergence of a replicating unit on a higher hierarchical level than individual cell replication. Note that this emergence of replicating cell aggregate is a natural consequence of a system with internal cellular dynamics with nonlinear oscillation, cell–cell interaction through media, and cell-type-dependent adhesion.

After the release of peripheral cells, the remnant core with type S and type Z cells stops cell divisions (Fig. 9.13(b)). This determines the lifetime of the replicating multicellular unit, given by its cell configurations and the deficiency of nutrition. This fact provides an interesting point of view with respect to the death of multicellular organism. As is well known, the death of a multicellular organism is not identical with the death of cells in the organism, but rather coincides with the death of the organism as a "system." For example, cells in a dead body often survive for a while. The emergence of multicellular organisms must be accompanied with such a "halting" state of the system. This halting state limits the size and the lifetime of an organism, which is required to complete a life cycle and to give rise to a new generation. In the present model, at some stage of development of cell aggregate, such a halting state is brought about by the lack of nutrition. At the first stage of multicellularity, where no special organ for transportation of nutrition exists, it is natural that there is a halting state in a cell cluster when it reaches a certain size.

At the first stage of multicellularity in evolution, two daughter cells fail to separate after division, and a cluster of identical cell types is formed first. To survive as a unit, differentiation of cells has to occur, and subsequently the multicellular cluster needs to release their active cells before the system falls into the halting state. Hence, germ-line segregation and a closed life cycle is expected to emerge simultaneously with a multicellular organism, as our simulation have demonstrated.

# 10

# Genetic Evolution with Phenotypic Fluctuations

## 10.1 Question to Be Addressed

Question: As discussed previously, the number of molecules in various cells of the same type can fluctuate strongly, even when each of the cells has identical genes. As molecules are synthesized and decomposed by chemical processes, the number fluctuations are an inevitable consequence of the laws of physics and chemistry, no matter how well the organism is designed. Indeed, the variation from cell to cell is quite large, as has already been discussed in Chap. 6. More generally, quantities characterizing the nature of an organism can fluctuate from individual to individual (Spudich & Koshland, 1976), even when they have identical genes and are put in the same environment, as has been confirmed quantitatively for bacteria in investigations involving the fluctuations of abundances of fluorescent proteins (Gardner et al., 2000; Ueda et al., 2001; Elowitz et al., 2002; Sato et al., 2003; Furusawa et al., 2005). Since living organisms are composed of many different types of molecules, such as proteins, RNA molecules, phospholipid molecules, and so on, their overall composition varies significantly even if they have identical genes. Thus, external features of organisms ("phenotypes"), including their behavior, vary from individual to individual also. The question that naturally arises is whether these fluctuations have some biological significance, and indeed, as evolution is one of the most important issues in biology, one should ask whether phenotypic fluctuations are relevant to evolution. At this point, it should be noted that most study of evolutionary genetics so far has focused on genetic variations because it is assumed that phenotypes are uniquely determined by genotypes. Our question of how (or whether) phenotypic fluctuations are related with (genetic) evolution has not been discussed. In short, are phenotypic fluctuations relevant to evolution?

To discuss the above question, let us first start by reviewing the basic viewpoints in evolution theory.

(i) *Existence of genotypes and phenotypes.* Phenotypes are characteristics of organisms that can be observed from the outside, and consequently that can influence other organisms. The size of an organism, the concentrations of chemicals, or enzyme activities that determine the characteristics of an organism, and behavior (such as motility) are examples of phenotypes. Genes neither directly influence other organisms, nor directly are observed by other organisms. Genes influence phenotypes in the ways they are expressed.

(ii) Fitness is a function of the phenotype and the environment. It gives an index of the average number of offspring (that survives for the next generation) per individual organism. Note that "environment" is used in a broad sense to include interactions with other individuals. Indeed, for an organism, other organisms surrounding it are a most important aspect of the environment, and hence populations of the same and other species need to be taken into account. (For example, how much of the available resources an organism can get depends crucially on the population density.) As long as organisms reproduce with the aim to increase their number while being constrained by finite resources, competition is inevitable. Thus selection works. As Darwin clearly pointed out, selection by fitness is a universal and inevitable consequence of a system consisting of replication units. To sum up, the fitness $F$ of an organism is a function of its phenotype and its environment, that is, we have $F$(phenotype, environment), and is not directly a function of its genes.

(iii) *Only the gene is transferred to the next generation.* Phenotypes are not directly transferred to offspring. Although a phenotype may change depending on the history of an organism or its environment, such a phenotype "acquired" over the life span of an individual is not transferred to its offspring. For example, even if one has engaged in vigorous exercise to strengthen one's muscles, such strong muscles are not transferred to one's children. Only the genes are transferred. This is called the "Weissman doctrine."

(iv) Mutations of genes occur randomly without any specific direction. (Note: this is closely related with assumption (iii).) As changes in genes occur at the molecular level of the DNA, and as such molecular changes occur through microscopic random processes which cannot "know" about the macroscopic state, this assumption of random change is rather natural.

(v) There is a flow only from gene to phenotype. This is what is called the central dogma of molecular biology. For example, through the developmental process the phenotype is determined depending on the genotype. Hence, the process can be summarized as *Gene* $\rightarrow$ *(Development)* $\rightarrow$ *Phenotype*.

Indeed, there is some controversy regarding assumption (iii). For example, in a unicellular organism, when a cell divides into two, some constituents other than DNA are also transferred, such as abundances of proteins, the structure of the membrane, and so forth. Examples of nongenetic inheritance have indeed been discussed (Jablonka & Lamb, 1995).[1] It is, however, generally believed that this kind of transfer to offspring constitutes a short-term memory which will decay within a few generations, while the genetic transfer to offspring is more precise thus constituting a long-term memory lasting over a large number of generations.

In multicellular organisms, usually only the germ cells are transferred to the offspring and changes in somatic cells cannot be transferred as shown by Weissmann. Still, there are possibilities that some maternal information may be transferred to offspring through aspects of an egg's composition other than DNA. However, it is again believed that such nongenetic inheritance is not important in most cases. Although it would be interesting to discuss the problem of this nongenetic or epigenetic inheritance, here we accept the standard assumption (iii), since epigenetice inheritance is thought not to play a major role in evolution in most cases.

As for assumption (iv), some controversy has arisen because of an experimental report of so-called "directed mutation" or "adaptive mutation" (Cairns et al., 1988; Shapiro, 1995). However, the result of the port is controversial and at least significance of the reported results is still unclear. We will therefore not investigate them further here and assume (iv) to be applicable following standard theory.[2]

In short, for the discussion on speciation, we adopt the standard assumptions above. It should be noted, however, that in evolutionary genetics, assumption (v) is generally replaced by the stronger form (v′): "a phenotype is a *single* valued function of a genotype." If this were always true, we could replace $F$ (phenotype, environment) in (i) by $F(f(\text{genotype}), \text{environment})$ and then we could discuss the evolutionary process in terms of the population dynamics only of genotypes (and environment).

Now, let us return to the original question. As we have confirmed that "gene $\rightarrow$ phenotype" involves nonnegligible fluctuations, assumption (v′) is not necessarily valid although such fluctuations could, of course, be irrelevant to evolution. However, we have to recall that selection does not work directly at the gene level, but at the phenotype level according to postulate (ii). Hence it is rather natural to expect that phenotypic fluctuations are relevant to evolution to some degree. On the basis of these notions, we conclude that

[1] Sonneborn gave an explicit example where the nature of the cilia of *Paramecium* is transferred to offspring without genetic change.

[2] Indeed, the theory to be discussed in Chaps. 10 and 11 is valid even if assumption (iii) or (iv) is violated, for example, under the epigenetic inheritance or under directed mutation.

it is important to investigate evolution while considering the role played by phenotypic fluctuations.

The question we address here is not just to satisfy the curiosity of theorists interested in fluctuations. It is also fundamentally related with general questions on evolution such as "Why do some organisms seem to evolve faster while the forms of some others have remained almost unchanged over huge time spans becoming so-called "living fossils"?" Besides selection pressure, is the mutation rate, that is, the change in the DNA sequence at the molecular level, the sole factor determining the tempo of evolution? Here we have to recall that the selection process works at the phenotype level, and not the gene level, and indeed our notion of interest – the speed in evolution – concerns the rate of change in the phenotype. Thus, the nature of the mapping from gene to phenotype is important for determining the speed of evolution. As the degree of phenotypic fluctuations decreases, this mapping becomes more rigid. Hence, it is relevant to consider the relationship between phenotypic fluctuations and evolution speed. If it is established, it will provide important clues to answer the above naive question on the tempo of evolution.

## 10.2 Logic

The simplest form of selection in evolution is regarded as a process in which the phenotype of a more advantageous form of some trait is chosen successively. In this simple case, evolution can be thought of as a process that moves the phenotype in a particular direction.

According to the central dogma of molecular biology (requisite (v)), gene controls phenotype. If we represent the phenotype as a variable in a dynamical system, then it is natural to assume that a gene controls the equations of the dynamical system, or, to be more specific, that it is a control parameter for a phenotype variable. Then, genetic mutations are represented as changes in the values of this control parameter, while evolution is modelled by selecting genes that gives rise to a higher fitness of the phenotype. Hence by the changes in the control parameter, a "force" that shifts the phenotype variable is assumed to be applied toward a certain direction, as a result of the selection. The evolution speed we are concerned with here is the rate of the change in the phenotype variable by the "force" that changes the genetic parameter.

On the other hand, even for organisms with identical gene parameters, the phenotype variable should vary among individuals to account for the phenotypic fluctuations. The question on the relationship between evolution speed and phenotypic fluctuations addressed here can, therefore, be rephrased as a question on the relationship between the rate of response of a variable against a force to change and the fluctuations of this variable. This interpretation is even clearer in the case of an artificial selection process where a given property, such as, for example, the enzyme activity, the concentration of some chemical, the body size, the running speed, etc., is enhanced (i.e., evolves) as a response to the applied selection condition. Consequently, the question investigated concerns the relationship between response and fluctuation, and it is natural to recall the fluctuation–response relation discussed in Chap. 3.

Before recasting the fluctuation–response relationship to adapt it to the present evolutionary context, it is useful to elucidate the terminology we adopt here. We refer to a measurable phenotype quantity (e.g., the concentration of a protein) in a biological system (say a cell or an organism) as a "variable" $x$ of the system, while the term "parameter" is adopted for a quantity that controls the changing of the variable. The parameter corresponding to $x$ is indicated as $a$. According to this definition, a gene given by a DNA sequence in a cell is regarded as one of the system's parameters in the artificial evolution experiment that we will discuss later. Now, even if the gene (the parameter) $a$ is identical for several individuals, there are fluctuations in the variable $x$. Then one can define a distribution $P(x;a)$ of the variable $x$ over biological systems (e.g., over cells) for a given parameter $a$ (i.e., gene). Here we apply the terms "average" and "variance" of $x$ with regards to the distribution $P(x;a)$ of the variable $x$.

With this terminology, we will state our theoretical proposition as follows: When we change the value of a parameter $a$ slightly so that $a \to a + \delta a$, the change in the average value of the variable $x$ will be proportional to its variance, that is,

$$\frac{\langle x \rangle_{a+\Delta a} - \langle x \rangle_a}{\Delta a} \propto \langle (\delta x)^2 \rangle, \tag{10.1}$$

where $\langle x \rangle_a$ and $\langle (\delta x)^2 \rangle$ are the average and variance of the variable $x$ for a given parameter value $a$, respectively. They are explicitly defined as $\langle x \rangle_a = \int x P(x;a) dx$ and $\langle (\delta x)^2 \rangle = \int (x - \langle x \rangle_a)^2 P(x;a) dx$, where $P(x;a)$ is the normalized distribution function of $x$ for the parameter value $a$. The integral is taken over the whole range of $x$, and the symbol $\langle \ldots \rangle$ denotes the average with respect to the distribution function $P(x;a)$.

Now we first derive a general relation, assuming that the distribution is approximately Gaussian. At $a = a_0$, we then have

$$P(x;a_0) = N_0 \exp\left(-\frac{(x - X_0)^2}{2\alpha_0}\right), \tag{10.2}$$

with $N_0$ a normalization constant so that $\int P(x:a)dx = 1$. Even when the control parameter $a$ changes somewhat, we assume that the distribution function remains approximately Gaussian, while its peak position shifts from $X_0$. Specifically, we consider the following form for the distribution:

$$P(x;a) = N \exp\left(-\frac{(x - X_0)^2}{2\alpha(a)} + v(x,a)\right). \tag{10.3}$$

As mentioned, $X_0$ is the peak position at $a = a_0$, while as $a$ changes, the peak position shifts because of the term $v(x,a)$. In general, we can take any function $v(x,a)$ as long as the condition $v(x,a_0)=0$ is fulfilled, by recalling that the distribution at $a=a_0$ is given by (10.2). As the $x$-independent term in $v(x,a)$ (i.e., the terms that depend only on $a$) can be incorporated into the normalization constant, $v(x,a)$ can be expanded as $v(x,a)=C(a-a_0)(x-X_0)+\cdots$, with $C$ as a constant, where ... is a higher order term in $(a-a_0)$ and $(x-X_0)$. Neglecting the higher order terms, the distribution function becomes

$$P(x:a)=N(a)\exp\left(-\frac{(x-X_0)^2}{2\alpha(a)}+C(a-a_0)(x-X_0)\right)\,, \tag{10.4}$$

with $N(a)$ a normalization constant so that $\int P(x:a)dx=1$.

Through the evolutionary process the peak position in the distribution of $x$ shifts because of the *generalized force* $C(a-a_0)(x-X_0)$. Assuming this distribution form, we study the change of the average values $\langle x\rangle$ against the change of the parameter from $a_0$ to $a_0+\Delta a$. Next we determine how the distribution $P(x;a+\Delta a)$ changes from $P(x;a_0)$ by expanding as

$$P(x;a_0+\Delta a)=N'\exp\left(-\frac{(x-X_0-C\Delta a\alpha(a_0+\Delta a))^2}{2\alpha(a_0+\Delta a)}\right)\,, \tag{10.5}$$

where $N$' is the normalization factor that absorbs all the $x$-independent terms in the exponential that depend on $a$. From the above equation, we obtain

$$\frac{\langle x\rangle_{a=a_0+\Delta a}-\langle x\rangle_{a=a_0}}{\Delta a}=C\alpha(a_0+\Delta a)\,. \tag{10.6}$$

Noting that $\alpha=\langle(\delta x)^2\rangle$

$$\frac{\langle x\rangle_{a=a_0+\Delta a}-\langle x\rangle_{a=a_0}}{\Delta a}=C\langle(\delta x)^2\rangle\,, \tag{10.7}$$

where the deviation between $\alpha(a_0+\Delta a)$ and $\alpha(a_0)$ is neglected as the above formulation is up to the lowest order in $\Delta a$. As long as $\Delta a$ is small, the $\Delta a$ dependence of $\alpha=\langle(\delta x)^2\rangle$ is negligible, and the variance in the above equation can be replaced by $\langle(\delta x)^2\rangle_a$. Equation (10.7) is a general result that will always hold for Gaussian-like distributions if higher order changes in the distribution can be neglected and if changes in parameters can be represented by a linear force that shifts its corresponding variable.

Now we reconsider the assumptions underlying this derivation by comparing it to the fluctuation–dissipation theorem in statistical physics (see, e.g., Kubo et al., 1985). In thermodynamics, the left-hand side is called a generalized force that produces a deviation of the variable from an equilibrium point. The magnitude of the deviation is small because the fluctuation–dissipation theorem in statistical physics is restricted to linear nonequilibrium thermodynamics. As the relation holds only near the equilibrium point, the condition for the smallness in the change of $a$ is assured.

Although our formula is formally similar to the fluctuation–dissipation theorem, the system in question is not near thermal equilibrium and hence, neglecting higher order $a$-dependence in $P(x; a)$ is an assumption that will need to be justified. The validity of this assumption and the plausibility of the correlation between the fluctuation and response should therefore be examined experimentally.

On the basis of the arguments given above, we propose that for a certain class of biological systems, the fluctuation–response relationship given by (10.1) holds for variables with Gaussian-like distributions under changes of the control parameter corresponding to the variable.

For example, consider the measurement of the growth rate of bacteria in a medium with some source of nutrition. The growth rate is influenced by the concentration of the nutrient, which works as a control parameter for the growth rate. Then, the fluctuation–response relationship suggests that the change in the growth rate $x$ by the change of the concentration of nutrient $a$ is proportional to the fluctuation of $x$.

Now let us apply this fluctuation–response relationship to evolution and represent a gene as a continuous (scalar) parameter. We would like to note the following two points with regard to this representation of a "gene as a parameter."

First, since genes are originally represented by a symbol sequence consisting of the letters A, G, C, T, a representation by a continuous parameter $a$ may look a bit strange. However, a gene is usually not represented by just a few letters but by a large number of letters. A mutation of a gene is due to a change in these letters. When introducing more changes in the letters successively, a larger influence on the phenotype can be expected. For example, in an evolution experiment often adopted in evolutionary engineering where the goal is to amplify some enzymatic activity (of bacteria), the enzymatic activity increases continuously from generation to generation, through a selection process applied after adding a mutation at each generation. Or, as a very simple theoretical illustration, one can consider the case in which the fitness decreases with the increase of the Hamming distance between the sequence under consideration and that of the fittest. Then, by starting from a nonfit sequence, the fitness increases continuously as the Hamming distance between the sequence and that of the fittest decreases. Consequently, it is, in general, appropriate to represent a gene as a continuous parameter, and indeed, this assumption is adopted in most theories of evolutionary processes.

Second, since genes are nothing but information expressed on DNA, they could, in principle, be included in the set of variables. However, according to the central dogma of molecular biology (requisite (v)), genes can affect phenotypes, the set of variables, but the phenotypes cannot change the code of the genes. During a life cycle, changes in genes are negligible compared with those of the phenotypic variables they control. Accordingly, we represent genes by a parameter.

Consider, for example, an artificial selection process to enhance a given quantity (e.g., enzyme activity or the abundance of some protein) which we call the fitness here. At each generation, by introducing mutations to organisms having identical genes, new organisms with different genes are produced. Among these, those with the highest fitness are selected for the next generation. Then, at the next generation, those with the highest fitness are selected again and so on.

On the other hand, this fitness, that is, the quantity used as the index for selection, fluctuates even among the clones, that is, isogenic individuals. This represents the phenotype fluctuations.

Here, we consider a particular gene that controls fitness, that is, the characteristic quantity under consideration. If the DNA sequence for the gene changes through mutation, there is an influence on fitness. As the mutations accumulate, so do the changes in fitness. In this case, the parameter associated with the characteristic quantity (phenotype) is the degree of substitution in the DNA sequence of the gene, and the rate of change of the parameter is given by the mutation rate.

In Fig. 10.1, we give a schematic representation of the above process. The peak of the distribution increases from generation to generation as a result of the selection process, while the evolution speed, that is, the increment between successive generations may change. In the above example, the mutation and selection processes at each generation introduce a force that changes the parameter $a$ toward the direction of increased fitness, whereas the "response" is the change (i.e., increase) in the phenotype (i.e., the measured quantity). Then, by adopting this interpretation, the following interpretation is possible for (10.1):

- change in $a$, genetic mutation,
- change in $\langle x \rangle$, change in the average phenotype due to the genetic change,
- $\langle (x - \langle x \rangle)^2 \rangle$, variance of the phenotypic distribution (fluctuations) of the clones.

With this interpretation (10.1) suggests that **the evolution speed of a phenotype (that is, the change of the average phenotype per generation) divided by the mutation rate is proportional to the phenotypic variance of the clone.**

Note that both the evolution speed, that is, the rate of change in the phenotype quantity per generation, and the variance of the quantity over the clones are experimentally measurable, so that the above relation between fluctuations and evolution speed is testable.

The above proposition can be interpreted as follows: As the degree of the fluctuations of the phenotype decreases, the change of the phenotype under a given rate of genetic change also decreases. In other words, the larger the phenotypic fluctuations are, the larger the evolution speed is. In this sense, we have a theoretical basis to characterize the degree of evolvability. This theory provides a first step in quantitatively characterizing our impression that organisms with larger phenotype fluctuations evolve faster.

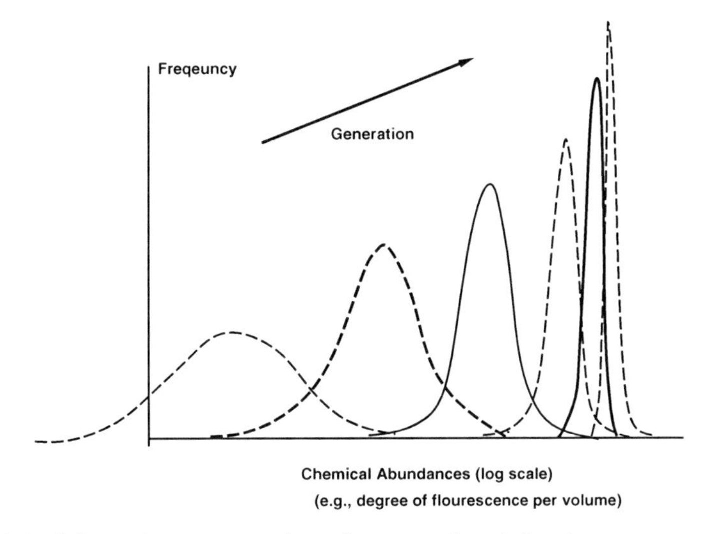


**Fig. 10.1.** Schematic representation of an experimental outcome concerning the fluctuation–response relation in an evolutionary context. At each generation, a given characteristic quantity is measured in each organism. This quantity varies depending on the organism's genes, but even among individuals with identical genes, the values are distributed. The combination of these two factors gives the initial distribution $P(x)$ as plotted. Then, the organisms with the gene that yields the largest value of the quantity $x$ are selected for the next generation. Hence the functions plotted in the graph for later generations are the distributions of this quantity for the clones, that is, individuals possessing the same genes (In other words, each of these graphs represents the distribution of phenotypes for a group of organisms that possess identical genes)

## 10.3 Model and Result of the Simulation

As an illustration of the above theory we consider a simple cell model with intra-cellular catalytic reactions, allowing for cell growth and division, that consist of a variety of chemicals whose concentrations are given by $(N_1, N_2, \ldots, N_K)$, for $K$ chemical species as was discussed in Sect. 6.3 (Furusawa & Kaneko, 2003a; Furusawa et al., 2005a). Depending on whether there is an enzymatic reaction from $i$ to $j$ catalyzed by some other chemical $\ell$ or not, a reaction path is connected in the network as $(i + \ell \rightarrow j + \ell)$. Here, nutrients (which have no catalytic activity) are transported through the cell membrane with the aid of some other chemicals that are "transporters."[3] Transported nutrients are successively transformed into other chemicals through catalytic reactions, including transporters. When these reactions

[3] In this chapter, we have revised the model of Sect. 6.3 to use transporters, but this revision is not essential for the overall behavior. All the results on Zipf's law and the log-normal distribution in Sect. 6.3 are still valid here.

progress because of the flow of nutrients, the number of molecules in a cell increases, until it goes beyond a given threshold $N$ and the cell divides into two.

Of course, how these reactions progress depends on the network, which is coded by genes. A change in a gene corresponds to a change in a network path. Here, we carry out evolution experiments where those reproducing cells are selected that have a higher concentration $c_{i_s}$ (or abundances $N_{i_s}$) of a given specific chemical $i_s$. To be specific, we first evolve cells so that they grow recursively, starting from a catalytic network chosen randomly. As discussed in Chap. 6, this is possible near a certain critical point that gives rise to efficient growth. After obtaining mother cells with recursive growth, we carry out an evolution experiment that increases the concentration of some chemical as follows. First, from each parent cell, $L$ mutant cells are generated by randomly adding or removing $m$ reaction paths to the reaction network of the parent cell. Thus the mutation rate $\mu$ is given by $m/M$, with $M$ the total number of paths. Then, for each of the networks, the reaction dynamics is simulated to identify those cells that can reproduce. (Indeed there are lethal mutants that cannot reproduce, which corresponds to the state with $D > D_c$ in Sect. 6.3.) Lastly, among those networks that can grow recursively, the top $n$ cells with regard to the abundances of the chemical species $i_s$ are selected for the next generation.

As the number of molecules is finite, for a given network, there are fluctuations in the abundances of each chemical as discussed in Sect. 6.3. Corresponding to the variable $x$, in the theory is the abundance $N_{i_s}$. To be precise, we choose $x = \log(N_{i_s})$ since the distribution of $N_i$ is close to the log-normal distribution as already mentioned, and since our theory is better suited for a variable $x$, whose distribution is close to a Gaussian distribution. As a measure of the phenotypic fluctuations, we compute the variance of $x$ for a network that gives the peak abundances at each generation.

The evolutionary changes in the phenotype distributions of the clones $P(x)$ are plotted in Fig. 10.2. Here, as the evolution progresses, both the speed of improvement (that is, increment of $x$ from one generation to the next) and the variance of the fluctuations decrease. To check the validity of the fluctuation–response relationship, we have plotted the increment of the phenotype $x$ (that is, the logarithm of the number of the molecules $N_{i_s}$) at each generation for the selected species successively. As shown in Fig. 10.3, the data plotted against the variance of $x$ are fitted well by a linear relationship, while the proportion coefficient increases with the mutation rate. The slope divided by the mutation rate is roughly constant. Hence the relationship we discussed in the last section

**evolution speed per mutation rate $\propto$ the phenotypic variance of clone**

is confirmed in this simulation.

Note that there is no a priori reason in the setup of the model that makes this relationship trivially hold. The relationship is confirmed for cells that

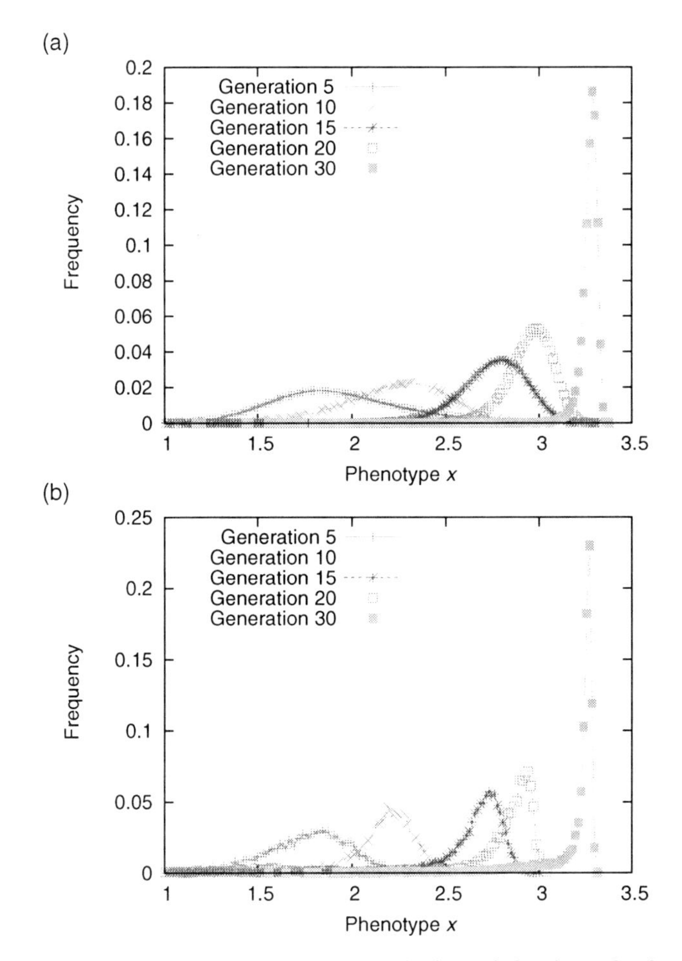


**Fig. 10.2.** Histogram of the phenotype $x$, which is defined as the logarithm of the number of molecules $n_{i_s}$ in the reaction-net cell model. Distributions for five generations in the course of evolution with a mutation rate $\mu = 0.01$ are plotted with changing colors. (**a**) Distribution $P(x)$ of the phenotype $x = \log(n_{i_s})$ of the selected clones at each generation. The distributions plotted here are obtained from 10,000 cells of the clone. (**b**) Distribution $Q(x)$ of the phenotype $x = \log(n_{i_s})$ over 10,000 mutants from the selected clones at each generation (see Sect. 10.5). For the simulations presented in this chapter, the parameters were set to $k = 1 \times 10^3$, $N_{max} = 1 \times 10^4$, and $\rho = 0.01$. The number of mutant networks $L$ is $1 \times 10^4$, and the top $n = 500$ cells with regard to the abundances of the chemical species $i_s$ are selected for the next generation. Reproduced from Kaneko and Furusawa (2006) with permission

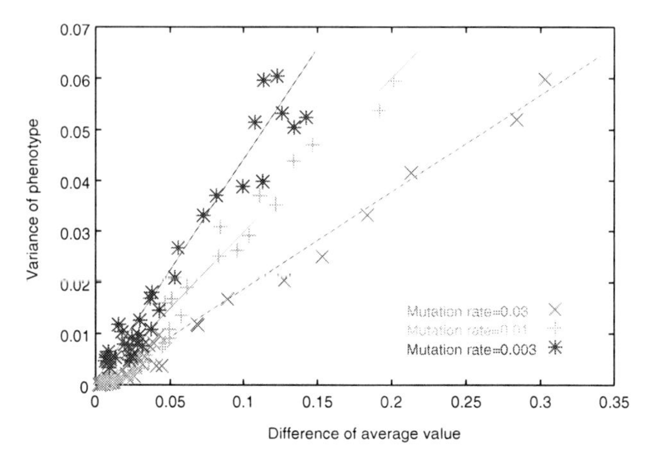


**Fig. 10.3.** Evolution speed versus phenotype fluctuations of clones. The abscissa gives the evolution speed that is measured as the difference between the average phenotypes $x$ of two successive generations, where the average is computed from the selected 500 cells. The ordinate gives the fluctuations as measured by the variance of $x$ of the clones of the selected cells, at each generation, computed from 200 clones per cell. Reproduced from Kaneko and Furusawa (2006) with permission

grow recursively, and indeed, such recursive growth is possible only near a critical state, as discussed in Sect. 6.3. If a phenotype changed drastically from generation to generation, it would be impossible to define an average phenotype for a given gene.

## 10.4 Experiment

To check whether the proposed relationship holds true in biological systems, we analyzed data from an *E. coli* evolution experiment. In our experiment, we use a fluorescent protein and select mutants with the highest fluorescence at each generation. The starting point for the experiment is a "weakly fluorescent protein," which was constructed by Ito et al. (2004).

Previously, Ito et al. (2004) succeeded in synthesizing a protein with a random sequence. By genetically attaching an arbitrarily chosen random sequence to the N terminus of a wild-type GFP (Green Fluorescent Protein) gene and transforming *E. coli* with the gene, a GFP mutant with low fluorescence was formed as the initial material for the evolution experiment. By applying random mutagenesis to only the attached fragment, a mutant pool with a diversity of $10^6$ cells was prepared. These mutants can have different degrees of fluorescence. One of the transformed *E. coli* cells was then selected on the basis of the level of its fluorescence intensity to be the parent for the next generation of clones. Successive generations were generated so that the (n+1)-th generation was generated from a parent clone selected from the $n$th generation, while always keeping diversity size and selection pressure the same as in the first generation.

To observe the diversity of the green fluorescence under identical experimental conditions, the clone selected at each generation was cultured in a liquid medium, and the distribution of the fluorescence intensity of the grown cells was measured with the help of flow cytometry. In this experiment, the fluorescence magnitude is the natural choice for the variable $x$. The fluorescence distributions obtained for each generation are shown in Fig. 10.4. It should be noted that each distribution has a finite width even though the *E. coli* cells are clones with the same genetic information living under the same conditions. The distribution here originates from the phenotypic fluctuation (i.e., the fluorescence variation) of the clones and does not relate to genetic variation.

Note that the distribution functions in Fig. 10.4 are plotted using the logarithm of the fluorescence intensity. The distribution of the fluorescence intensity among the individual cells is close to a log-normal form, as has been discussed in Sect. 6.5 (at least, it is close to being symmetric after taking the logarithm, while without taking it, the distribution has a larger tail on the higher intensity side). Therefore, as the quantity representing the phenotype, it is convenient to use the logarithm of the fluorescence. (Indeed, the results of flow-cytometry measurements are usually displayed on a log scale.)

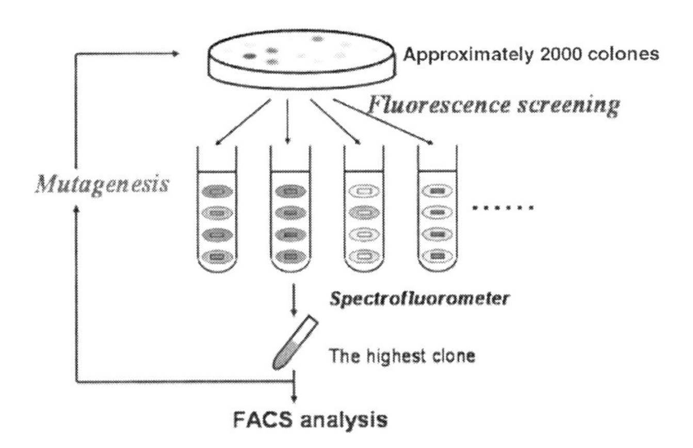


**Fig. 10.4.** Schematic representation of the selection process in the experiment; Reproduced from [Yomo et al. 2005] with permission

The genetic mutation at each generation changes the protein structure, leading to an increase in the average fluorescence through the selection process. Hence the mutation, that is, amino-acid substitution of the polypeptide sequence attached to the GFP corresponds to change in the scalar parameter $a$ in our theory, which controls the variable $x$, the fluorescence intensity.

Shown in Fig. 10.5 is the average value of the distribution and its variance plotted against the generation number. Note that both the change in the average value between two successive generations and the variance decreases from generation to generation. Thus this result shows a positive correlation

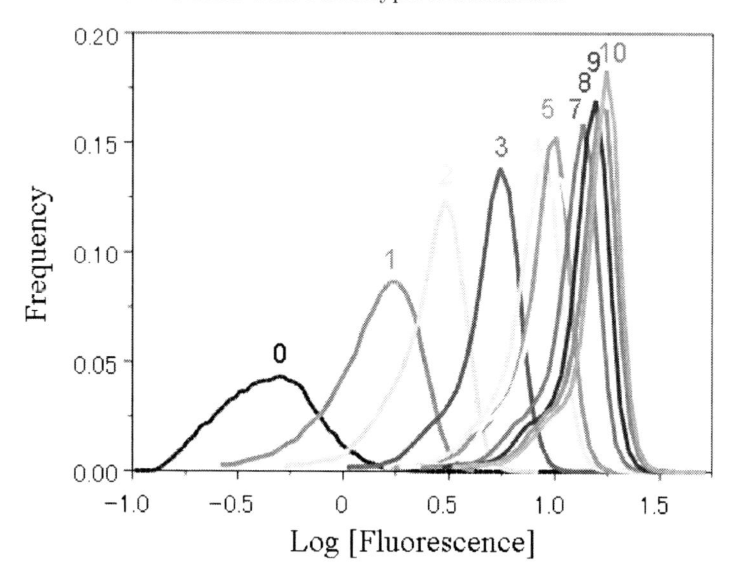


**Fig. 10.5.** Histogram of the logarithm of the fluorescence intensity for each generation, for the experiment described in the text. The number above the peak of each distribution indicates its generation number. We evaluated the fluorescence intensity of each *E. coli* in each generation by dividing its FF value by its FS (forward scatter) value measured by cytometry. This is because the FS value roughly indicates the size of *E. coli*, while the FF value is usually proportional to the FS value. Reproduced from Sato et al. (2003)

between the response and the fluctuation that is roughly consistent with our result (10.1) which predicts that the change of average value should be proportional to its variance.

To check the validity of relationship (10.1) more quantitatively from the experimental results, we plot in Fig. 10.6 the change in the average value per generation against the variance multiplied by the parameter change, where the parameter change is given by the synonymous mutation rate, also shown there. If the proportionality in (10.1) holds perfectly, all these points should lie on a straight line that passes through the origin. Also, in Fig. 10.7, to serve as a comparison, we plot the change in the average value per generation directly against the parameter, that is, without multiplying by the variance. The correlation shown by the results where we multiply by the variance is clearly much better than that shown by the results when we do not. Indeed, the correlation coefficient of the variance-multiplied results is 0.8, while the unmultiplied one is only 0.2. From this result it is reasonable to state that the change of fluorescence is more highly correlated in the case where the mutation rate is multiplied by the variance than in the case where only the mutation rate is considered. It should also be noted that this correlation appears irrespective of the sudden change in the mutation rate that we introduced in this experiment

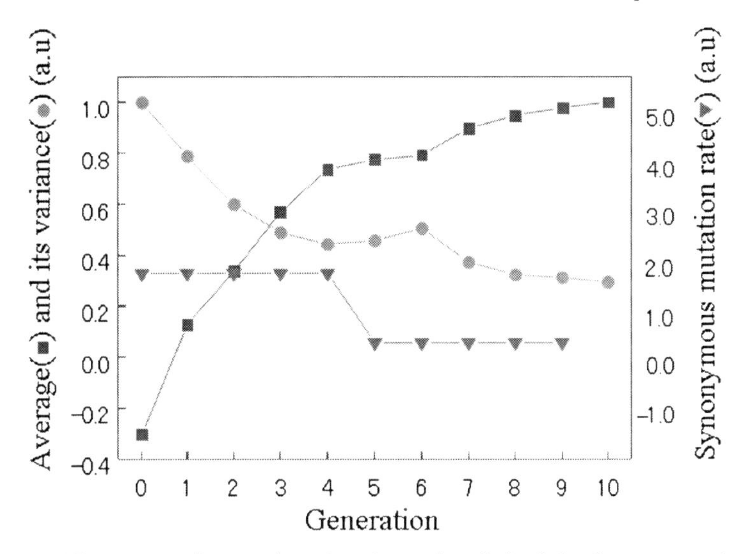


**Fig. 10.6.** The average (*squares*) and variance (*circles*) of the fluorescence intensity for the clone cells selected at each generation, plotted versus the generation number. The average values and variances are computed from the Gaussian-like distributions as their peak positions and half widths. The synonymous mutation rate for each generation is also plotted as triangles. Reproduced from Sato et al. (2003)

at the fifth generation, as can be seen in Fig. 10.6. This observed correlation is not a trivial result because the variance has decreased by about one third in later generations compared to the first generation. In other words, explicit $a$-dependence in $\alpha(a)$ in (10.5), if it exists at all, is not very strong.

To close this subsection, we note again that the variable under consideration here is not the fluorescence intensity itself but its logarithm. The averages in Figs. 10.6 and 10.7 are also computed from the logarithm of the fluorescence. In other words, we adopted the logarithm of the fluorescence as the variable $a$ of our theory. The reason for this choice is as follows. First, the distribution is nearly Gaussian in the logarithmic scale, as discussed in Chap. 6. Since our theory is based on Gaussian-like distributions, it is natural to use the logarithm of the abundances as the variable. Indeed, the observed linear relationship between the response and variance is true only with this choice of variable. Second, it should also be noted that most chemical characteristics depend on the logarithm of chemical abundances (recall that in chemical thermodynamics equations, the logarithm of the concentration generally appears as the relevant term, as for example is seen in the definition of the pH or the expression for the chemical potential in dilute solutions). Finally, the use of the logarithm as a measure of phenotypic quantities is often adopted in evolution theory, as also proposed by Haldane (1949) (see also Kimura, 1983).

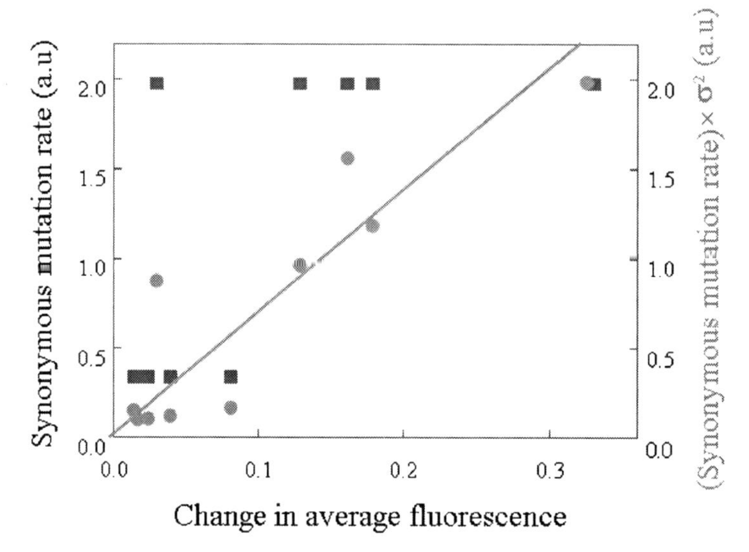


**Fig. 10.7.** The variance of the fluorescence intensity multiplied by the synonymous mutation rate are plotted versus the change of average fluorescence intensity as circles, using the data from Fig. 10.6. The line is a linear fit to the data, which turns out to pass through the origin. For reference, the synonymous mutation rate versus the change of average fluorescence intensity value is also plotted as squares. The correlation coefficient for the linear fit for the variance mutiplied by the mutation rate (circles) is 0.79, while that for just the synonymous mutation rate (squares) is 0.21. We see that the former shows a better linear correlation than the latter. Reproduced from Sato et al. (2003)

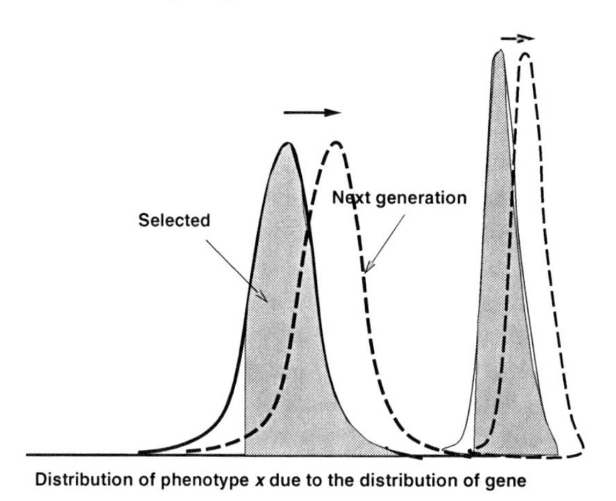


**Fig. 10.8.** Schematic representation of Fisher's theorem

## 10.5 Relevance to Biology

### 10.5.1 Phenotype–Genotype Relationship at the Fluctuation Level

Fluctuations are, of course, discussed in the standard theory of population genetics. In that theory, however, the phenotypic fluctuations of organisms with identical genes is not explicitly discussed, as assumption (v′) in Sect. 10.1 is adopted. Instead, the distributions of genes and their relevance to evolution speed are considered in detail. Recall that following the postulate, the selection process works through fitness, which is a function of the phenotype. With a genetic change, the phenotype, and accordingly the fitness, of an organism changes in turn influencing the rate at which offspring can be produced. Consequently, it is expected that a larger fitness variance due to a wider genetic distribution results in a larger change in fitness from one generation to the next (see Fig. 10.8).

Indeed, the proportionality between evolution speed and the variance of fitness due to the genetic distribution was established by Fisher and is called the fundamental theorem of natural selection (Fisher, 1930; Edwards, 2000).

In contrast, the relationship we have proposed here concerns the evolution speed and the phenotypic fluctuations of clones (i.e., organisms having identical genes). In this case, the detailed mechanism producing the phenotypic fluctuations has not yet been elucidated, and no established formula is available for the degree of phenotypic fluctuations. Since Fisher's theorem states the proportionality between evolution speed and phenotypic variance due to genetic variance and the study presented here demonstrates the proportionality between evolution speed and intrinsic phenotypic fluctuations of clones, it is reasonable to conclude that there should be some relationship between phenotypic variance due to the genetic variance and the phenotypic variance of clones.

In general, however, one would not expect a straightforward relationship between the two since the genetic fluctuations generally depend on the mutation rate and the population distribution of organisms with different genes, while the phenotypic fluctuations of clones do not. The existence of a relationship, if there is one at all, must therefore originate in some evolutionary constraints concerning stability.

Quite recently, a possible relationship between the two was found by means of a stability analysis of the statistical distribution of phenotypes and genotypes (Kaneko & Furusawa, 2006). In this analysis, the assumption is made that there is a two-variable distribution in gene and phenotype. With this assumption, an inequality between the two fluctuations is first derived, and then, it is shown that for the mutation rate that achieves the highest evolution speed, this inequality is replaced by an equality. For a general mutation rate, the genetic variance increases roughly proportionally to the mutation rate. Accordingly, the proportionality between the phenotypic variance of different genes (genetic fluctuations measured by a phenotype) and the phenotypic variance of the clones of the most dominant genotype is derived. If this argument is valid, the fluctuation–response relationship in evolution discussed above and Fisher's theorem can be shown to consistently be related with each other. We will now explain this in some more detail (Kaneko & Furusawa, 2006).[4]

In considering evolution, we also need to take into account the distribution of the genotype $a$, instead of regarding it as a given parameter. Through the evolutionary process, the dominant genotype $a$ changes, and the dominant phenotype $x_0(a)$ changes accordingly. Now, to investigate evolution with regards both to the distribution of phenotype and gene, we introduce a two-variable distribution $P(x, a; h)$ with $h$, a given environmental condition. Then by changing the environment $h$ (or the selection pressure in artificial evolution experiment), the most dominant genotype $a$ and the phenotype change in accordance with the peak in this two-variable distribution $P(x, a; h)$. The existence of such a distribution function is the first assumption here. The second assumption we make is that at each stage of the evolutionary process, the distribution has a single peak in $(x, a)$ space. Through the evolutionary course, the dominant gene $a$ changes, which is given by the peak position of $P(x, a : h)$, which in turn changes depending on the environmental condition $h$.

This second assumption is based on the notion that some kind of evolutionary stability exits, that is, that at each stage in the evolutionary process, neither the phenotype nor the genotype distribution is spread out, and that the distribution has a clear peak around a dominant gene and phenotype. This stability condition might be a bit strong to postulate for every evolutionary process, but for a gradual evolution process (without speciation), it would be a reasonable assumption. In the artificial selection experiment, a gradual change in the phenotype could be reinterpreted as a gradual increase of the selection pressure to increase the parameter $h$ (e.g., by increasing the threshold value beyond which a cell [or an organism] is selected so that it has a higher concentration of some protein or function). Or, by borrowing a concept from thermodynamics, it may be formulated as "quasi-static process", where a transition process is modelled as a series of successive (slightly different) equilibrium states.

We now modify the distribution $P(x; a)$ in Sect. 10.3 to become a two-variable distribution $P(x, a)$ as follows:

$$P(x, a) = N \exp\left[-\frac{(x - X_0)^2}{2\alpha(a)} + C(a - a_0)(x - X_0) - \frac{1}{2\mu}(a - a_0)^2\right], \quad (10.8)$$

[4] The theory outlined below follows the spirit of Einstien in his theory of Brownian motion, which is the original form of fluctuation–response relationship (Einstein, 1905, 1906). In the theory, consistentcy between microscopic fluctuation and macroscopic motion leads to the relationship, while we study the relationship between microscopic phenotypic fluctuation and genetic change here.

where $N$ is a normalization constant. Here the Gaussian distribution $\exp(-\frac{1}{2\mu}(a-a_0)^2)$ is introduced to represent the distribution of genes around $a=a_0$. The variance here is (in a suitable unit) nothing but the mutation rate $\mu$. The above equation can then be rewritten as

$$P(x,a) = N\exp\left[-\frac{(x-X_0-C(a-a_0)\alpha)^2}{2\alpha(a)} + \left(\frac{\alpha C^2}{2} - \frac{1}{2\mu}\right)(a-a_0)^2\right]. \tag{10.9}$$

In order for this distribution to have a single peak (i.e., not to be flattened along the direction of $a$) the following condition (besides $\alpha > 0$) should be satisfied: $\frac{\alpha(a_0)C^2}{2} - \frac{1}{2\mu} \leq 0$, i.e.,

$$\mu \leq \frac{1}{C^2\alpha(a_0)} \equiv \mu_c \tag{10.10}$$

This means that the mutation rate has an upper bound beyond which the distribution does not have a peak in the gene-phenotype space.

Indeed, such a limit in the mutation rate for a single-peaked distribution has already been discussed for a fitness landscape where one type is the fittest among other neutral types (Eigen & Schuster, 1979) (see also Sect. 4.1). Eigen and Schuster have shown that there is a critical mutation rate beyond which the fittest organisms cannot maintain their dominance in the population, and error catastrophe discussed in Sect. 4.1 occurs. Since our formulation assumes a continuous change of genes by mutation, and explicitly takes into account phenotypic fluctuations, it differs from Eigen's approach, but the two are similar with regards to the limit on the mutation rate required to maintain the stability of the single-peaked distribution.

Now we investigate the phenotypic variance due to the genetic distribution. First, we consider the average $\overline{x}_a$ over the distribution $P(x,a)$ for a given fixed $a$, and then we consider the distributions of $\overline{x}_a$ for changes in $a$ around $a_0$. The variance $V_g$ of this distribution of $\overline{x}_a$, that is,

$$V_g = \langle(\overline{x}_a - \overline{x}_{a_0})^2\rangle \tag{10.11}$$

is that discussed in Fisher's theorem. Then noting that

$$\overline{x}_a \equiv \int x\exp(-V(x,a))dx = X_0 + C(a-a_0)\ , \tag{10.12}$$

we obtain

$$V_g = \langle(\overline{x}_a - \overline{x}_{a_0})^2\rangle \ = \ C^2\langle(\delta a)^2\rangle = C^2\mu\ . \tag{10.13}$$

Accordingly, the inequality $\mu < 1/(C^2\alpha(a_0))$ is rewritten as

$$V_g \leq \alpha(a_0). \tag{10.14}$$

Recall that $\alpha$ is the phenotypic variance $\langle \delta x^2 \rangle$ of organisms with identical genes. To distinguish $\alpha$ from $V_g$, let us call it "intrinsic phenotypic fluctuation" (variance) and denote it as $V_{ip}$. Then the above inequality is written as

$$V_g \leq V_{ip}. \tag{10.15}$$

This means that the phenotypic fluctuation over clones (organisms with identical genes) is larger than the phenotypic variance due to the genetic variance. For a more general derivation of the above inequality (see Kaneko & Furusawa, 2005b).

Now, the genetic variance in the population $\langle (\delta a)^2 \rangle$ is proportional to the mutation rate, and the evolution speed increases with it. Still there is a limit given by the inequality, beyond which the evolutionary process yielding successively better types no longer works since an error catastrophe occurs that leads to a breakdown of the single peak distribution. Hence the highest evolution speed is achieved just before this catastrophe occurs. This condition is given by the above critical mutation rate, where the inequality (10.15) is replaced by the equality $V_g = V_{ip}$. At the critical mutation rate, the evolutionary path passes through marginally stable states so that the phenotypic fluctuations due to the distribution of genes equals the phenotypic variance of the clones.

Recall that Fisher's fundamental theorem of natural selection states that the evolution speed is proportional to $V_g$, the phenotypic variance by genetic variance. Then, at the state with optimal evolution speed, the above relationship means that the phenotypic variance of the clones is proportional to the evolution speed. When the mutation rate is smaller than this optimal value, the original inequality is satisfied.

Note, in the above formulation, $\langle (\delta a)^2 \rangle = \mu$ and $V_g \propto \mu$. Recalling that $V_g$ at $\mu_c$ equals $V_{ip}$, we get

$$V_g = \frac{\mu}{\mu_c} V_{ip} \,. \tag{10.16}$$

(Note that $V_{ip}$ is independent of $\mu$.) Since $V_g$ is proportional to the evolution speed according to Fisher's theorem, $V_{ip}\mu \propto V_g \propto$ (*evolution speed*) follows. Hence, the evolution speed is proportional to $\mu V_{ip}$, and the fluctuation–response relationship in the earlier sections is shown to be consistent with Fisher's theorem.

It should be stressed that the derivation of our expression uses only the stability condition, and does not depend on specific mechanisms of evolution or selection pressure. The assumption we made is the existence of a distribution $P(x,a) = \exp(-V(x,a))$ (or in other words, the existence of the "potential function" $V(x,a)$) and evolutionary stability, where we assume that at each generation in the evolutionary course, the distribution has a single peak, that is, stability in both genetic and phenotypic space. This is not obvious at all. Since a phenotype is a function of a gene, assuming the existence of a two-variable distribution is a tricky matter as genetic and phenotypic fluctuations are treated in the same way. Hence, we need to check our theory by models and experiments.

### 10.5.2 Confirmation of the Relationship by the Model Presented in Sect. 10.3

To confirm the validity of our theory, we have carried out several simulations of the simple cell model studied in Sect. 10.3. Here, to investigate the distributions of phenotypes due to genetic variation, we compute the phenotypes $x$ (the logarithm of the concentration of the chemical in concern) over 200 clonal cells to obtain the peak position of the phenotype distribution. This peak position of the phenotype $x$ differs from mutant to mutant, and when considering $L$ mutants we can obtain the distribution $Q(x)$ of these peaks. The variance of the peak positions over all mutant networks $V_g$ is nothing but the variance of $Q(x)$. $Q(x)$ is plotted in Fig. 10.2b, while the phenotype distribution of the clones $P(x)$ is shown in Fig. 10.2a. As can clearly be seen, the distributions evolve jointly, satisfying $V_{ip} > V_g$ as is expected from the theory.

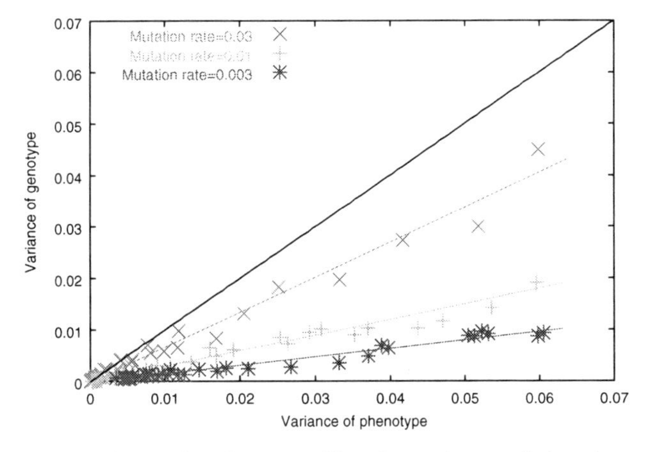


**Fig. 10.9.** The relationship between $V_{ip}$, the variance of the phenotype $x$ of the clone as measured in Fig. 10.2, and $V_g$, the variance of the phenotype $x$ over 10,000 mutants of the selected cell. Plotted with identical symbols are the relationships over courses of evolution with fixed mutation rates as indicated in the figure. (Reproduced from Kaneko & Furusawa, 2005)

We have plotted $V_g$ versus $V_{ip}$ in Fig. 10.9, and find that the expected inequality is indeed satisfied, and also that $V_{ip} \propto V_g$ holds for each evolutionary process with a fixed mutation rate.[5] As the mutation rate $\mu$ increases, the

[5] One may wonder whether in this model the phenotypic variance $V_{ip}$ decreases as $1/N$ when increasing of the total number molecules $N$ in a cell, while $V_g$ does

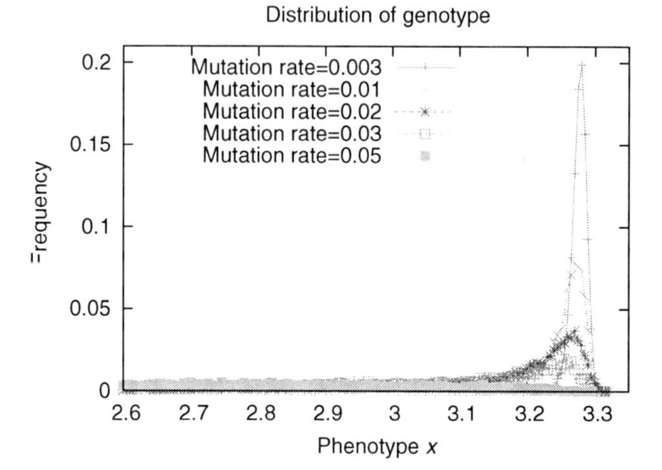


**Fig. 10.10.** Distribution of the phenotypes $x$ over 10,000 mutants, generated with the mutation rates 0.003, 0.01, 0.02, 0.03, and 0.05. Around the mutation rate 0.03, the distribution flattens, and the peak position starts to shift downward. (Reproduced from Kaneko & Furusawa, 2006)

slope of $V_g/V_{ip}$ increases approaching the diagonal line $V_g = V_{ip}$. On the other hand, with the increase of $\mu$, mutant populations exhibiting very low values of $x$ increase, a situation that corresponds to the collapse of the catalytic reaction process for cell growth. As shown in Fig. 10.10, beyond some mutation rate $\mu_{thr}$, the distribution becomes flat, and the peak in the distribution starts to shift downward. By investigating the $V_g - V_{ip}$ relationship, we have confirmed that around $\mu \approx \mu_c$, $V_g$ approaches $V_g \approx V_{ip}$. Indeed, for mutation rates beyond this threshold mutation rate, evolution does not progress, as the distribution is almost flat, and the value of $x$ after selection cannot increase from generation to generation. Thus the evolution speed is optimal around $\mu \approx \mu_c$. Summing up, there is a threshold mutation rate $\mu$ beyond which the evolutionary process does not progress (that is, an error catastrophe occurs), where $V_g$ approaches $V_{ip}$. All of these numerical results support the theoretical prediction described earlier. It should also be stressed that the results here do not depend on the specific algorithm for evolution, such as the ratio of selected networks for the next generation. The selection pressure, given by the fraction of selected networks, influences only the proportionality coefficient between the evolution speed and $V_g$ as predicted by Fisher's theorem, but it does not influence the inequality and proportionality between $V_g$ and $V_{ip}$.

---

not. Here we have to recall that a cell with recursive and efficient growth is at a critical state represented by a power law distribution (Zipf's law), as shown in Chap. 6. At this critical state $V_{ip}$ does not decrease with an increase in $N$, and the results here are therefore valid independently of $N$ as long as it is sufficiently large.

We have now confirmed our theoretical predictions. However, we have not yet completely clarified why the theory is actually valid, although one possible reason is that certain genetic changes can give rise to the same kind of effects on the reaction dynamics as can phenotypic fluctuations. For example, consider a reaction process $i$ to $j$ catalyzed by $\ell$. If the concentration $c_\ell$ changes because of fluctuations, then the rate of the reaction will change as well. Still, such a change in the rate could equally well be induced by changing a path in the network or the catalytic activity of the reaction, which is coded by a gene. In this sense there is a correspondence between the phenotype change induced by genetic change and that induced by the phenotypic fluctuations.

Of course it is very important to verify our proposition experimentally, for example by using the example discussed in Sect. 10.4. Indeed, preliminary data from the experiments seem to suggest the validity of our proposition.

### 10.5.3 Biological Significance

#### Genetic Assimilation

Waddington, in his pioneering study, proposed the concept of "genetic assimilation" (Waddington, 1957), in which a phenotype change by an environmental change is later "assimilated" into a genetic change. Indeed, such assimilation is possible under the standard Darwinian framework satisfying postulates (i)–(v) in Sect. 10.1. Here we should note that the degree of such a phenotype change is also correlated with the fluctuations of the phenotype according to the fluctuation–response relationship described in this chapter. Then, by also noting the linear relationship between evolution speed and phenotypic fluctuations, it is expected that the evolution speed is correlated with the response rate of the phenotype against environmental change. This is often called *phenotypic plasticity* (Callahan et al., 1997; West-Eberhard, 2003). The relationship between the phenotype change due to environmental factors and evolution will be established through phenotypic fluctuations. Hence, on the basis of our present study, it will be possible to reformulate Waddington's notion of genetic assimilation or the Baldwin effect (see next chapter) (Baldwin, 1896; Bonner, 1980; Ancel & Fontana, 2002) quantitatively, and study the evolutionary process in terms of the plasticity measured through phenotypic fluctuations (de Visser et al., 2003).

#### Evolution Speed

According to our theory, evolution speed is highly correlated with phenotypic fluctuations that depend on intracellular reaction processes, developmental processes, and so forth. Even for identical mutation rates, the evolution speed can differ depending on phenotypic fluctuations. This viewpoint will be important for understanding the tempo of evolution, for example, why in some organisms called *living fossils* the phenotype has changed only little over many generations.

Furthermore, by considering that the phenotype fluctuations can also depend on environmental conditions, it is expected that the evolution speed can be enhanced under some environmental conditions that amplify the phenotypic fluctuations. Such a dependence of the evolution speed on the environment was also discussed in the context of the concept of adaptive mutation, and controversy has continued for over a decade about its interpretation and biological relevance (Cairns et al., 1988; Shapiro, 1995). The present study on evolution speed may shed a new light on this controversy.

### Heritability

The inequality $V_{ip} > V_g$ concerns phenotypic variances due to genetic and nongenetic origins. If these variances are independent and added naively, the total phenotypic variance we observe from the wild-type population could be represented as $V_{tot} = V_{ip} + V_g$ (or $V_{ip} + V_g + V_e$, when the phenotypic variance due to environmental fluctuation $V_e$ is also added independently). In this simple representation, the inequality we propose implies that phenotypic fluctuations by genetic variations are smaller than those of nongenetic origin, that is, $V_g/V_{tot}$ is less than half. Indeed this quantity $V_g/V_{tot}$ is called heritability $h^2$ (Futuyma, 1986; Maynard-Smith, 1989). In standard population genetics, $V_{tot} - V_g$ is attributed to the phenotypic variance $V_e$ of nongenetic origin, while it is not clear how much of it is attributed to $V_{ip}$ or $V_e$. Indeed "intrinsic" phenotypic fluctuations through "developmental noise" (Spudich & Koshland, 1976; West-Eberhard, 2003; de Visser et al., 2003), as discussed here, contribute to the total phenotypic fluctuation as well. Although one needs to be careful when considering the correlations among the variances and the quantitative estimate of $V_g$, it is interesting to note that this heritability $h^2$ often takes a value around 1/3 (Maynard-Smith, 1989), and less than half for most cases. It will be important to discuss the heritability in the light of our inequality with regards to intrinsic phenotype fluctuations.

### Decrease in Phenotypic Fluctuation Through Evolution

In both the model and the experiment, the phenotypic fluctuations decrease while the evolutionary process progresses. This suggests the empirical rule that the more optimized a certain gene is, the less likely is the appearance of mutants with higher fitness. This rule is also confirmed by many evolution experiments. In the evolution experiment of Sect. 10.4, the change in the fluorescence intensity at each generation decreases, while the fluorescence intensity becomes larger as the optimization process progresses. This decrease in phenotypic fluctuations is consistent with the increase in robustness through evolution as often suggested (West-Eberhard, 2003; de Visser et al., 2003), and is also consistent with the decline in phenotypic fluctuations in "closed systems," as proposed in Chap. 3.

### Relevance of the Phenotypic Fluctuations to Evolution Revisited

Although we have succeeded in demonstrating the relevance of phenotypic fluctuations, the results of our model and experiment suggest that the fluctuations decrease during the course of the evolution. Hence the evolution speed will decline accordingly, and one might wonder whether evolution may finally come to a halt.

Here we have to be cautious in applying the above arguments to evolution in nature. First, for both the experiment and the model, we have assumed a fixed fitness criterium for selecting a surviving phenotype. In nature where organisms interact with many other organisms, such fixed fitness criteria cannot generally be defined. Fitness can change over time, depending on environmental fluctuations and populations of surrounding organisms. In such cases, if the phenotypic fluctuations have declined, the organisms lose the ability to adapt to environmental change. Hence only those that maintain sufficiently large fluctuations will survive.

Another topic associated with phenotypic fluctuations is the error catastrophe that occurs for $V_g \approx V_{ip}$. Results of our model in Sect. 10.3 show a decrease in $V_{ip}$ while evolution progresses. Then when $V_{ip}$ is small, an error catastrophe occurs around $V_g \sim V_{ip}$ if the mutation rate is increased a little bit. When this catastrophe occurs, the phenotype distribution is flattened as shown in Fig. 10.10, and the fluctuations are suddenly enhanced. In other words, the plasticity in the phenotype is regained. Indeed, we have studied several other models, and found that such a sudden increase in phenotype fluctuations always follows an error catastrophe, reached through evolution. Hence recoveries of phenotypic fluctuations (plasticity) can occur, leading to the regaining of evolvability.

Note that although we have chosen a one-dimensional parameter $a$ in our theory (say, as a Hamming distance from the fittest DNA sequence), evolution in nature occurs in a high-dimensional space, represented by a set of symbol sequences. Up to some stage of artificial selection in a given direction, the one-dimensional representation may be valid, but as the phenotypic fluctuations decrease, the residual directions in the gene cannot be negligible, and the instability along that direction has to be taken into account.

Indeed, such an increase in the instability through evolution is also discussed by several biologists (Visser et al., 2003) as due to the influence of "hidden genes," together with the increase in robustness (i.e., the decrease in fluctuations) mentioned above. We will come back to this problem in Chap. 12.

# 11

# Speciation as a Fixation of Phenotypic Differentiation

## 11.1 Question to Be Addressed

**Question: Characteristics of biological species are not distributed continuously, but are separated into discrete types, that form species. Indeed, Darwin asked why organisms are separated into distinct groups, rather than their character being continuously distributed (Darwin, 1859). Consider the problem of "origin of species," that is, how two (or more) species are formed from single identical species. In the beginning of this speciation process, all individuals have similar genes. If two organisms with minor genetic change only slightly differ in phenotype, then they share the common niche and compete for survival. Then it is hard that two groups with only slight genetic difference coexist. Only individuals having phenotype that are fitted better at the environment should survive. Then, how can the speciation occur when organisms live in the same space interacting each other?**

On the earth, there are a huge variety of species. To consider the origin of such diversity, it is crucial to understand the process how a single species differentiates into two or more. This is the problem of speciation.

Of course there has been debates on the speciation over a century. Still, as will be discussed now, one could say that in spite of progress in the understanding of evolution ever since Darwin (1859), the speciation is not yet fully understood. In the recent book, Maynard-Smith and Szathmary (1995) wrote that *we are not aware of any explicit model demonstrating the instability of a sexual continuum.*

Let us first recall the discussion in Chap. 10. As stated therein, the basic standpoints in evolution theory are as follows:

(i) Existence of genotype and phenotype.
(ii) Fitness is given as a function of the phenotype and the environment.
(iii) Only the gene is transferred to the next generation.

(iv) The mutation of genes occur randomly without any specific direction.
(v) There is flow only from gene to phenotype. The process is summarized as *Gene → Development → Phenotype.*

Here again, we adopt these standard assumptions, for the discussion on speciation. Still as in Chap. 10, we do not take the assumption (v′) "phenotype is uniquley determined from gene". In other words, phenotype is not necessarily uniquley determined from gene, i.e., it is not a single valued function of genotype, in contrast to the often-adopted assumption in the theory of evolutionary genetics.

Indeed, explanation of the speciation under the assumption of (v′) is not so easy. The reason is as follows: If slight genetic change leads to slight phenotype change, then individuals arising from mutation from the same genetic group differ only slightly according to this picture. Then, these individuals compete each other for the same niche. Unless the phenotype in concern is neutral, it is generally difficult that two (or more) groups coexist. Those with a higher fitness would survive.

One possible way to get out of this difficulty is to assume that two groups are "effectively" isolated, so that they do not compete. Some candidates for such isolation have been searched for, where the most well-known example is spatial segregation. This is the mechanism, so-called allopatric speciation, which we briefly discuss now.

In the speciation, there are two classes, roughly speaking; that is, sympatric speciation and allopatric speciation. The former is the speciation sharing the same space, that is, the former individuals live in the same space, interact, and mate each other during the course of speciation. The latter is the speciation when individuals are separated in space so that they neither interact nor mate during the speciation. Indeed, often two species that are derived from a single species live separately in space. It sounds also natural to assume that the organisms that live in different environmental conditions tend to evolve different characteristics. Accordingly, allopatric speciation is often believed to be the basic and most frequent mechanism.

Still, is it obvious that speciation progresses when individuals live distant in space? For example, even if individuals live either at the top or foot of the mountain, it does not necessarily mean that they will be different species. Characteristics of individuals may change continuously according to their location, but they do not necessarily differentiate into distinct groups. Two organisms from distinct space, isolated from each other, cannot mate, but when these two are put at the same place, they may mate to produce offspring which again reproduce. Furthermore, the fact that different species live in distinct space at present does not necessarily mean that the spatial isolation is the "cause" of speciation. It may be a result of speciation, instead. Indeed, we will show such example later.

On the other hand, sympatric speciation, if a general theory for it is established, can be more fundamental, since it does not require spatial segregation in the beginning. As will also be discussed, once sympatric speciation is established, then, later spatial segregation can evolve naturally. Furthermore, there are several examples that suggest sympatric speciation. In some lakes or in isolated islands, speciation is observed. For example, in Lake Victoria, which is known to have dried out about 12 thousands years ago, about few hundred species of specific fish, *Cichlid*, coexist. Similar examples are known for Tanganica or Malawi lakes. Since there are no spatially separated regions in the lakes so that fish can go everywhere in the lake, the diversification into many species is considered to be progressed sympatrically. The diversification to a great many species in tropical rain forest is also suggested to be due to sympatric speciation.

Hence mechanisms for sympatric speciation have been sought for over decades. However, sympatric speciation is often regarded to be rather difficult theoretically, since it confronts with the question already raised. If the organisms with only slight difference have similar phenotype, they compete each other for the same niche at the same space. Then genes with better phenotypes are selected, and coexistence of two separated groups sound difficult.

Again, one possible way to resolve the question is that the two groups are "effectively" isolated even though they live in the same space. Accordingly, people searched for some mechanisms how two groups do not mix and survive independently. The most well known and thoroughly studied mechanism is the sexual isolation by mating preference.

For such study, one assumes two groups that are almost identical with regards to the survivability, and then search for a mechanism that these two groups do not mutually interact with regards to (sexual) reproduction (Maynard-Smith, 1966; Felsenstein, 1981; Lande, 1981; Doebeli, 1996; Coyne & Orr, 1998; Howard & Berlocher, 1998). As a simplified illustration, consider that individuals have color from blue to red, and bluish individuals prefer to mating bluish, and reddish prefer to mate reddish ones. Then, even if these individuals live in the same space, they somehow separate in the "color" space. In this way, the interactions are eliminated even if they live in the same space. In general, if one considers a mechanism of mating preference with regards to some character, and assumes that there is a tendency that each does not mate with those with different characters, then the separation would be possible. In the above example of color, the red and blue groups will be separated from each other, to form blue and red species. Of course, this description on the speciation by mating preference is too simplified, and one needs theoretical elaboration to give a condition for speciation, but in the mainstream theory of sympatric speciation there underlies this type of argument. In fact, there have recently appeared some models showing the instability of sexual continuum, without assuming the existence of discrete groups in the beginning. (Probably, the argument based on the runaway is most persuasive [Lande, 1981; Turner & Burrows, 1995; Howard & Barlocher, 1998].)

Even though two groups coexist at the same spatial location, they can be genetically separated if two groups do not mate each other. Hence, the mating preference is now established as a mechanism for sympatric speciation. Indeed, often, between two close species, the individuals from different species avoid mating. However, here again, we cannot conclude that the mating preference that exists at present is the "cause" of speciation. Such mating preference could be a "result" of speciation. Furthermore, in this theory, why there is such mating preference itself is not answered. In this sense, as a theory it is not completely self-contained.

In addition, what one can show in the mating preference theory is only that the two groups may exist, but not that two groups should exist, because the blue group can exist by itself and the red by itself. Hence, if during the course of speciation one group is extinct by fluctuations, there is no reason that the lost group is recovered. In this sense one cannot answer the "necessity of speciation" in the theory.[1]

Then if one does not assume some kind of mating preference in advance, isn't it too possible to consider any plausible, general mechanism for sympatric speciation? Recall that in most studies of speciation, the interaction between individuals leads to competition for their survival. Difficulty in stable sympatric speciation without mating preference lies in the lack of a known clear mechanism how two groups, which have just started to be separated, coexist in the presence of mutual interaction. Of course, if the two groups were in a symbiotic state, the coexistence could help the survival of each. Then the coexistence of two groups would be possible. However, the two groups have little difference in genotype in the beginning of speciation process, according to the assumption (v)'. Then, it would be quite difficult to imagine such a "symbiotic" mechanism.

Note that the above difficulty comes from the assumption that the phenotype is a single-valued function of genotype (v)', a stronger version of (v). Is this single-valued-ness always true? To address this question, we reconsider the genotype–phenotype relationship. Indeed, there are three reasons that we doubt this single-valued-ness.

---

[1] Another recent proposal is the introduction of (almost) neutral fitness landscape and exclusion of individuals with similar phenotypes, as was discussed as evolutionary branching in adaptive dynamics (Geritz et al., 1998). If the phenotype is almost neutral, then similar phenotypes coexist at a rather broad range. On the other hand, if there is exclusion for similar phenotypes, it is natural that the phenotypes split into discrete bands separated with some distance. For example, Dieckmann and Doebeli (1999) have succeeded in showing that two groups are formed and coexist, to avoid the competition among organisms with similar phenotypes, assuming a rather flat fitness-landscape. A similar idea has been extended to several models recently (Dieckmann & Doebeli, 1999; Kondrashov & Kondrashov, 1999; Kawata & Yoshimura, 2001). This provides one explanation and can be relevant to some sympatric speciation. However, it is not so clear how the phenotype that is not so important as a fitness works strongly as a factor for exclusion for a closer value. Furthermore, we are more interested in the differentiation of phenotypes that are functionally different and not neutral.

First, as discussed in Chap. 7, Yomo and his colleagues have reported that specific mutants of *E. coli* show (at least) two distinct types of enzyme activity, although they have identical genes (Ko et al., 1994). These different types coexist in an unstructured environment of a chemostat, and this coexistence is not due to spatial localization. Coexistence of each type is supported by each other. Indeed, when one type of *E. coli* is removed externally, the remained type starts differentiation again to recover the coexistence of the two types. The experiment demonstrates that the enzyme activity of these *E. coli* are differentiated into two (or more) groups, because of the interaction with each other, even though they have identical genes.

Second, some organisms are known to show various phenotype from a single genotype. This phenomenon is often related to malfunctions of a mutant (Holmes, 1979), and is called as low or incomplete penetrance (Opitz, 1981).

Third, we have explained a theoretical mechanism of phenotypic diversification in Chaps. 7 and 8, as the isologous diversification for cell differentiation (Kaneko & Yomo, 1994, 1997, 1999; Furusawa & Kaneko, 1998). As already stated, phenotypic diversity will arise from a single genotype and develop dynamically through intracellular complexity and intercellular connection. Although we have explained the theory for the behavior of each cell, the theory itself is applied to any unit with replication. Then, when organisms with plastic developmental dynamics interact with each other, the dynamics of each unit can be stabilized by forming distinct groups with differentiated states in the pheno-space. Here the two differentiated groups are necessary to stabilize the states of each other. Otherwise, the developmental process is unstable, and because of the interaction the two types will be formed again (provided there is a sufficient number of individuals). In this sense the two groups are in a "symbiotic" state, as requested. This theoretical mechanism is demonstrated by several models and shown to be a general consequence of coupled dynamical systems.

Now, the problem we address here is as follows: If we do not assume (v)' (but by assuming (i)–(v)), is there any mechanism that two groups mutually require each other for the survival in the beginning of the separation of the two groups. Here, we seriously take this problem of this interaction-induced phenotypic differentiation from the same genotype, and discuss its relevance to evolution. We will show that this phenotypic differentiation is later fixed to genotypes through mutation to genes and gives a general mechanism for the sympatric speciation, in spite of the fact that we have assumed only the flow from gene to phenotype.

From a different point of view, we will study a possible relationship between plasticity in phenotype introduced by interaction and evolution, as also studied in Chap. 10. In contrast to the phenotypic fluctuation discussed there, we are interested in differentiation of phenotype through developmental process. The isologous diversification theory shows that there can be developmental "flexibility," in which different phenotypes arise from identical gene sets, as in the incomplete penetrance aforementioned. Here we study how this theory is relevant to evolution. Indeed, the question how developmental process and evolution are related has been addressed over decades (Maynard-Smith et al., 1985). The so-called Evo-Devo study has attracted much attention these days following the accumulation of genome information including discover of homeobox (Gehring, 1998) and progresses in developmental biology. So far, most of the study remains at the re-consideration of accumulated data, but theoretical study to relate with developmental plasticity and evolvability is strongly requested.

We consider correspondence between genotype and phenotype seriously, by introducing a developmental process with which a given initial condition is led to some phenotype according to a given genotype. "Development" here means a dynamic process from an initial state to a matured state through rules associated with genes. (In this sense, it is not necessarily restricted to multicellular organisms.) With this study, we hope to establish a novel viewpoint for the EvoDevo study.

## 11.2 Logic: Interaction-Based Speciation

Consider an organism with internal phenotype states that reproduce. As the number of such organisms is increased, they start to interact strongly. Here following the isologous diversification theory in Chaps. 7–9, we assume that phenotypes of individuals with identical genotypes differentiate into two groups. This can occur through developmental dynamics associated with the interaction among individuals. At this moment, the two groups have identical (or almost identical) genes, but they take a different phenotype. Of course, such differentiation does not always occur, but as long as the homogeneous phenotype state over individuals is destabilized through the interaction, such differentiation is possible as shown in Chaps. 7 and 8. Now we consider how this differentiation introduces differentiation also in gene over a long-term span through mutation of genes and selection process, assuming the standard evolution theory (i)–(v).

Following the above setup, there are two groups with distinct phenotypes, which we call "upper" and "lower" groups as in Fig. 11.1. For example, these two groups have different chemicals and different way in the use of resources. Here the existence of two groups is mutually necessary. At this moment, they have same genotypes, while the offspring of each group can either take the same phenotype with its mother or take the other phenotype, since the two have the same genes. The phenotype differentiation is not fixed.

In reproduction, there occurs some mutational change to genes randomly. Of course, this mutational change occurs in the same way for the upper and lower groups. However, the selection can work in a different way. For example, consider that the upper group uses a (metabolic) reaction path A for growth, and the lower group uses a different path B. Then if the genetic change to enhance the enzyme activity for the reaction A occurs for the upper group,

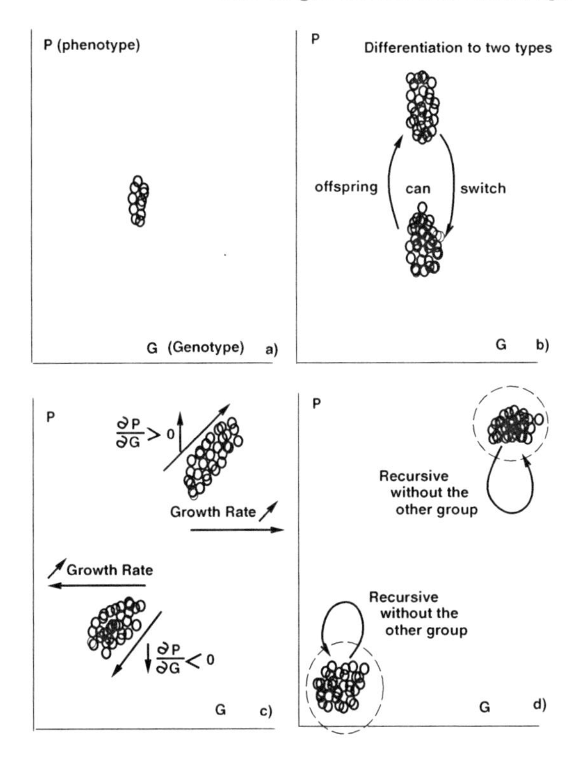


**Fig. 11.1.** Schematic representation of the speciation scenario obtained from our simulation and theory. A pair (phenotype, genotype) is plotted successively with time: (**a**) the stage of interaction-induced phenotypic separation, (**b**) the stage of genotype–phenotype feedback amplification, (**c**) the stage of genetic fixation, and (**d**) speciation completed. Reproduced from Kaneko and Yomo (2002b)

the organism can grow faster, and has a higher fitness. On the other hand, for the lower group the change to suppress the reaction A and enhance the reaction B will be favored. Hence the opposite direction in the genetic change between the upper and lower groups are favored for the reproduction. Now the opposite change for some gene is selected (see Fig. 11.1c). This change in gene will enhance the difference in phenotypes between the two groups, that is, the upper group becomes shows a further "upper" phenotype and vice versa.

After this process the two groups start to have a different genotypes also. When the difference in genes is so large that each group can take only a single phenotype. For example, if the enzymatic reaction for A is enhanced and that for B is highly suppressed, then the organism can take only the phenotype with higher A and lower B, but not that with lower A and higher B. At this stage, each group recovers a one-to-one correspondence between the phenotype and the genotype, and the two have different phenotypes and genotypes. Through competition for reproduction and mutational change of genes, the phenotypic differences are fixed to genes, until the groups ("species") are completely separated in genes as well as in phenotypes. Now the speciation process is completed.

When we view only the starting stage (i) and the final stage (iv), it looks like nothing but the speciation by genetic change. However, if we view whole the process (i)–(iii), it is hard to conclude that the speciation is caused by genetic change. Rather, the phenotypic change exists first, and later it is consolidated to gene.

Note this process is rather stable against external and internal disturbances. Indeed, it will also be demonstrated that the proposed theory for speciation works also under sexual recombination. Indeed, the hybrid offspring of the two groups becomes sterile, and also provides a basis for mating preference, a major mechanism in sympatric speciation. The hybrid offspring of the two groups becomes streile, which also provides a basis for mating preference, a major mechanism in sympatric speciation.

Here we again recall that the two groups are "symbiotic" at the stage (i). As already seen in Chaps. 7 and 8, each of the two groups reproduces under the existence of the other. Hence the interaction is essential to the differentiation. In our theory, interaction is not an obstacle to speciation but promotes it. Thus the "sympatric" speciation is favored rather than "allopatric." Also, the speciation process is robust with respect to fluctuations, because if one group disappears by fluctuations, it is recovered at the stage (i), as discussed in Chaps. 7 and 8.

## 11.3 Model

### 11.3.1 Basic Strategy

Since we have to study the correspondence between genotype and phenotype, we need to introduce a developmental process that leads a given initial condition to some phenotype according to a given genotype. "Development" here means a dynamic process from an initial state to a matured state through rules associated with genes, in its general sense. (In this definition, it is not necessarily restricted to multicellular organisms.) To consider this process, we assume that the state of each individual organism is characterized by a set of variables and parameters that govern the dynamics of the variables. As in Chap. 10, it is appropriate to represent phenotype by a set of state variables. For example, each individual $i$ has variables $(X_t^1(i),X_t^2(i), \dots, X_t^k(i))$, which defines the phenotype. This set of variables can be regarded as concentrations of chemicals, rates of metabolic processes, or some quantity corresponding to a higher function characterizing the behavior of the organism. The state is not fixed in time, but develops from the initial state at birth to a matured state when the organism is ready to produce its offspring. The dynamics of the state variables $(X_t^1(i),X_t^2(i), \dots, X_t^k(i))$ is given by a set of equations with some parameters.

Genes, since they are nothing but information expressed on DNA, could in principle be included in the set of variables. However, following the argument in Chap. 10 it is reasonable to represent it by a set of parameters in the dynamics governing the phenotype. In terms of dynamical systems, the set corresponding to genes can be represented by parameters $\{g^1(i), g^2(i), \ldots g^m(i)\}$ that govern the dynamics of phenotypes, since the parameters in an equation are not changed through the developmental process, while the parameters control the dynamics of phenotypic variables. Only when an individual organism is reproduced, this set of parameters changes slightly by mutation. For example, when $\{X_t^\ell(j)\}$ represents the concentrations of metabolic chemicals, $\{g^1(i), g^2(i), \ldots g^m(i)\}$ is the catalytic activity of enzymes that controls the corresponding chemical reaction.

Now, our model is set up as follows:

(1) *Dynamical change of states giving a phenotype*: The temporal evolution of the state variables $(X_t^1(i), X_t^2(i), \ldots, X_t^k(i))$ is given by a set of deterministic equations, which are described by the state of the individual, and parameters $\{g^1(i), g^2(i), \ldots g^m(i)\}$ (gene), and the interaction with other individuals. This temporal evolution of the state consists of internal dynamics and interaction.
   (1-i) The internal dynamics (say metabolic process in an organism) are represented by the equation governed only of $(X_t^1(i), X_t^2(i), \ldots, X_t^k(i))$ (without dependence on $\{(X_t^\ell(j)\}$ $(j \neq i)$), and are controlled by the parameter sets $\{g^1(i), g^2(i), \ldots g^m(i)\}$.
   (1-ii) Interaction between the individuals: The interaction is given through the set of variables $(X_t^1(i), X_t^2(i), \ldots, X_t^k(i))$. For example, we consider such interaction form that the individuals interact with all others through competition for some "resources." The resources are taken by all the individuals, giving competition among all the individuals. Since we are interested in sympatric speciation, we take this extreme all-to-all interaction, by taking a well-stirred soup of resources, without including any spatially localized interaction.

(2) *Reproduction and death*: Each individual gives offspring (or splits into two) when a given "maturity condition" for growth is satisfied. This condition is given by a set of variables $(X_t^1(i), X_t^2(i), \ldots, X_t^k(i))$. For example, if $(X_t^1(i), X_t^2(i), \ldots, X_t^k(i))$ represents cyclic process corresponding to a metabolic, genetic, or other process that is required for reproduction, we assume that the unit replicates when the accumulated number of cyclic processes goes beyond some threshold, or one can use a cell division condition adopted through Chaps. 4–9 where the variables $\{X_k^i(t)\}$ are chemical concentrations.

(3) *Mutation*: When each organism reproduces, the set of parameters $\{g^j(i)\}$ changes slightly by mutation, by adding a random number with a small amplitude $\delta$, corresponding to the mutation rate. The values of variables $(X_t^1(i), X_t^2(i), \ldots, X_t^k(i))$ are not transferred but are reset to initial

conditions. If one wants to include some factor of epigenetic inheritance, one could assume that some of the values of state variables are transferred. Indeed we have carried out this simulation also, but the results to be discussed are not altered (or confirmed more strongly).

(4) *Competition*: To introduce competition for survival, death is included both by random removal of organisms at some rate and by a given death condition based on their state.

We have carried out numerical experiments of several models with the above conditions (1)–(5). For example, we have chosen a model with reaction dynamics, where the phenotypic variables are chemical concentrations and genetic parameters are reaction coefficients showing the catalytic activities of enzymes. We have also chosen coupled dynamical systems with dynamic variables for phenotypes and genetic parameters giving the equation of motion. The results to be discussed are commonly observed, as long as the variables show differentiation with the interplay between internal dynamics and interaction.

## 11.4 Result

### 11.4.1 Process for Genetic Diversification

From several simulations satisfying the condition of the model in Sect. 11.3, we have obtained a scenario for a sympatric speciation process. We have studied several models (Kaneko & Yomo, 2000, 2002b), including that with a catalytic reaction network (Takagi et al., 2000), and same scenario is confirmed, with the speciation process we observed. This is schematically shown in Fig. 11.1, where the change of the correspondence between a phenotypic variable ("P") and a genetic parameter ("G") is plotted at every reproduction event.

First, we sketch our interaction-induced process of genetic diversification triggered by isologous diversification, obtained from simulations of several models. This scenario is summarized as follows.

In the beginning, there is a single species, with one-to-one correspondence between phenotype and genotype. Here, there are little genetic and phenotypic diversity that are continuously distributed (see Fig. 11.1a). We assume that the isologous diversification starts because of developmental plasticity with interaction, when the number of these organisms increase. Indeed, the existence of such phenotypic differentiation is described by isologous diversification as is supported by several numerical experiments. This gives the following stage I.

#### Stage I: Interaction-Induced Phenotypic Differentiation

When there are many individuals interacting for finite resources, their phenotypes start to differentiate even though the genotypes are identical or differ only slightly. Phenotypic variables split into two (or more) types (see Fig. 11.2). This interaction-induced differentiation is an outcome of the mechanism aforementioned. Slight phenotypic difference between individuals is amplified by the internal dynamics, while the phenotypes tend to be clustered into two (or more) types, through the interaction between organisms. The two distinct phenotype groups (brought about by interaction) are called "upper" and "lower" groups, tentatively.

This differentiation is brought about, since the population consisting of individuals taking identical phenotypes is destabilized by the interaction. Such instability is, for example, caused by the increase of population or decrease of resources, leading to strong competition. Of course, if the phenotype $X_t^j(i)$ at a matured state is rigidly determined by developmental dynamics, such differentiation does not occur. It is the assumption we make in the present theory that permits this – that there exists such developmental plasticity in the internal dynamics, when the interaction is strong. Recall again that this assumption is theoretically supported.

Note that the difference is fixed at this stage at neither the genetic nor phenotypic level. After reproduction, an individual's phenotype can switch to another type.

### Stage II: Amplification of the Difference Through Genotype–Phenotype Relationship

At the second stage the difference between the two groups is amplified at both the genetic and the phenotypic levels. This is realized by a positive feedback process between the change of geno- and phenotypes.

First the genetic parameter(s) separate as a result of the phenotypic change. This occurs if the parameter dependence of the growth rate is different between the two phenotypes. Generally, there are one or several parameter(s) $g^\ell$, such that the growth rate increases with $g^\ell$ for the upper group and decreases for the lower group (or the other way round) (see Figs. 11.2 and 11.3).

Certainly, such a parameter dependence is not exceptional. As a simple illustration, assume that the use of metabolic processes is different between the two phenotypic groups. If the upper group uses one metabolic cycle more, then the mutational change of the parameter $g^\ell$ to enhance the cycle is in favor for the upper group, while the change to reduce it may be in favor for the lower group. Indeed all numerical results support the existence of such parameter(s). This dependence of growth rate on the genotypes leads to the genetic separation of the two groups, as long as there is competition for survival, to keep the population numbers limited. Hence, this second stage is *always* observed in our model simulation when the phenotypic differentiation at the first stage occurred.

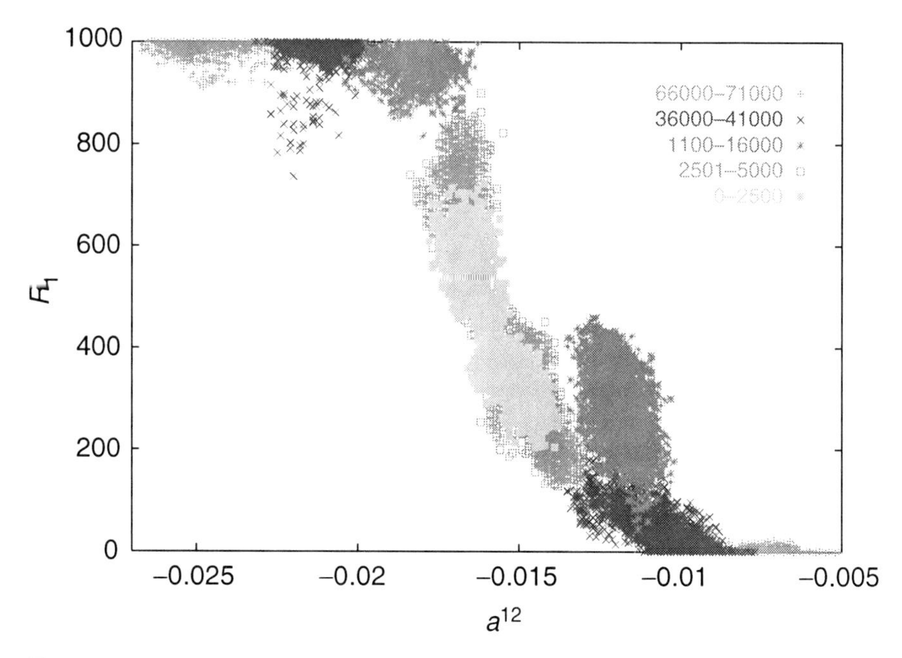


**Fig. 11.2.** Evolution of genotype–phenotype relationship. In the model adopted here, there are several cyclic processes, and when the sum of the times of each cyclic process $R^{\ell}(i)$ completed is larger than some threshold, the element divides. Here phenotype is represented by $R^1, R^2, \ldots R^k$ (we take $K = 3$). These cyclic processes mutually interact through some dynamical processes controlled by parameters $a^{\ell j}$. Here $(R^1, a^{12})$ is plotted for every division of individuals. Different division steps (first 2500, 2501–5000,. . . , 66,000–71,000) are plotted with different symbols as shown. Initially, phenotypes are separated, even though the genotypes are identical (or only slightly differ), as shown in light blue. Later, the genotypes are also separated, according to the difference in phenotypes. In the simulation, the population size fluctuates around 300, after an initial transient. (Hence the generation number is given roughly by dividing this division number by 300.) Reproduced from Kaneko and Yomo (2000), with permission

The genetic separation is often accompanied by a second process, the amplification of the phenotypic difference by the genetic difference. In the situation of Fig. 11.1c, as a parameter $G$ increases, a phenotype $P$ (i.e., a characteristic quantity for the phenotype) increases for the upper group, and decreases (or remains the same) for the lower group.

## Stage III: Genetic Fixation

After the separation of two groups has progressed, each phenotype (and genotype) starts to be preserved by the offspring, in contrast to the situation at the first stage. However, up to the second stage, the two groups with different

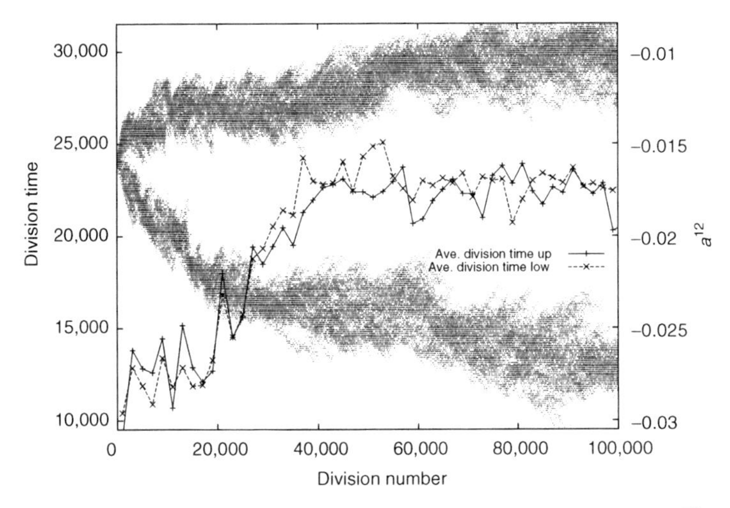


**Fig. 11.3.** The evolution of the genetic parameter. The parameter $g = a^{12}(i)$ is plotted as a dot at every division (reproduction) event, with the abscissa as the division number. The average time necessary for division (reproduction) is plotted for the upper and lower groups, where the average is taken over 2,000 division events (6th–8th generation). As the two groups are formed around the 2,000th division event, the population size becomes twice the initial, and each division time is also approximately doubled. Note that the two average division speeds of the two groups remain of the same order, even when the genetic parameter evolves in time. Reproduced from Kaneko & Yomo (2000)

phenotypes cannot exist in isolation by itself. When isolated, offspring with the phenotype of the other group starts to appear. The two groups coexist depending on each other (see Fig. 11.1d).

Only at this third stage, each group starts to exist by its own. Even if one group of units is isolated, no offspring with the phenotype of the other group appears. Now the two groups exist on their own. Such a fixation of phenotypes is possible through the evolution of genes (parameters). In other words, the differentiation is fixed into the genes (parameters). Now each group exists as an independent "species," separated both genetically and phenotypically. The initial phenotypic change introduced by interaction is fixed to genes, and the "speciation process" is completed.

To check the third stage of our scenario, it is straightforward to study the further evolutionary process if only one isolated group exists. To do this, we pick out some population of units only of one type after the genetic fixation is completed and both the geno- and phenotypes are separated into two groups. Then we start the simulation again. When the groups are picked from later generations after the genetic fixation process, the offspring keep the same phenotype and genotype. Now, only one of the two groups can exist in isolation. Here, the other group is no longer necessary to maintain stability.

This recursive production by each group characterizes the third stage of our scenario.

In contrast, at the second stage, the separation is not fixed rigidly. Units selected from one group at this earlier stage again start to show phenotypic differentiation, followed by genetic separation, as demonstrated by several simulations. After some generations, one of the differentiated groups recovers the geno- and phenotype that had existed before the transplant experiment. This means that there remains some plasticity at this stage, which is in a sharp contrast with the third stage.

At the third stage, two groups with distinct genotypes and phenotypes are formed, each of which has one-to-one mapping from genotype to phenotype. This stage now is regarded as speciation. (In the next section we will show that this separation satisfies hybrid sterility in sexual reproduction and is appropriate to be called speciation.) Summarizing the stages I-III, phenotrype differentiation drives the genetic separation, in spite of the flow only from gene to phenotype. Phenotype differentiation is consolidated to genotype, and then the offspring take the same phenotype as their ancestor.

## 11.5 Further Remarks on the Differentiation Scenario

### 11.5.1 Condition for Genetic Diversification

Now, we show in our model that interaction-induced phenotypic differentiation is a necessary (and sufficient) condition for the formation of genetically distinct groups.

**Phenotype differentiation is necessary and sufficient for the sympatric speciation in our theory.**

#### Sufficient

If phenotypic differentiation at stage I occurs in our model, then the genetic differentiation of the later stages *always* follows, in spite of the random mutation process included. How long it takes to reach the third stage can depend on the mutation rate, but the speciation process itself does not depend on the mutation rate. However small the mutation rate may be, the speciation (genetic fixation) always occurs.

Once the initial parameters of the model are chosen, it is already determined whether the interaction-induced phenotype differentiation will occur or not. If it occurs, then always the genetic differentiation follows.

#### Necessary

On the other hand, in our setting, if the interaction-induced differentiation does not exist initially, there is no later genetic diversification process. If the

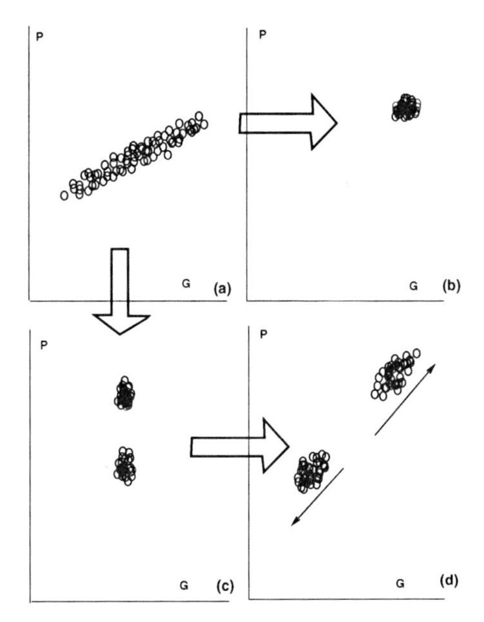


**Fig. 11.4.** Schematic representation of the evolution starting from large genetic variance. (**a**) Initial genetic and phenotypic distribution, (**b**) without the phenotypic differentiation, no speciation follows, and (**c**) if phenotypic differentiation occurs, then the speciation follows later (**d**)

initial parameters characterizing nonlinear internal dynamics or the coupling parameters characterizing interaction are small, no phenotypic differentiation occurs. Also, the larger the resource per individual is, the smaller the effective interaction is. Then, phenotypic differentiation does not occur. In these cases, even if we take a large mutation rate, there does not appear differentiation into distinct genetic groups, although the distribution of genes (parameters) is broader. We have also made several simulations starting from a population of units with widely distributed parameters (i.e., genotypes). However, unless the phenotypic separation into distinct groups is formed, the genetic differentiation does not follow (Fig. 11.4a, b). The genetic diversification process is not driven by imposing distributed genotypes initially. Only if the phenotype differentiation occurs, the genetic differentiation follows (Fig.11.4c, d).

It is also important to note that the separation into *discrete* phenotypic groups is necessary to have differentiation into discrete genetic groups. Indeed when the phenotype is broadly distributed without forming discrete groups, the evolution does not lead to form discrete groups in genotype and phenotype, even though the phenotype variation is large. For some other models with many variables and parameters, the phenotypes are often distributed broadly, but continuously without making distinct groups. In this case, distinct genetic groups do not appear through the mutations, although the genotypes are broadly distributed (see Fig. 11.5).

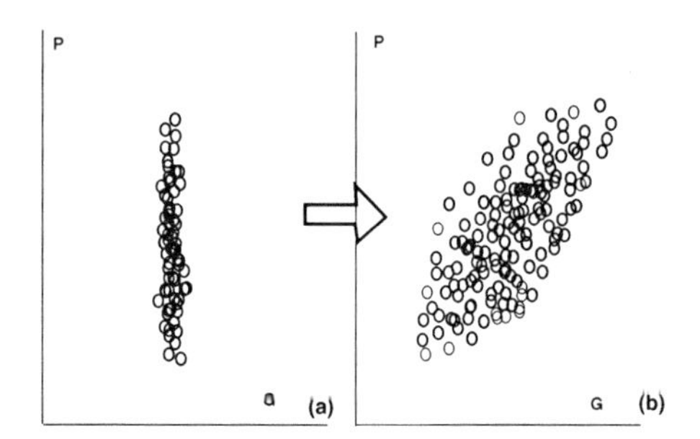


**Fig. 11.5.** Schematic representation of the evolution when phenotype is distributed without clear differentiation to discrete states. (**a**) Initial distribution of phenotypes and genotypes. (**b**) After the evolution, genes are also distributed broadly, but no speciation follows

### 11.5.2 Coevolution of Differentiated Groups

Note that each of the two groups forms a niche for the other group's survival mutually, and each of the groups is specialized in this created niche. For example, some chemicals secreted out by one group are used as resources for the other, and vice versa.

Hence the evolution of two groups is mutually related. At the first and second stages of the evolution, the speed for reproduction is not so much different between the two groups. Indeed, at these stages, the reproduction of each group is strongly dependent on the other group, and the "fitness" as a reproduction speed of each group by itself alone cannot be defined. At stage II, the reproduction of each group is balanced through the interaction so that one group cannot dominate in the population. This is why the growth speeds of the two groups are almost same, as shown in Fig. 11.2.

Later at the third stage, the division speeds start to be different. Here, the degree of the interaction term between the two groups gets weaker than that of the internal dynamics term. Now the fitness as a reproduction speed can be defined, and a group with a higher reproduction rate surpasses in the population.

### 11.5.3 Robust Speciation

Note that the speciation process of ours occurs under strong interaction. At the second stage, these two groups form symbiotic relationship. As a result, the speciation is robust in the following sense. If one group is eliminated externally, or extinct accidentally at the first or second stage, the remaining group forms the other phenotype group again, and then the genetic differentiation is started again. The speciation process here is robust against perturbations.

### 11.5.4 Deterministic Nature of Evolutionary Process

As mentioned, if phenotypic differentiation (stage I) occurs in our model, then the genetic differentiation of the second stage *always* follows, in spite of the random mutation process included. Once the initial parameters of the model are chosen, it is already determined whether such differentiation will occur or not. Indeed, we have numerically studied the differentiation and evolution process of the two groups by taking different random number sequence for mutation. As far as we checked more than a dozen runs, the differentiation and evolution of the two groups always occur. The phenotypes and genotypes of each of the two groups are almost identical for each run. Of course, the existence of mutation is required, but the genetic separation is not mutation driven. The evolution to distinct genetic groups is rather deterministic in nature, in this sense.

Note also that the evolution process is rather fast. In the simulations, the phenotypic separation is completed around 50 generations after it is started. The speed of the change of the parameters (genes), of course, depends on the mutation rate. The velocity of the parameter change per generation is found to be proportional to the mutation rate. Although this velocity goes smaller with the mutation rate, the time required for the split of the two groups, indeed, does not increase so much: If the mutation rate is smaller, the difference in the parameter values (for gene) between the two groups is smaller, but the phenotype is separated. Also the genotype distribution is more tightly focused around the two groups as the mutation rate is lower. Thus, the separation of the two groups in genes is easily established. To sum up, a fast separation process is a characteristic feature of our mechanism.

### 11.5.5 Speciation: Reproductive Isolation Under Sexual Recombination

The speciation process is defined both by genetic differentiation and by reproductive isolation (Dobzhansky, 1937). Although the evolution through stages I–III leads to genetically isolated reproductive units, one might still say that it should not be called "speciation" unless the process shows isolated reproductive groups under the sexual recombination. In fact, it is not trivial if the present process works under sexual recombination, since the genes from parents are mixed by each recombination. To check this problem, we have considered some models so that the sexual recombination occurs to mix genes. To be specific, the reproduction occurs when two individuals $i_1$ and $i_2$ satisfy the maturity condition, and then the two genotypes are mixed. As an example we consider two offspring $j = j_1$ and $j_2$, from the individuals $i_1$ and $i_2$ as

$$g^\ell(j) = g^\ell(i_1)r_j^\ell + g^\ell(i_2)(1 - r_j^\ell) + \delta \tag{11.1}$$

with a random number $0 < r_j^\ell < 1$ to mix the parents' genotypes.

In spite of this strong mixing of genotype parameters, the two distinct groups are again formed. Of course, the mating between the two groups can produce an individual with the parameters in the middle of the two groups. When parameters of an individual take intermediate values between those of the two groups, at whatever phenotypes it can take, it either does not reach the condition for reproduction, or even if it does, it takes much longer time than those of the two groups. Before the maturity condition for reproduction is satisfied, the individual will be removed by death with a high probability. As the separation process to the two groups further progresses, an individual with intermediate parameter values never reaches the condition for the reproduction before it dies.

This process for increasing sterility is demonstrated clearly by measuring the average offspring number of individuals over given parameter (genotype) ranges and over some time span. An example of this average offspring number is plotted in Fig. 11.6, with the progress of the speciation process. As the two groups with distinct values of parameters are formed, the average offspring number of an individual having the parameter between those of the two groups starts to decrease. Soon the number goes to zero, implying that the hybrid between the two groups is sterile.

In this sense, sterility (or low reproduction) of the hybrid appears as a *result, without any assumption on mating preference.* Now genetic differentiation and reproductive isolation are satisfied. Hence it is proper to call the process through stages I–III as speciation.

Before closing this subsection, it should be noted that our mechanism for the speciation works in asexual and sexual reproduction in the same way. The phenotype separates into two groups first, also in the present case with sexual recombination. Later the change is mapped onto the parameters $g^m$. The speciation process progresses following the three stages given in Sect. 10.4. Indeed the stability of the speciation against sexual recombination is naturally expected, since the coexistence of two distinct phenotype groups is supported by isologous diversification, that is, differentiation to distinct phenotypes under the same genotypes. Even though the genes are mixed, the phenotypes tend to be separated into distinct groups, because of the interaction. Hence the separation into distinct groups is not blurred by the recombination. Here, sexual recombination is just another factor to fix the differentiation, and is not the essence of the present process. In short, the speciation process is initiated by phenotype difference, which is later fixed to genotypes by mutation and/or sexual recombination.

### 11.5.6 Evolution of Mating Preference[2]

So far we have not assumed any preference in mating choice. The mating to produce a sterile hybrid continues, and the split into two groups occurs

[2] This and the next subsetcions can be skipped unless one is interested especially in these topics.

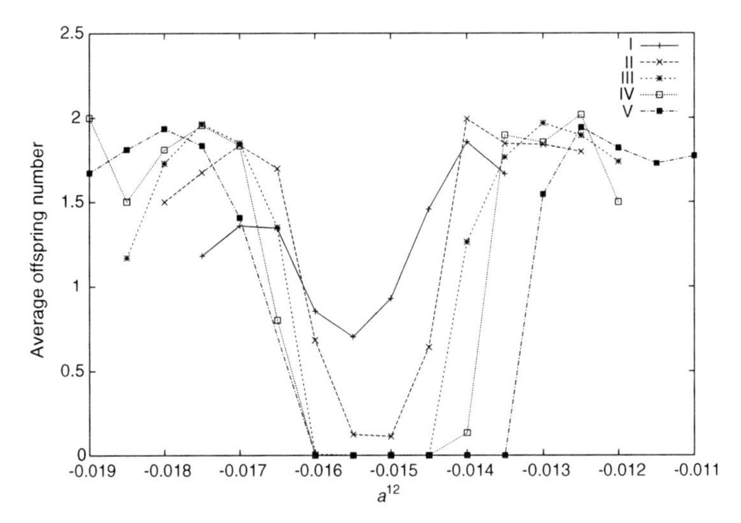


**Fig. 11.6.** The average offspring number before death is plotted as a function of the parameter (genotype), for simulations with sexual recombination. As an extension to include sexual recombination, we have also studied a model in which two organisms satisfying the above threshold condition mate to reproduce two offspring. When they mate, the offspring have parameter values that are randomly weighted average of those of the parents, as given in the text. We have measured the number of offspring for each individual during its life span. By taking a bin width 0.005 for the genotype parameter $g = a^{12}$, the average offspring number over a given time span is measured to give a histogram. The histogram over the first 7,500 divisions (about 20 generations) is plotted by the *solid line* (I), and the histogram for later divisions is overlaid with a different line, as given by II (over 7,500–15,000 divisions), III (15,000–22,500), IV (22,500–30,000), and V (37,500–45,000). As shown, a hybrid offspring will be sterile after some generations. Here we have used the model for Fig. 11.2 and the same initial condition and imposed recombination. In the run, the population fluctuates around 340 (reproduced from Kaneko & Yomo, 2000)

after mating, and thus can be called *postmating isolation*. However, it is then natural to expect that some kind of mating preference evolves to reduce the probability to produce a sterile hybrid. Here we study how mating preference evolves as a result of postmating isolation.

As a simple example, we include loci that can act as a gene for determining mating preference. We assume another set of genetic parameters that controls the mating behavior. For example, each individual $i$ has a set of mating threshold parameters $(\rho^1(i), \rho^2(i), \dots \rho^k(i))$, corresponding to the phenotype $(X^1(i), X^2(i), \dots, X^k(i))$. If $\rho^\ell(i_1) > X^\ell(i_2)$ for some $\ell$, the individual $i_1$ denies the mating with $i_2$ even if $i_1$ and $i_2$ satisfy the maturity condition. In simulations with mating thresholds $\{\rho^m(i)\}$, we choose a pair of individuals who satisfy the maturity condition, and check if one does not deny the other. Only if neither denies mating with the other, the mating occurs to produce offspring, when the genes from parents are mixed in the same way as in the previous section. If these conditions are not satisfied, the individuals $i_1$ and $i_2$ wait for the next step to find a partner again.

Here the set of $\{\rho^m\}$ is regarded as a set of (genetic) parameters, and changes by mutation and recombination. The mutation is given by addition of a random value to $\{\rho^m\}$. Initially all of $\{\rho^m\}$ (for $m = 1, \ldots, k$) are smaller than the minimal value of $(X^1(i), X^2(i), \ldots, X^k(i))$, so that any mating preference does not exist. If some $\rho^\ell(i)$ gets larger than some of $X^\ell(i')$, the mating preference appears. Hence we do not assume any mating preference in advance and examine if it evolves or not.

An example of numerical results is shown in Fig. 11.7, where the change of phenotype $X^m$ and some of the parameters $g^j$ are plotted. Here, by the phenotype differentiation, one group (to be called "up" group) has a large $X^m$

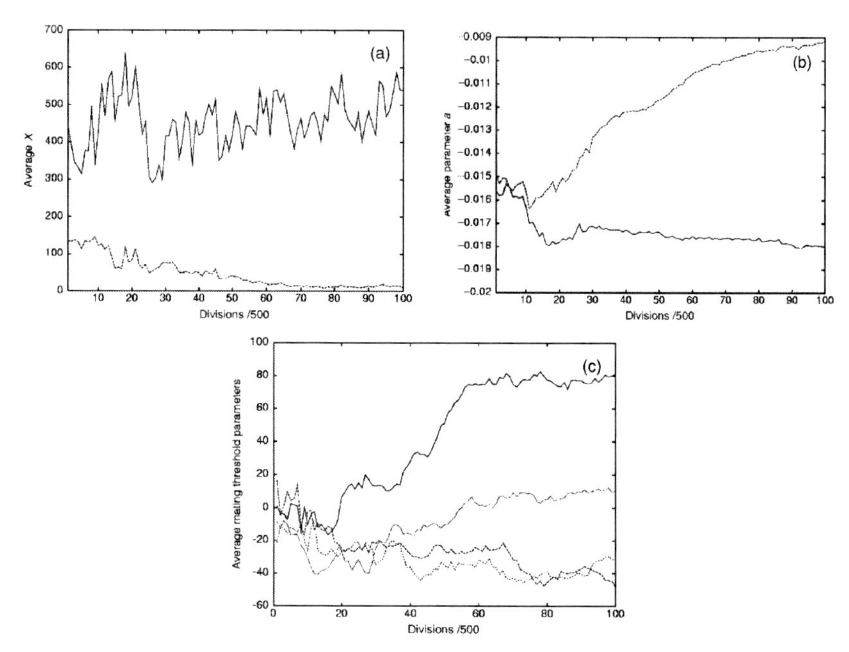


**Fig. 11.7.** An example of the speciation process with sexual recombination and the evolution of mating preference, Here two groups of distinct phenotype (large $X^1$, small $X^2$) and (small $X^1$, large $X^2$) are formed at the first few generation, which we call "up" and "down" groups. We have measured the average $\overline{X^j}$ at reproduction events, $\overline{a^{\ell m}}$, $\overline{\rho^j}$ for each group per 500 divisions. (The population here is roughly 500, and thus the average is roughly over one generation.) Change of the average $\overline{X^j}$, $\overline{a^{\ell m}}$, and $\overline{\rho^j}$ is plotted with divisions (generations). (**a**) $\overline{X^1}$ (up group; *solid line*), $\overline{X^1}$ (down group; *broken line*), (**b**) $\overline{a^{12}}$ (up group; *solid line*), $\overline{a^{12}}$ (down group; *broken line*), (**c**) $\overline{\rho^1}$ (up group; *solid line*), $\overline{\rho^1}$ (down group; *broken line*), $\overline{\rho^2}$ (up group; *broken line*), $\overline{\rho^2}$ (down group; *thin broken line*). Adapted from Kaneko & Yomo (2002b)

value for some $m = \ell$ and almost null values for some other $m = \ell'$. Hence, sufficiently large positive $\rho^{\ell'}$ gives a candidate for mating preference.

Right after the formation of two genetically distinct groups following the phenotype separation, one of the mating threshold parameters ($\rho^1(i_1)$) starts to increase for one group. In the example of the figure, "up" group has phenotype with (large $X^1$, small $X^2$) and the other ("down") group with (small $X^1$, large $X^2$). There the "up" group starts to increase $\rho^1(i_{up})$, and $\rho^1(i_{up}) > X^1(i_{down})$ is satisfied for an individual $i_{down}$ of the "down" group. Now the mating between the two groups is no longer allowed, and the mating occurs only within each group. Thus the mating preference evolves to prohibit intergroup mating producing sterile hybrids (Fig. 11.7).

Although the evolution of the mating preference here is a direct consequence of the postmating isolation, it is interesting to note that the coexistence of the two species is further stabilized with the establishment of mating preference. Without this establishment, there are some cases that one of the species disappears because of the fluctuation after very long time in the simulation. With the establishment, the two species coexist much longer (at least within our time of numerical simulation).

### 11.5.7 Formation of Allele–Allele Correlation

In diploid organisms, there are two alleles, and the two alleles do not equally contribute to the phenotype. For example, often only one allele contributes to the control of the phenotype. If by recombination, the loci from two alleles are randomly mixed, then the correlation between genotype and phenotype achieved by the mechanism so far discussed might be destroyed. Indeed, this problem was pointed out by Felsenstein (1981) as one difficulty for sympatric speciation.

Of course, this problem is resolved if genotypes from two alleles establish high correlation. To check whether this correlation is generated, we have extended our model to have two alleles and examined whether the two alleles become correlated. Here, we adopted the model studied so far, and added two alleles further. In mating, the alleles from the parents are randomly shuffled for each locus. In other words, each organism $i$ has two sets of parameters $\{g^{(+)\ell}(i)\}$ and $\{g^{(-)\ell}(i)\}$. Each $g^{(+)m}(i)$ is inherited from either $g^{(+)m}$ or $g^{(-)m}$ of one of the parents, and the other $g^{(-)m}(i)$ is inherited from either $g^{(+)m}$ of $g^{(-)m}$ of the other parents. Here parameters at only one of the alleles work as a control parameter for the developmental dynamics of phenotype.

We have carried out some simulations of this version of our model (Kaneko, 2002c). Here again, the speciation proceeds in the same way, through stages I, II, and III. Hence our speciation scenario works well in the presence of alleles. In this model, the genotype–phenotype correspondence achieved at stage III could be destroyed if there were no correlation between two alleles. Hence we have plotted the correlation between two alleles by showing two-dimensional pattern $(g^{(+)1}(i), g^{(-)1}(i))$ in Fig. 11.8. Initially there was no correlation, but

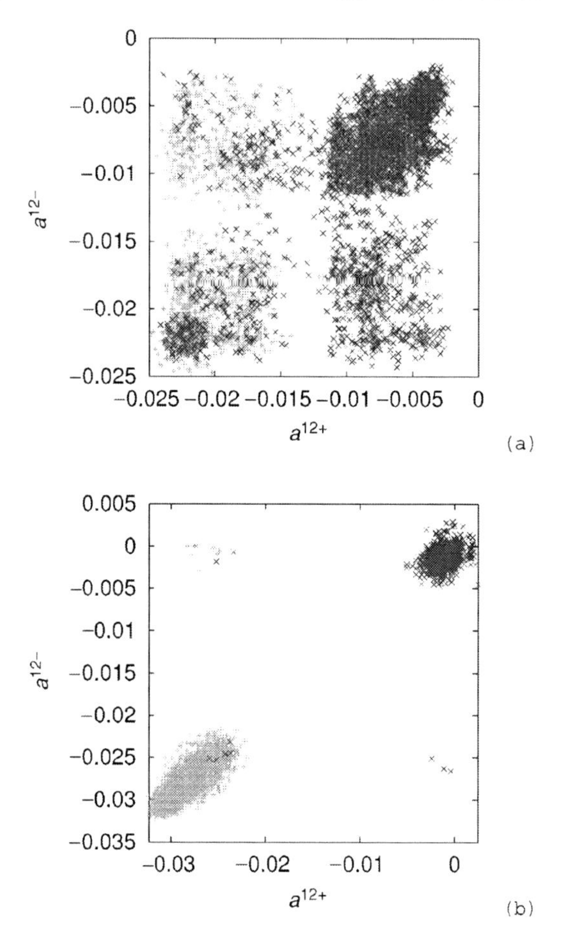


**Fig. 11.8.** Correlation in alleles. The point $(a^{(+)12}(i), a^{(-)12}(i))$ is plotted at every reproduction event, Two alternate groups with distinct phenotypes are plotted with alternate gray scales. (**a**) Plots at the division from 20,000th to 36,000th. Here phenotypic differentiation already occurred but the genetic separation is not completed. (**b**) Plots at the division at a much later stage (from 50,000th to 63,000th) where genetic separation already occurred. Adapted from Kaneko (2002c)

through temporal evolution, the correlation is established. In other words, the speciation in phenotype is consolidated to genes, and later is consolidated to the correlation between two alleles.

### 11.5.8 Allopatric Speciation as a Result of Sympatric Speciation

By extending our theory so far, we can show that spatial separation of two species is resulted from the sympatric speciation discussed here. To study this problem, we have extended our model by allocating to each organism a resident position in a two-dimensional space. Each organism can move around the space randomly but slowly, while resources leading to the competitive interaction diffuse throughout space much faster. If the two organisms that satisfy the maturation condition meet in the space (i.e., they are located within a given distance), then they mate each other to produce offspring.

In this model, we have confirmed that the sympatric speciation first occurs through stages I–III in Sect. 10.2. Later these two differentiated groups start to be spatially segregated, as shown in Fig. 11.9. Now sympatric speciation is shown to be consolidated to spatial segregation (Kaneko, 2002c).

The spatial segregation here is observed when the range of interaction is larger than the typical range of mating. For example, if mobility of resources causing competitive interaction is larger than the mobility of organisms, spatial segregation of sympatrically formed species results.

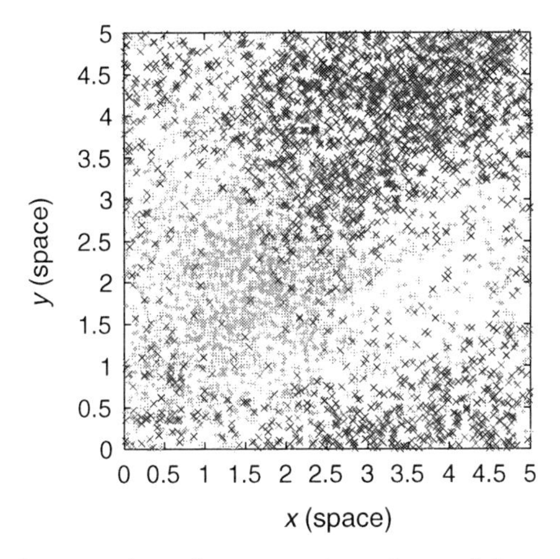


**Fig. 11.9.** Spatial separation of two species, observed in a numerical experiment. Model is with spatially local mating. Each organism moves the position with Brownian motion, given as the random number over $[-\delta_f, \delta_f] = [-0.0025, 0.0025]$, while it is located within a two-dimensional square of size $5 \times 5$, with periodic boundary condition. Mating is possible if two organisms satisfying maturing condition are located within the distance 0.25. Position of the organisms at every division is plotted, at each division event from 45,000 to 50,000. Two alternate groups with distinct pheno- and geno-types are plotted with alternate gray scale. Adapted from Kaneko (2002c)

Instead of spatially local mating process, one can assume slight gradient of environmental condition, for example, a gradient in resources. In this case again, it is expected that sympatric speciation takes place first, and later spatial separation occurs accordingly.

To sum up, we have pointed out here the possibility that speciation events considered to be allopatric when viewed only from the present pattern can be a result of the sympatric speciation of our mechanism. The sympatric speciation is later consolidated to spatial segregation of organisms.

## 11.6 Constructive Experiment

Discussion on the actual mechanism of evolution, however, often remains anyone's guess. Most importantly in our scenario, in contrast, is experimentally verifiable. For example, the evolution of *E. coli* is observed in the laboratory, as has been demonstrated by Kashiwagi et al. (1998, 2001) and Xu et al. (1996). Since the strength of interaction can be controlled by the resources and the population density, one can check whether or not the evolution at a genetic level is accelerated through interaction-induced phenotypic diversification.

Although the confirmation has not yet been completed, there are already some experiments supporting the importance of interaction to the genetic diversification (Kashiwagi et al., 2001). Here we will briefly discuss the results. We again use the bacteria *E. coli* for this experiment. They are cultured in liquid media, while mutations to genes for Glutamine synthetase are introduced by mutagenesis, at three times each separated by many generations. This enzyme is essential to nitrogen metabolic process and is essential to the survival of the bacteria in the medium we set in the experiment. In the mutagenesis, we amplify DNA by the PCR method and introduce random change in the DNA sequence of the gene that codes the enzyme Glutamine synthetase. By using these bacteria after mutagenesis, we get a population of bacteria whose genes are almost identical except the genes for the Glutamine synthetase. With this experiment, we can study the change of bacterial population distribution with different genes for the enzyme.

In the liquid media, populations of the bacteria are continuously cultured, where some nutrition chemicals are supplied at constant rate while the liquid including the bacteria is poured out with the same rate. After some time, the culture reaches a steady state so that the environment and the total population of bacteria are constant (with some fluctuations). In this culture, Glutamine acid is included as an important nutrient, which the bacteria have to transform into the glutamine that is essential for survival, with the aid of the enzyme Glutamine synthetase. Depending on the catalytic activity of the enzyme, the efficiency for the Glutamine synthesis differs, and hence the growth speeds also differ. Since there are mutations to the genes for this enzyme, bacteria with different genes have different growth speed with regards to the glutamine synthesis.

After the cultivation, we extract out the bacteria and perform mutagenesis as mentioned above and repeat the cultivation. We have repeated this cycle three times. Now we found that within the medium, several types of bacteria coexist with different genes for the enzyme. Each type has a different enzyme activity. In Fig. 11.10 plotted are the population distribution of each type of bacteria. Recall that in this medium, as a single cell, the bacteria with the highest catalytic activity of glutamine is the fittest. This is indeed the wild-type W2. However, we have found here that W2 does not dominate in the population but other types with much lower catalytic activity coexist, as shown in Fig. 11.10. Furthermore, the population distribution in each type changes over the generations. This suggests that the growth speed of each type is not solely determined by its own, but depends strongly on the interaction with other types.

To confirm the importance of the interaction, Kashiwagi et al. have carried out two sets of experiments. First they changed the population density in the tank, which is controlled by changing the flow rate into and out of it. In

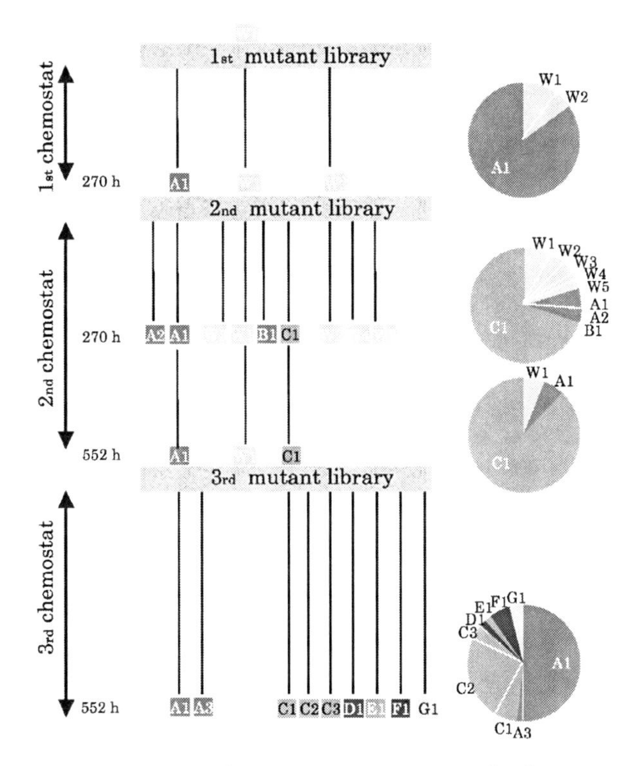


**There is a coexistence of some mutants at each chemostat culture even under a strong selection pressure.**

**Fig. 11.10.** Evolution experiment in a high density of bacteria, *E. coli.* After cultured over some time span, the cells are under mutagenesis, to produce the next stage. The three stages are repeated. The population density of different strains, that is, with different genotypes is shown, with corresponding photogenic tree measured from the genetic change. Based on Kashiwagi et al. (2001) under the courtesy of Kashwagi and Yomo

the above experiment, there are $10^{10}$ bacteria cells per 75 $m\ell$. As another experiment, they studied the case of much lower density $10^7$ bacteria cells per 75 $m\ell$. In Fig. 11.11, the fractions of the population of W2 type are plotted for three examples of $10^{10}$ cells per 75 *ml*, and also $10^7$ cells per 75 *ml*. In the former cases, the fraction of W2 cells is less than half, which fluctuates in time, but does not depend much on each experiment. On the other hand, in the latter case with lower density, only the W2 population exists and occupies the whole population. In a low density only the type with the highest fitness survives, while in a higher density the interaction allows for the coexistence of several types. The result suggests that the coexistence of diverse types is facilitated by the interaction. In the case with weak interaction due to low-density populations, only the fittest survives. On the other hand, in a high-density case the fitness is determined through interactions, and depends on the population distribution. Hence it is not possible to assign a single fittest type. To sum up, their experimental results confirm the viewpoint schematically displayed in Fig. 11.12.

The interaction here is mediated by some chemicals between cells. Hence as a candidate of important chemical, we choose glutamine. Indeed, when a variety of cells coexist, the concentration of glutamine in the medium is $\sim$1.7 μM, even though it is not introduced (note that supplied are glutamine acid, not glutamine). This suggests that the glutamine synthesized in some cells is diffused out, which influence all other cells. This interaction is nothing but we have introduced in the models of Chaps. 7–9 and 11. In these models, some chemicals that are synthesized within a cell diffuse out, through which the cells interact.

As a control of interaction, it is natural to cut off the interchange of this glutamine among cells. For this, we introduce an enzyme into the medium to transform glutamine back into glutamine acid that already exists in the medium. This is carried out by supplying Glutaminase into the medium (see

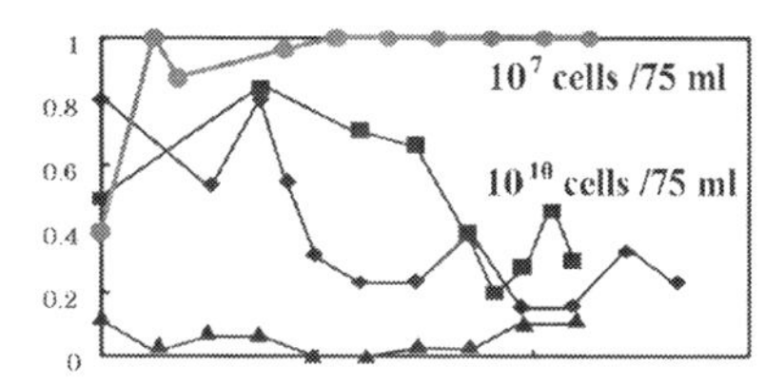


Crowded condition rendering sufficient interaction among the cells can cause fitness change and lead to the coexistence of closely related mutants.

**Fig. 11.11.** Changes in the population fraction of W2 strain, which is the "fittest" with regards to the Glutaminase catalytic activity. The *solid line* shows three sets of experiment by using the total population density of $10^{10}$ cells per 75 $m\ell$, showing coexistence of diverse cell strains. The *dotted line* shows a result from the density of $10^7$ cells per per 75 $m\ell$, where the population fraction of W2 strains reach 100%. Based on Kashiwagi et al. (2001) under the courtesy of Kashwagi and Yomo

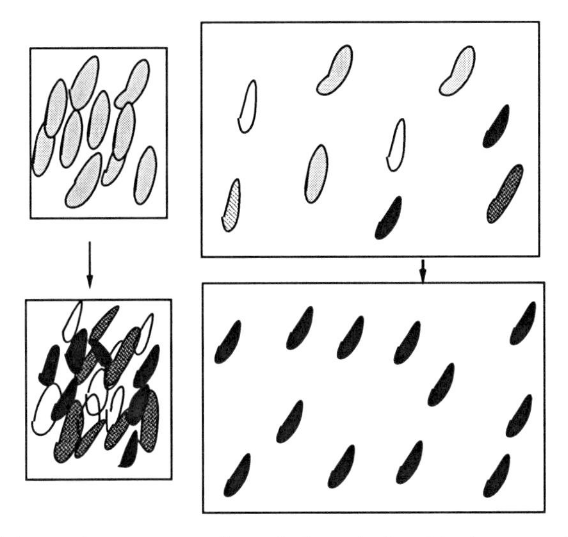

**Fig. 11.12.** Schematic representation of the relationship between the population density and diversity of cell types. When crowded diverse types of *E. coli* coexist, while in low density the fittest one is selected

Fig. 11.13). The temporal evolution of the population fraction of cell types other than W2 are plotted in Fig. 11.13. As a result of this "cutting-off-glutamine" experiment, only the "fittest" W2 type survives. This indicates that coexistence of a variety of cell types is mediated by the cell–cell interaction due to glutamine.

So far we have confirmed the relevance of interaction to diversification of cell types. Next, we discuss the phenotypic plasticity. A quantitative study with the use of flow cytometry is under progress, but here we will briefly mention change in the morphology of the colony of the bacteria. To avoid too much complexification, let us focus only on two types of strains, W1 and H. Here the strain H has a higher catalytic activity of Glutaminase, but these two coexist through the interaction.

In this case when the strain H is cultured in the agar medium, they show two different patterns, one "crumbled" and the other "flower-like." Furthermore, to confirm that these changes are not due to the genetic change, cells are taken from the flower-like colony, and cultured. Again two types of colonies are formed with some fraction. To sum up, the phenotypes (the colony shape) of H strain are differentiated, and these two phenotypes coexist.

So far, we have not completely clarified the relationship between this phenotypic plasticity and the genetic diversification into different types. Still, through the long-term experiment to change the concentration of the H strain cells, it is suggested that together with a drastic change of population ratio of H to W1 strain cells, the fractions of two colonies of H strains are also changed. Still, we have not yet confirmed which is first, but at least the change in pheno-

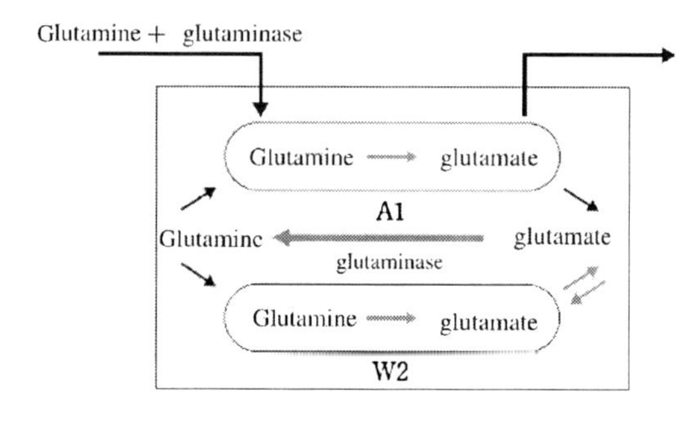


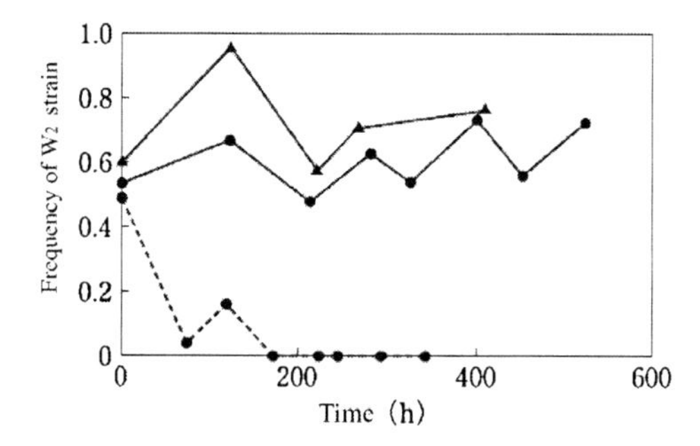


**Fig. 11.13.** Result of an experiment to cut off the interaction by adding Glutaminase in the medium. Schematic representation (above), and the population fraction of cell types other than W2 strains (below), where the data from the cutoff experiment are plotted as the *dotted line*. By cutting off the interaction the fraction of bacteria other than W2 vanishes, and the coexistence of different types is no longer possible

typic plasticity and the genetic diversification at a population level are highly correlated.

## 11.7 Relevance to Biology

The conclusion in this chapter is that sympatric speciation can generally occur under strong interaction if the condition for interaction-induced phenotype differentiation is satisfied. We briefly discuss relevance of our theory to biological evolution.

### 11.7.1 Tempo in the Evolution

Since the present speciation is triggered by interaction, the process is not so much random but rather deterministic. Once the interaction among individuals brings about phenotypic diversification, speciation always proceeds directionally without waiting for a rare, specific mutation. The evolution in our scenario has a "deterministic" nature and a fast tempo for speciation, which is different from a typical 'stochastic' view of mutation-driven evolution.

Some of the phenotypic explosions in the history of evolution have been recorded as having occurred within short geologic periods. Following these observations, punctuated equilibrium was proposed (Gould & Eldegridge, 1977). Our speciation scenario possibly gives an interpretation of this punctuated equilibrium. It may have followed the deterministic and fast way of interaction-induced speciation.

### 11.7.2 Decrease in the Phenotypic Plasticity

In the process of speciation we outlined, the potentiality of a single gene to produce several phenotypes that existed initially is consumed and decline. After the phenotypic diversification of a single genotype, each gene through mutation takes only one of the diversified phenotypes in the population. Thus, the original one-to-many correspondence between the genotype and phenotypes is consumed. Through the present process of speciation, the potentiality of single genotypes to produce various phenotypes decreases unless the new genotypes introduce another positive feedback process to amplify the small difference.

As a result, one may see single genotypes expressing only one (or a small number of) phenotypes in nature. Since most organisms at the present time have gone through several speciation processes, they may have reduced their potentiality to produce various phenotypes. According to our theory, if the organisms have a high potentiality, they will undergo a speciation process before long and the potentiality will decrease. In other words, natural organisms tend to lose the potentiality to produce various phenotypes in the course of evolution. As a reflection on the evolutionary decline of the potentiality, one can expect that mutant genes tend to have a higher potentiality than the wild-type gene. As mentioned in Sect. 10.1, the low or incomplete penetrance (Opitz, 1981), i.e., higher potentiality to produce diverse phenotypes from clones, is known to often occur in mutants, compared with higher penetrance (i.e., lower variety in phenotypes) in a wild type. Our result is consistent with these observations, since wild types are in most cases a consequence of evolution, and so in them the one-to-one correspondence is recovered. However, in mutants, there is a greater potential to have phenotypic plasticity.

### 11.7.3 Relevance of Developmental Plasticity to Speciation

The relationship between development and evolution has been discussed extensively. Our theory states the relevance of developmental plasticity to speciation. Taking our results and experimental facts into account, one can predict that organisms emerging as a new species have a high potential to produce a variety of phenotypes. It is interesting to discuss why insects, for example, have a higher potential to speciation from this viewpoint. Also examining if living fossils, such as *Latimeria chalumnae*, *Limulus*, and so forth, have a stable expression of a small number of phenotypes.

Note that through the speciation process discussed here, plasticity in phenotypes declines as mentioned. Through the speciation process, the original plasticity is "consumed" so that a single phenotype is generated from each genotype. In this sense, the process agrees with the hypothesis in Chap. 3 stating that in a closed system the plasticity decreases. Then, how does the speciation process continue on the Earth?

Here we should consider the meaning of "closed system." Indeed, there should be some occasions when the potentiality is regained, so that the evolution continues. For example, changes in the environment may influence the developmental dynamics to regain loose correspondence, or introduction of novel degrees of freedom or genes may provide such looseness. If, by external environmental change, the population distribution is drastically changed, some interactions are added or eliminated. If by geographical change (say by the collision of continents) species that were not interacting before start to meet, a new interaction term is added.

It should also be noted that developmental processes are often sensitive to environmental conditions, as recent studies on environmental endocrine disruptors have suggested. The developmental plasticity, once lost by the completion of speciation, may be regained by some change in the environmental condition or by interactions with other organisms.

Another source in regaining the plasticity is endosymbiosis. In endosymbiosis, two organisms that were independent join and have tight interaction. In terms of dynamical systems, some terms or some variables in the equations are added or eliminated. In these cases, the recovery of plasticity may be possible. We will briefly discuss an example in Chap. 12.

#### Remark: Phenotypic Plasticity

Note that the so-called "phenotypic plasticity" usually means such property that a single genotype produces **alternative phenotypes** in **alternative environments** (Callahan et al., 1987; Spitze & Sadler, 1996; Weinig, 2000). In contrast, in our phenotypic differentiation, distinct phenotypes from a single genotype are formed **under the same environment**. In fact, in our model, this phenotypic differentiation is necessary to show the later genetic differentiation. Without this differentiation, even if distinct phenotypes appear for different environmental conditions as in the case of "phenotypic plasticity," genetic differentiation does not follow. In spite of this difference, it is true that both are concerned with flexibility in phenotypes. Some of phenotypic plasticity studied so far may bring about developmental flexibility of ours, by slightly changing the environmental condition.

### 11.7.4 Unified Theory for Speciation in Sexual and Asexual (and Unicellular) Organisms

One important point in our theory is that the speciation in asexual and sexual organisms are explained within the same theory. Of course, the standard definition of species using hybrid sterility is applied only for sexual organisms. However, it is true that the asexual organisms, or even bacteria, exhibit discrete geno- and pheno-types. It is suggested that "species," that is, discrete types with reproductive isolation, may exist in asexual organisms (Roberts & Cohan, 1995; Holman, 1987). There are also discussions that the potentiality of speciation in asexual organisms is not lower than the sexual organisms. In this sense, the present theory sheds a new light to the problem of speciation in asexual organisms as well.

In relation to this, it should be more natural to define "species," without assuming sexual reproduction. Our proposition is as follows: Consider a state space (see Chap. 3) for phenotype and genotype. Each organism is represented as a point in this space. When a huge variety of organisms are plotted in the space, all of them are not continuously distributed, but the points are concentrated onto several clusters. The points form clusters ("clouds") that are separated from each other. When successive descendants of an organism in a given cluster continue to belong in the same cluster, this cluster continues to exist, separated from others, through temporal evolution. Then this cluster is called "species," which naturally satisfies reproductive isolation.

### 11.7.5 Adaptive Radiation

When a new environment is opened, a successive speciation process often follows – this is called adaptive radiation. By choosing a model with many cyclic processes, we can observe successive speciation into several groups from a single genotype. With the increase of population, the phenotypes first split into two groups, each of which is specialized in some processes. They survive depending on each other. With evolution, they form distinct genetic groups. With the further increase of population, new instability appears resulting in further separation into more groups from (each) group, which is later fixed to genes. This process can continue successively. Accordingly we can study adaptive radiation on the basis of our theory.

### 11.7.6 On the Interaction

Our theory for speciation crucially depends on the interaction. Indeed there is an established theory, called "frequency-dependent selections" which considers the interaction. In the frequency-dependent selection, fitness is modulated by the population distribution of other species. As genetically (and accordingly phenotypically) different groups interact with each other, the fitness changes according to the population of each group (Futuyma, 1986). At the third stage of our theory, the condition for this frequency-dependent selection is satisfied, and the evolution progresses with the frequency-dependent selection. However, the important point in our theory lies in the earlier stages where a single genotype leads to different phenotypes. Indeed this intrinsic nature of differentiation is the reason why the speciation process here works at any (small) mutation rate and also under sexual recombination, without any other ad hoc assumptions.

Another theory that takes advantage of interaction is resource competition. There, coexistence of two (or more) species after the completion of the speciation is discussed (Tilman, 1976, 1981). This theory gives an explanation how two species are specialized in their own niche, by separating out their resource use. Still, the speciation process itself is not discussed, because two individuals with a slight genetic difference can have only a slight difference there, in contrast to our theory. Rather, our theory provides a basis for resource competition also.

### 11.7.7 Allopatric Speciation

It is often believed that allopatric speciation is more common than sympatric speciation. However, we have to be careful in judging if the speciation is sympatric or allopatric, because in a biological system, we often tend to assume **causal relationship** between two factors, from the observation of just **correlation** of the two factors. For example, when the resident area of two species, which share a common ancestor species, is spatially separated, we often guess that the spatial separation is a cause for the speciation. Indeed, allopatric speciation is often adopted for the explanation of speciation in nature. However, in many cases, what we observed in field is just correlation between spatial separation and speciation. Which is the cause is not necessarily proved. Rather, spatial segregation can be a result of (sympatric) speciation. Consider, for example, the segregation of resident area in a city between rich and poor people. Most of us do not assume that people in the "rich area" are rich because they live there. Rather, in most cases, the spatial separation is a *result* of differentiation in wealth, but not a *cause*. In the same way, it is sometimes dangerous to assume allopatric speciation even if the residence of two species are separated.

Of course, geographical isolation by sudden change of environment may lead to allopatric speciation. Still, some of the data regarded as an evidence of allopatric speciation could be interpreted otherwise. It is possible to consider otherwise: After sympatric speciation has taken place and is established, the niche of the two species becomes different. Then the two groups may segregate in space according to the difference in the environment. After this process is completed, the two species are spatially separated, as we have discussed in Sect. 11.5.8. This might be regarded as a demonstration of allopatric speciation, but in this case sympatric mechanism is the trigger to this speciation.

### 11.7.8 Baldwin's Effect

In our theory phenotype change is later consolidated to genotype. Indeed, genetic "takeover" of phenotype change was also discussed as Baldwin's effect

(Baldwin, 1896; Bonner, 1980), where the displacement of phenotypic character is fixed to genes. Genetic assimilation of phenotype is then discussed by Waddingtom (1957) as mentioned in Chap. 10. The fixation to genotype at the second stage in our theory is understood as an example of this Baldwin's effect, while the phenotypic differentiation in the first stage is also essential to the speciation mechanism proposed here.

Here, we have demonstrated the relevance of the Baldwin's effect to speciation explicitly by considering the developmental process, using a model with (metabolic) reaction dynamics. The phenotype differentiation is formed through the developmental process to generate different characters due to the interaction. Distinct characters are stabilized by each other through the interaction. With this interaction dependence, the two groups are necessary for each other, and robust speciation progresses. The stabilization of development by genetic assimilation is also discussed by Newman (1994).

The Baldwin's effect was often discussed in the context of transmission of a novel type of behavior. Such behavior is sometimes related with the morphology of an organism. For example, consider sympatric speciation of *cichlid* in African lakes. It is known that two species coexist that eat the gills of other fish from the left or right side, respectively. This difference in behavior is also reflected in the morphology of the mouth of the fish. The jaw of each species is curved to the right or to the left, respectively, so that it is easier for them to attack other fish from one side. In other words, distinct types of behavior and developmental process are tightly coupled. Our theory suggests that developmental plasticity induced by the interaction with other fish first leads to differentiation of phenotypes into "right-side" and "left-side" types, with regards to both behavior and development. Later this difference is assimilated to genes, as shown by the second stage of our theory that is given by Baldwin's effect. We suggest that the speciation into these two types thus progressed rapidly in these lakes.

### 11.7.9 Reversing the Order

According to our theory, sympatric speciation under sexual reproduction starts first from phenotypic differentiation. Developmental plasticity induced by interaction leads to phenotypic differentiation, and then genetic diversification takes place. This leads to hybrid sterility, and later mating preference evolves. Further later, this differentiation can be fixed to correlation in alleles or to spatial segregation.

This order may be different from most commonly adopted studies (while the order from phenotype to genotype was discussed as Baldwin's effect, as mentioned). Hence, our theory will be verified by confirming this chronic order in the field. One difficulty here, however, lies in that the process from phenotypic differentiation to the last stage is rather fast according to the theory, and it may not be so easy to observe the intermediate process in the field. Still, there may be some hope for the observation, by first searching for phenotypic differentiation of organisms with identical genotype under the identical environment. In this respect, the *cichlid* data of Lake Nicaraguan may be promising (Wilson et al., 2000), since phenotypic differentiation corresponding to different ecological niche is observed even though clear genetic difference is not observed yet.

# 12

# Conclusion

## 12.1 Summary

In this book, we have presented an understanding of living systems in which they are regarded not as finely tuned machines but, rather, as realizations of the universal structure characteristic of reproductive systems arising from complex dynamical systems. From this point of view, we have proposed a theoretical framework to answer questions like the following. (1) In a cell, there exist many chemical components that reproduce through mutually catalytic processes. Given this kind of reaction network, why is it the case that the component responsible for carrying genetic information is separated from the rest? Also, how do the genotype and phenotype come to diverge, and how is their relation formed? (2) Why are such catalytic reaction networks possessing so many variables able to reproduce recursively on the one hand, while keeping some variability that makes evolution possible on the other? What statistical properties must be possessed by the state of the constituent chemical components to realize such behavior? (3) From interactions among cells, there arises differentiation into distinct cell types. How, in the presence of large fluctuations of chemical concentrations, can this generate a stable developmental process? Furthermore, how does each cell contained within some group "know" the properties of the group? (4) In the developmental process, the omnipotence of the original embryonic stem cells is gradually lost, as they differentiate into stem cells and then into committed cells. How does this irreversibility arise? In general, the plasticity of cell states decreases through this irreversible differentiation process. How is this plasticity characterized? (5) During the process of cell differentiation, how do cells create and read positional information and with this carry out a stable process of pattern formation? (6) How is the plasticity of phenotypes related with evolution, that is given by genetic changes? Does the differentiation in phenotype caused by interactions among individual organisms bring about a divergence of genotype and lead to speciation?

In Chaps. 4–11, we have provided answers to these questions from the results of both theoretical and experimental studies. These answers are given not as a result of fine control in a biological system, but as the universal properties of dynamical systems possessing the capacity of reproduction. As we have seen, these properties arise (to a certain extent) in appropriately constructed experimental systems also. We do not propose to understand life in terms of the "rigid," precisely determined behavior of logical systems. Rather, the picture we present is based on complex, "loose" dynamics from which recursive "types" are formed. These types then come to be fixed in the relatively rigid system represented by the genes. From this point of view, we have treated the stability of the developmental process, the (ir)reversibility of differentiation, and the possibility for evolution.

Allow us here to summarize the understanding we have realized into the following four concepts, which lie at its heart.

### 12.1.1 Isologous Diversification – General Tendency Toward Differentiation from Identical Units Through Interaction

The phenomenon in which two originally identical states become distinct through the amplification of small fluctuations is known in physics as "symmetry breaking." A state is said to possess *symmetry* if there exists a transformation of the system under which this state does not change. Such a transformation could be, for example, the exchange of left and right, the spatial translation of the system as a whole, etc. Symmetry breaking is the loss of a symmetry property that a state once possessed. Note that the Turing theory discussed in Chap. 9 represents one type of symmetry breaking. In that case, an initially spatially uniform state (i.e., a state that is invariant under spatial translation) becomes nonuniform when a periodic pattern appears. As a result of this change, the system loses its symmetry with respect to arbitrary spatial translations, and now possesses symmetry with respect to only the special translations that shift the pattern by integer multiples of its wavelength.

Now, let us consider cell differentiation. Here, initially, all cells are the same. Then, as the cells increase in number, they begin to differentiate, and cells possessing different internal states appear. The process of differentiation thus can be interpreted as a type of symmetry breaking. This symmetry breaking results from the dynamics of the internal states of the cells and the interactions among cells. In this process, the states of the cells that were once all the same are caused to diverge, as small fluctuations become amplified. Then, through interactions among cells, the respective states consisting of these now diverse individuals are stabilized, and the differentiated state is thereby maintained.

In general, the systems we consider consist of units (e.g., individual cells) characterized by internal dynamics (e.g., catalytic reaction systems) that are nonlinear in nature. Then, because of these internal dynamics and the interactions among units, the state of the system in which all cells are in the same state becomes unstable. As a result, the units undergo differentiation and come to exist in a variety of states.

One of the important results of our theory is the stability of the process of development. Given a system with some distribution of cell types, we have found that if the state of just one cell changes from the original, then, because this new state is not stable, this cell will be drawn back to the original state. This can be understood from consideration of the stability of states in terms of attractors, as discussed in Chap. 3. Consider the situation in which we fix the states of all but one cell. Then, the original state of the one "test" cell corresponds to the attractor determined by the fixed cells. In this situation, if the state of this cell is perturbed slightly, it will be pulled back to this attractor. Also, the stability of the system can survive such a "macroscopic" perturbation as the removal of many cells. The reason for this is that the distributions of the states and types of the cells are interrelated, and if the proportion of each cell type is not maintained within a certain range, the stabilizing mechanism mentioned above ceases to exist, and the system as a whole becomes unstable. For this reason, the states of some of the cells will begin to change, and eventually, the system will recover a distribution of cell types that is close to the original distribution. The stabilization mechanism acting in this case thus becomes clear.

In the cell differentiation model presented in Chaps. 7–9, the destabilization that results from the increase in cell number is stabilized through differentiation. However, in general, instability is not triggered only by the increase of the number of cells. It can also result from changes in the environmental conditions, for example, because the strengths of interactions depend on these conditions. Actually, it is known that if some bacteria (such as *E. coli*) are cultivated in an environment lacking nutrients, their cellular states split into two types. (This can be observed by using a cell sorter to separate cells according to the quantity of a certain protein that they contain. When this is done, the distribution of this quantity is found to exhibit a two-peak form.) One example of such an experiment is discussed in Chap. 7. If this type of differentiation becomes genetically fixed, it represents a step in evolution. This is an example of a general type of phenomenon in which effects at different levels are mutually strengthened and thereby fixed. Below we discuss this type of phenomenon in more detail.

### 12.1.2 Dynamic Consolidation – a Process Through which Plastic Differentiation is Fixed at a Separate Level with the Aid of Dynamic Interference Among Processes

The general idea of the differentiation discussed above is that a system of originally identical cells comes to possess a variety of cell types through the amplification of fluctuations, and this differentiated state is stabilized by the interactions among cells. We now discuss how differentiation at one level is transferred to a different level. Indeed, in living systems, it is often the case that a differentiation at one level leads to a differentiation at another level, and, through the mutual strengthening of these effects at different levels, they become fixed. We refer to this phenomenon as "dynamic consolidation."

As an example of dynamic consolidation, in Chap. 9, we saw how the processes of cellular differentiation and pattern formation can have a mutually strengthening interaction. There, the differentiation of cell states causes the appearance of a spatial pattern and, then, this pattern promotes differentiation. This is an example of positive reciprocal feedback through which both differentiation and pattern formation become stabilized. In this way, each cell acquires positional information, and this allows for stable pattern formation. Indeed, a pattern formed in this manner is able to restore itself when subjected to many kinds of disturbances, including the removal of multiple cells.

As another example, we discussed in Chap. 11 how the differentiation of phenotype can be transferred into a differentiation in genotype, and how this process thus becomes stabilized in the form of new species. In the first stage of this process, with the differentiation of phenotype, the type of the descendants diverges somewhat from the original. Then, when this change becomes fixed in the genotype, and the descendants thus come to possess the same genotypes and phenotypes, the original differentiation becomes strengthened. Although the change appears first in the phenotype, when the genotype also differentiates, the differences among phenotypes become amplified. Thus, through this positive feedback, the difference in phenotype causes a difference in genotype, which furthers the difference in phenotype. In addition, the differentiation in genotype leads to a change in mating preference, and mating thereby becomes limited to pairs of the same type. In this way, the difference between the two (or more) types is amplified, and they become stabilized as distinct species. Then, in the case that the individual organisms are motile, there is a tendency for them to become spatially separated into species-specific groups. This further strengthens their distinct existences.

This type of behavior, in which the divergence of types on one level spreads to another level, was also considered in Chap. 4, where we discussed the phenomenon of the division of roles of molecules in a catalytic reaction network. In this case, the degrees of enzymatic action among the molecule species and, accordingly, their replication rates vary. These differences then come to represent the differences among molecule species with respect to the degrees to which they control cell properties and the degrees to which their concentrations are preserved in cell reproduction. Although no such model calculation has yet been carried out, we conjecture that the evolutionary process that follows this kind of differentiation also involves some kind of positive feedback mechanism, as a clear division of roles among the elements in charge of genetic information and those in charge of metabolic processes emerges (see Sect. 6 of Chap. 4).

With the above examples, we have elucidated the role of dynamic consolidation in the origin of life, the process of development and the process of speciation.

### 12.1.3 Itinerancy – Long-Term Change of Several Quasi-Steady States with Self-Organized Transition Rules

The different types that emerge from differentiation through isologous diversification represent stable states. By contrast, from a plastic state, several types of states will be differentiated. In the case of a plastic state, the system will make repeated transitions, approaching one metastable state and then moving away, then approaching another metastable state, and moving away again.... This kind of dynamic process, in which a system jumps among metastable states is known as "chaotic itinerancy." If we consider such a system on a long-time scale, we can regard this behavior of jumping among states itself as constituting a single global state corresponding to a single attractor. (Of course, this would be an attractor in an extremely high-dimensional state space.) However, here instead of representing this entire temporal trajectory of the system as a single attractor, we focus on the transitions among the "lower level" states that this trajectory visits. There are many variables involved in these transitions, and for this reason they appear to be somewhat erratic, but they are not completely random. For example, it is not the case that a transition can happen from any state to any other state. Indeed, there exist rules governing the probabilities for transitions among the various states, and given that the system is in a particular state, there are some states to which it is likely to make a transition and some states to which it is very unlikely to make a transition. In addition, these transitions depend on the internal states of the individual elements, and thus the actual transitions that occur depend on the circumstances under which the system exists. We discussed an example of this type of behavior in Chap. 8. There, in a model of differentiation from stem cells, while in the original plastic state, the system exhibits chaotic itinerancy globally, over the entire state space. Then, as the development process progresses, the system becomes fixed at one state. In this way, the chaotic itinerancy ceases, and a differentiated state is realized.

In the mutually catalytic reaction network presented in Chap. 6, the system makes transitions among a number of states exhibiting recursive production. These dynamics allow for the evolution of so-called "protocells." In this situation, after the system remains for some time in a state in which recursive production is carried out, the original combination of cellular chemical compositions becomes unstable, and as a result, the number of chemical species present decreases. Following this decrease in the number of chemical species existing within a cell, there is great variation among the states of cells appearing in successive generations, until there is realized a new cell state in which, again, nearly recursive production is carried out in a stable manner, with the variety of chemical species maintained. In this sense, the process consisting of the decrease in the the number of effective degrees of freedom (number of variables) and the resulting destabilization of the system and subsequent transition can also be understood in terms of the chaotic itinerancy considered in Chap. 3.

In the transition processes that appear in living systems, it is often the case that unless the system passes through one particular state it cannot proceed to the next one, and in this way the range of the states that can be visited becomes irreversibly limited. The existence of such behavior may explain why the changes undergone by cellular states are often described as chains of *if–then* relations. Although the internal changes experienced by a cell are realized as the result of many chemical reactions and therefore, strictly speaking, cannot be expressed in terms of logical operations, the fact that they sometimes do behave approximately like *if–then* systems may be attributable to the existence of such underlying itinerant dynamics.

### 12.1.4 Minority Control – the Tendency for Replicators with Minority Populations to Control the Behavior of the System

There is a fundamentally important question of why within the cell only a few special molecules are responsible for carrying the so-called genetic information. (This is related to the topic of Sect. 12.1.2.) With the hope of answering this question, in Chaps. 4 and 5, we investigated reaction networks modeling systems of multiple mutually catalytic replicating molecules. There, because the numbers of individual molecules of the various species within an actual cell will not necessarily be large, we represented these by discrete (0, 1, 2 . . . ) rather then continuous variables. Doing this, we found that in this type of system, some species for which there are just a few molecules take on the important role of controlling cellular reproduction. If an element exhibits reproduction and selection, and the constituent species composing its internal states are represented by discrete variables, as is the case for cells and the chemicals they contain, the selected states differ from those in the case that the species are represented as continuous-valued variables. This behavior can be understood as follows. In a system consisting of a number of constituent species that all are essential in the synthesis of the others, if one species were to disappear completely, the system as a whole would be affected strongly. For this reason, those species whose numbers are small play a relatively important role in determining the behavior of the system. In fact, we may consider this to be the reason that genotype and phenotype are separate characterizations.

We believe that the situation in which control is exercised by constituent elements that are minority in the population is a general characteristic of reproductive systems. In particular, as we saw in Chap. 6, in the itinerancy discussed above, it is often the case that constituent species with minority populations play important roles in opening the narrow path followed by the system trajectory. Then, because the behavior comes to be determined by a sort of *on-off* condition – that is, whether molecules of these species are present or not – the system can appear to be functioning according to logical operations. The apparent control of cell states by the *on* and *off* of genes may be related to this mechanism of minority control.

As discussed in Sect. 12.1.1, cell differentiation occurs when a state becomes unstable. This change is initiated by minority molecular species. Also, considering this phenomenon of minority control on a more macroscopic level, it may be possible to understand the role of germ line segregation in multicellular organisms. In the reproduction of multicellular organisms, none of the large number of somatic cells are passed on to the next generation. Instead, in the reproductive process, only a small number of germ cells are passed on to the next generation, and these cells control the entire process. Indeed, with this minority control by germ cells, the capacity for evolution is maintained, as in the experiment discussed in Sect. 4.4. It is necessary to study this point both theoretically and with constructive experiments.

### 12.1.5 The Universal Properties of Reproductive Systems

The four concepts summarized above are fundamental in the understanding of living systems presented in this book. The other fundamental concept is that of a system composed of reproductive elements possessing internal states. One very important characteristic property of living systems is the capacity for reproduction at multiple levels. Within a cell, there is replication of proteins and DNA. At a higher level, cells, and again at a higher level, individual organisms also possess the ability to reproduce. To this time, the dynamics of systems of elements possessing internal states have been widely studied, with much attention given to the role of interactions among elements. In addition, the phenomenon of selection in reproductive systems has been investigated as a fundamental theory of evolution. However, in such studies, the point of view of studying reproductive systems with the goal of elucidating universal properties has not played a prominent role.

In Chaps. 5 and 6, we saw that the dynamics of systems exhibiting reproduction are intrinsically governed by equations of the form $dx/dt = ax$, and in such situations, the distribution of the concentrations of various chemical components generally takes a log normal form. The reason that distributions of this form appear is that the size of the fluctuations in the concentration of each component grows at a rate that is essentially proportional to the average of this concentration. In general, with fluctuations of this form, even if the distribution is not exactly log normal, the distribution of concentrations will not be symmetric (in particular, will not be Gaussian). Rather, it will be weighed to the side of larger population.

In replicating chemical reaction systems, distributions of concentrations usually take a log-normal form, and, in general, such systems display large fluctuations. However, in living systems, precise control may often be necessary. How is it possible to reconcile these two apparently opposing factors? Although we cannot yet give a definitive answer to this question, we believe that the phenomenon of minority control, discussed above, is universal in replicating systems, and thus it would appear that this is one source of the control that is exercised in these strongly fluctuating systems.

In the type of systems we now consider, the size of the fluctuations of a given constituent depends on the quantity of that constituent present. This implies that there is a tendency for the system to change in a particular direction. Also, in the case that there are many coupled variables, the randomness introduced by noise will be largely canceled out, and hence fluctuations will be suppressed. In addition, in a complex reaction network, the strengths of the fluctuations vary among the variables. In systems that include autocatalytic processes, treated in Sect. 5.4 and Chaps. 7 and 8, the strength of the fluctuations and the nature of the temporal variation differ among the variables. This type of state dependence of the time evolution and fluctuations of the variables leads to the itinerancy summarized in Sect. 12.1.3.

Systems of reproducing elements possessing internal states can, through interactions among elements, exhibit isologous diversification, as discussed in Sect. 12.1.1. In this way, differentiation and diversity of states emerge. As described in Chap. 7, in the process of isologous diversification, the type of instability peculiar to reproductive systems is removed through interactions, and a stable process of differentiation results. Thus isologous diversification is the foundation of the stability exhibited by living systems.

As discussed above, states of reproductive systems, in general, can exhibit large fluctuations and temporal variation. Indeed, each time a new element is produced, the internal state changes. In a system exhibiting such constantly changing states, if there happens to appear a state that changes only little in time, it will probably be easily maintained as it continues to carry out reproduction. In this type of system, transitions from very plastic states that exhibit large changes to less plastic states that are largely fixed occur quite generally. Dynamics of this kind are easily expressed in terms of the itinerancy discussed in Sect. 12.1.3, but if the relations among different processes become fixed, these dynamics will become those of dynamic consolidation, touched upon in Sect. 12.1.2.

From the point of view of physics and dynamical systems, in the last decade, the universal properties of systems of interacting elements with internal states have become clear (Kaneko & Tsuda, 2000). Analogously, formulating the universal laws governing "reproductive systems" will represent an important step in realizing an understanding of living systems.

## 12.2 Machine Versus Life Revisited

In Chap. 1, we emphasized the difference between machines and life. With the understanding obtained through consideration of results presented in this book, it is worthwhile returning to this point.

- *Fluctuations.* In our standard machine, we exercise a control to reduce the fluctuations in the output. A negative feedback mechanism is sometimes adopted for this purpose. In a living system with the potentiality for reproduction, however, a positive feedback process is unavoidable. The log-normal distribution discussed in Chap. 6 is a result of such a process, which leads to large fluctuations. Living systems generally function in the presence of such large fluctuations.

Of course, at some stage it is important to reduce fluctuations. In a living system, however, a process for reducing fluctuations itself will exhibit fluctuations. As there exists no "supervisor" that can reduce the fluctuations in a living system, if such a process of reduction exists, it must operate within the system. The minority control mechanism discussed in Chap. 4 may provide one solution, while parallel mutual feedback loops that appear in the intermingled hypercycle networks considered in Chap. 5 offers another possible mechanism for the reduction of fluctuations.

However, notwithstanding the possible existence of such fluctuation reduction mechanisms, fluctuations are unavoidable, and biological systems often take advantage of them. The cellular differentiation process discussed in Chaps. 7–9 is triggered by fluctuations, while it was shown there that coupled dynamical systems provide models of developmental process robust against fluctuations.

In Chap. 10, we studied the problem of evolution as another phenomenon in which fluctuations play a role. There, it was shown that the rate of evolution is proportional to (or strongly correlated with) the strength of phenotypic fluctuations. On the basis of an experiment performed on an embedded gene network described in Sect. 8.5 (Kashiwagi et al., 2005), we have recently shown that spontaneous adaptation (without the use of a signaling network) can be realized by taking advantage of the fluctuations in gene expression.

- *Strong interference among parts.* We usually design our standard machine in such a manner that interference among its parts is reduced. In a living system also, "modular" structure is often adopted to separate out one process from all other complicated processes. However, such separation is usually incomplete, and therefore interaction among processes and parts is almost always unavoidable. However, rather than posing a problem, this interaction is often useful, as biological systems often take advantage of such strong interactions. For example, in the developmental process discussed in Chaps. 7–9, intracellular reaction processes and cell–cell interactions yield contributions to the dynamics of cells that are of the same order, and therefore neither can be treated as a small perturbation. Indeed such an interplay between intracellular and intercellular dynamics leads to cell differentiation and robust development in the theory discussed in Chaps. 7–9. The speciation process described in Chap. 11 also requires such strong interactions among individuals.

  Although we have not discussed it in the present book, this type of interplay between intracellular and intercellular dynamics (i.e., the condition in which interactions within cells and among cells exercise influences of similar magnitudes on the cellular dynamics) is also important in studying molecular processes, as the molecules (i.e., proteins) are very densely crowded in a cell, as shown in Sect. 1.5, and therefore molecule–molecule interactions cannot be treated as perturbations of the intramolecular dynamics. (See Nakagawa & Kaneko [2004] and Kaneko [2005b] for examples of such inter/intramolecular dynamics.)

- *The interrelationship between rules and dynamics.* In a machine, a designer is outside of the system, while in a living system, there is no external designer. Of course, a "gene" is regarded as providing a blueprint for such a design, and it acts as a rule to control the behavior of a cell. On the other hand, a gene exists within a cell system, and therefore the manner in which it is expressed and the manner in which it was generated through evolution are related to the cellular state, that is, the phenotype. Indeed, even though there is information flow only from gene to phenotype, the phenotype dynamics constrain the evolution of the gene, as discussed in Chaps. 10 and 11. Indeed, the relationship between the intrinsic phenotypic fluctuations ($V_{ip}$), and the fluctuations due to genetic variance, ($V_g$), studied in Chap. 10, is a result of such an interrelationship. The inequality $V_{ip} \geq V_g$ is conjectured on the basis of evolutionary stability. In contrast, there is no such constraint in a machine, and thus in that case, $V_{ip}$ can be controlled in such a way that it is much smaller than $V_g$. In this sense, the rate $\frac{V_{ip}}{V_g}$ may serve as an index for biological plasticity, or deviation from a "machine-like" behavior.

## 12.3 Fluctuations, Response, and Stability

In this book, we have investigated what we call the "softness" or plasticity of cellular systems. In Chap. 3, we presented the hypothesis that this softness can be expressed in terms of the size of the fluctuations displayed by a system. We also consider the property of softness to be that which is responsible for the ease with which a system changes in response to an external operation. Thus, we believe there is a connection between the size of fluctuations and the response exhibited by the system. In Chap. 3, we presented the idea that the fluctuations in a system and the response coefficient corresponding to an externally applied operation are proportional, as a generalization of the Fluctuation–Dissipation Theorem of thermodynamics.[1]

The idea of a relation between fluctuations and response is not new in physics. This idea was originally formulated in Einstein's theory of Brownian motion [Einstein 1905, 1906]. In particular, for systems near equilibrium,[2]

[1] Although "fluctuation–dissipation" is the standard terminology in physics (Kubo et al., 1972), in its generalized form in the context of living systems, rather than "dissipation," the term "response" is more easily understood, and we use it for this reason.

[2] The states to which this statement applies are those for which the deviation from equilibrium can be expressed and treated linearly.

their proportionality has been clearly established. Here, as an extension of the concept of the equilibrium state in thermodynamics, we consider in a general context a stable system and the fluctuations about some values of the variables used to express the state of this system. We suppose that the distribution for each of these variables possesses a single peak. Now, we study the manner in which the system changes when the parameters controlling the state of the system (e.g., those representing the external environment, such as the concentrations of the various chemical species surrounding the system, those corresponding to the genes controlling the state of a cell, etc.) are changed slightly. Then, it is shown that the response (i.e., the changes in the peak positions of the distributions) is proportional to the strength of the fluctuations (i.e., the widths of the distributions). Mathematically, there is nothing surprising about this. What is important here is that we propose to describe a living system in terms of a **state**, $(x_1, x_2 \ldots)$, a distribution, $P(x_1, x_2 \ldots; a_1, a_2 \ldots)$, and **parameters**, $(a_1, a_2 \ldots)$. It is our fundamental hypothesis that such a point of view is effective for the purpose of understanding the phenomena exhibited by living systems. Here, we regard the application of a "generalized force" as the alteration of the parameter $a$, and we consider small such changes (i.e., those in the linear response regime). Here, we stress that the feasibility of such a treatment is a hypothesis, and the fluctuation–response relation in biological contexts should not be considered in strict mathematical terms. Rather, this is something whose rigorous formulation will require extensive experimental study.

In this approach, we can consider the "generalized force" applied to the system as corresponding to, for example, the change in the environmental conditions or the combined process of mutation and selection. With the premises stated here, the response coefficient and the size of the fluctuations are proportional.

In our treatment, care must be taken in the choice of the variables used to express the system. As mentioned in Chap. 6, in most reproduction systems, log-normal distributions are the fundamental form in which the system behavior appears. By contrast, in equilibrium statistical physics, Gaussian distributions appear most commonly, and indeed the fluctuation–response relationship is formulated assuming Gaussian distributions. Thus, in situations that the distributions of variables of interest take log-normal forms, as in the case of cell reproduction systems, in order to apply relationships of a fluctuation–response type, we need to employ the logarithms of the original variables.

Now, let us reconsider the softness of living systems treated in this book from the point of view of the relation between the response coefficient and fluctuations.

- *Fluctuation–response relations in the development process.* As seen in Chaps. 7 and 8, as the number of cells in a system increases and the process of development progresses, cell differentiation begins to take place.

In this process, the plasticity of the cells, that is, the ability they possess to change into different types, decreases. Now, note that from the point of view of a single cell, the change in the number of cells that occurs during development appears as a change in the environmental conditions. Thus, the change in the state of a cell during the development process can be considered a response to changes in the environment. With this manner of thinking, in consideration of the hypothesized fluctuation–response relation, it is natural to conjecture that there is a correlation between the change in the fluctuations in the state of a cell and that cell's plasticity. Indeed, the theory studied in Chaps. 7 and 8 predicts that as the embryonic stem cells change, eventually yielding committed cells, and the omnipotence of these original cells is lost, the size of the fluctuations exhibited by the cell states decreases. In other words, the plasticity and fluctuation strength are reduced together.[3]

It is vitally important to conduct experiments investigating the changes exhibited by fluctuations in cell states. These studies could examine, for example, (well-controlled) experimental systems of differentiating cells beginning from stem cells, or the changes undergone by the fluctuations in intestinal bacteria (*E. coli*) as they experience differentiation (see Chap. 6). Such systems would provide the opportunity to elucidate the nature of fluctuation–response relations in differentiation processes.

It is a fundamental concept in the theory of thermodynamics that fluctuations diverge at phase transitions of certain kinds (i.e., second-order phase transitions). From this point of view, then, it is natural to hypothesize that at the beginning of the differentiation process, in which stem cells produce cells of various types, the states of the stem cells exhibit large fluctuations. Again in physics, it is also known that in a first-order transition, for example, from a solid to a liquid or gas, the temperature can remain constant despite the input of heat, because of the latent heat of the transition. Extending this idea to the case of cellular differentiation, we can imagine the situation in which a cell does not immediately respond to an effect that would alter its state. For example, suppose a cell exists in some differentiated state A. Then, suppose that an external operation whose effect is to push the cell into some state B is applied. For example, such an operation could consist of altering the concentrations of the chemical species in the cell's environment. Then, in analogy to the first-order thermodynamic phase transition, it may be the case that the cell remains in the original state for some significant period of time before making the transition to state B. By carrying out experiments that investigate such behavior, it should be possible to make quantitative determinations of

[3] The theory discussed here does not provide predictions about the direction of the change but only that the plasticity and fluctuations change together. Separate considerations are necessary to demonstrate that these in fact decrease during the development process. Discussion of this point is given in the following section.

the ease or difficulty with which cell states can be changed. Experimental results of this kind would certainly be useful in the development of a phenomenological theory that can describe the phase transitions exhibited by cellular systems.

- *The fluctuation–response relation in evolution.* As discussed in Chap. 10, the phenomenon of selection in evolution can be regarded as a process in which the fittest phenotype with respect to some trait is chosen. With this interpretation, evolution can be thought of as a process that is pulled along in a particular direction. It is natural, then, to consider a fluctuation–response relation in the context of evolution and to consider evolution itself as a response to some externally applied operations. In this case, the response in question is that of the phenotype, and thus if there are fluctuations of the phenotype and these can be measured, the idea of a fluctuation–response relation can be applied.
  As discussed in Chap. 10, since living systems necessarily experience fluctuations, even for a fixed set of genes, there are fluctuations in the phenotype. With these considerations, we have investigated evolution in Chap. 10 while considering the role played by phenotypic fluctuations. The validity of the fluctuation–response relationship has been demonstrated in numerical simulations of a cell consisting of a reaction network, while the positive correlation of phenotypic fluctuations and the rate of evolution was also observed in an artificial selection experiment that employed bacteria to enhance fluorescence of the proteins within the investigated cells.
  Note that the relationship studied here regards the phenotypic fluctuations exhibited by clones, that is, individuals who possess identical genes. By contrast, most study in evolutionary genetics carried out to this time concerns phenotypic fluctuations that result from genetic variation. In Chap. 10, we proposed a relationship between these two types of phenotypic fluctuations, that is, those exhibited by clones and those resulting from genetic variations.
- *The fluctuation–response relation corresponding to operations.* Perhaps the most direct way of investigating the fluctuation–response relation is to make measurements of the change undergone by the state of a cell as its environmental conditions (e.g., the concentrations of certain chemical substances) are altered. For example, we could study the relation between the fluctuations in the cellular state of bacteria and the amount of change displayed by this state as the nutritional level of the environment is changed little by little.
- *Development and evolution from the operational point of view.* Elucidating the relation between evolution and development is one of the fundamental problems in biology, and it has been the subject of investigation for many years. Häckel asserted that ontogeny repeats phylogeny. In other words, each stage in the process of development of an individual corresponds to

a stage in the evolution of the species to which that individual belongs. The relation described by Häckel, however, is no longer considered a "law" in any sense, because of its overly broad and vague expression. However, there are probably very few biologists who would deny that there is some relationship between the processes of development and evolution. This "sense" of a connection, however, has not yet been successfully formulated theoretically.

Operations capable of bringing about evolution have been a main topic in the realm of biotechnology from ancient times, when humans began developing techniques of artificial selection to create domesticated plants and animals to their liking, to the present, in the context of evolutionary engineering. Also, today, operations that can affect development, for example, with regard to somatic clone technology and regenerative engineering, are of great interest. It may be the case that the relationship between development and evolution can be understood by considering the types of operations that can influence either.

The fluctuation–response relation enters the discussion here because the change in development and evolution caused by such operations are regarded as responses to external "forces" and thus are connected to fluctuations of some kind. Now, if the fluctuations relevant to the processes of development and evolution are not independent, their relationship would represent an interesting connection between development and evolution. Indeed, if the "responses" to operations affecting evolution and development possess proportionality relations with the *same* fluctuations, then, through these fluctuations, clearly there must be a relation between the two different types of operations. We believe that if experiments are carried out from this point of view, the relation between the processes of development and evolution and their relation to fluctuations can be clarified.

- *Problems to address.* Although we believe that such an approach will be effective, there are several fundamental problems concerning the application of the fluctuation – response relation to living systems. First, in normal thermodynamic contexts, the coefficient in the proportionality relation between the fluctuations and response is expressed in terms of a universal quantity, the temperature. However, for the situations in which we are interested, whether we are considering the states of a cell or an organism, the coefficient of proportionality in the fluctuation – response relation cannot be the thermodynamic temperature. Furthermore, even if we were able to construct fluctuation–response relations for a number of systems, it is quite possible that the proportionality coefficient would not be a universal quantity, but vary from case to case. If, on the other hand, we could identify coefficients that are each universal for at least a certain class of cells, then it may be possible to formulate a theoretical description of the "activity" of cells that experimentalists know on an intuitive level.

Next, we must recognize that the formulation of the fluctuation–response relation involves a number of conditions. First, this is a relation that applies to responses to "small" perturbations. Thus, in order for it to be of use in any given case, there must exist a realm of small yet meaningful perturbations of the system. It should be noted, however, that even for perturbations that are somewhat too large for a strict proportionality relation to hold, it will likely be the case that there remains a strong correlation between fluctuations and response, and therefore in this situation too, it should be possible to formulate some kind of relation between the two. Nevertheless, in any attempt to formulate for application to living systems a fluctuation – response relation that goes beyond the usual proportionality form, we must first treat the following points. (The discussion below is of a somewhat speculative nature, and the reader may wish to skip it.)

- First, in the usual situation, the fluctuation–response relation applies to a response to the change in a system parameter that results from an externally applied operation. However, in the cases of the evolution and development of organisms, the parameter change in question represents autonomous behavior of the system. For example, the change in cell number that occurs during the process of development is due to the reproduction carried out by the cells themselves. Thus, here, the change undergone by the parameter of interest is determined by the properties of the system of cells under consideration. In situations like this, there arise phenomena of reciprocal influences between the properties of the whole and the properties of the parts – something of a circulation of effects between different scales, which is characteristic of so-called complex systems. In Chaps. 7–9, we saw how through this circulation, stable states of individual cells and the properties of the cell aggregate are formed. To obtain an understanding of living systems, it is necessary to elucidate the relation between the fluctuations exhibited by the states of individual cells and the fluctuations in the distribution of cells in the structure to which they belong. For example, in the models studied in Chaps. 7–9, initially, each cell possesses a variety of components, and the fluctuations in their concentrations are large. Then, as the cells differentiate and their properties become more fixed, the fluctuations in the cell states are attenuated, but in the system as a whole, there comes to appear a distribution of various cell types. It is important to obtain a general theory with which this reciprocal relationship between the "parts" and the "whole" can be understood.
- *How do we separate the "state" from the "noise" (fluctuations)*? According to the equilibrium fluctuation-dissipation theorem in thermodynamics, fluctuations in the state of a system consist of unbiased noise, that is, thermal noise and molecular fluctuations. In this case, it is not difficult to clearly distinguish the state from the random fluctuations. However, in the study of life, the systems of interest are quite generally in nonequilibrium states, and for this reason, we must consider the validity of the

approximation we make when we separate the system into state variables and fluctuations.

First, if the dynamics of the system state themselves possess some directionality, what we wish to consider as noise are the deviations from this directional change. However, if we cannot make a clear distinction between the dynamics of the state and the fluctuations, it will no longer be possible to regard the fluctuations as random. Then, we will have situations in which, for example, when some state variable realizes a particular value, the noise will become more intense, or if we move in a specific direction in state space, the size of the fluctuations will increase, and in this way, fluctuations will come to possess directionality and selectivity. In general, as the number of state variables increases, it becomes more likely for this type of situation to be realized. Indeed, it has been proposed that chaotic itinerancy should be understood in terms of such behavior as a general phenomenon characterizing state spaces of many variables. In that interpretation, when the system is near one attractor, its state can be described by the dynamics of a small number of state variables and fluctuations about these dynamics. Then, as those fluctuations gradually increase in intensity, it eventually becomes necessary to use a larger number of variables in order to express the state of the system. Again, after some time, the system will settle into the neighborhood of another attractor, and it will again be possible to describe its state in terms of a (different) small set of variables and fluctuations. Considering this kind of behavior, it becomes clear that for application to living systems, we must formulate generalized fluctuation–response relations that can account for the role played by fluctuations in the selection of transitions between states.

- *How do we distinguish between variables and parameters?* In the case of physical phenomena, the concept of an externally applied operation and the response to it can usually be clearly defined. A typical example is an externally controlled magnetic field and the effect this introduces through a magnetic interaction. In the case of biological systems, of course, there are similar situations. For example, we can alter the concentrations of the chemical species in the environment of a cell and observe the resulting change in the cell state. However, in the cases of development and evolution, the situation would seem to be different. Here, the processes to which the systems respond – the increase in cell number and the selection of individuals according to the fitness – exist within the systems themselves. Thus here, to apply the idea of the system changing in response to an external operation, it is necessary to somehow take these processes outside the system, treating them as changes undergone by external parameters. For example, in the experiments discussed in Chap. 10 investigating the evolution of bacteria, it is necessary to regard the change undergone by the genes as a change in the parameters controlling the cell state. Implicit in this manner of description are the assumptions that the genes change much more slowly than the phenotypes of the cell states and, of course, that genes can in fact be treated as parameters that control the cell state. As discussed in Chaps. 10–11, this point of view and its assumptions underlie present-day evolutionary biology. To verify the validity of this approach, or to make progress toward a more effective approach, it is necessary to determine to what extent such a separation of state variables and control parameters is possible and what the limit on this extent implies about the nature of the fluctuation-response relation that can be applied to such a system.

- *Memory.* If the present behavior of a system depends on the states in which it existed to this time, then this history must be accounted for in the state variables used to describe it. For example, the behavior of a cell depends on whether or not at some time in its past it was kept in a culture lacking in nutritional components. In this case, the memory of the past experience would be reflected in the cell state. In the dynamical systems approach that we employ, the effects of such historical conditions must be accounted for by the values of the variables describing the present cell state. Now, if it were possible to introduce a set of variables that correspond to this system "memory," we could use the type of approach considered to this point, that is, that of treating a state and the fluctuations around it. However, if we did attempt to account for memory in this manner, it would be necessary to prepare a very large number of variables. In consideration of this fact, it seems that a more effective approach would be to employ a smaller number of state variables and regard everything that is not described by these as noise. Then, as time goes on and the system "acquires history," certain components of the fluctuations due to noise can be successively reinterpreted and treated as new variables. Obviously, the problems inherent in such an approach are closely related to those of distinguishing the state from the fluctuations.

## 12.4 The "Law" of Decreasing Plasticity in a Closed System

In the previous section, we discussed the representation of plasticity in terms of fluctuations and its relation to response. Now, it is an important concept that even in the absence of an external operation, the plasticity of a system can decrease in time. This is seen, for example, in the processes of development and evolution. Experimental and numerical results obtained from the models considered in this book have led us to conjecture the "law" that the plasticity of a closed system decreases in time (see also Chap. 3). Below we present some specific examples of such behavior.

- The decrease in plasticity of cell states during the process of development: As cells differentiate and become committed, they lose plasticity. According to theoretical models, this is realized as a decrease of the variation in

the chemical constituents of the cell (in other words, a decrease of the variation in gene expression) and transitions to states that are more resistant to change.
- In the evolution process discussed in Chap. 10, we considered a fixed condition of selection to increase some chemical concentration or the degree of fluorescence. In this case, as evolution progresses, the phenotypic fluctuations always decrease in successive generations. In this sense, the plasticity decreases through evolution resulting from a given fixed selection condition.
- Loss of phenotypic plasticity in the speciation process through fixing in genotype: According to the models of speciation considered in Chap. 11, in the case that there exist several possible phenotype states, as these become fixed in the genes, the phenotypes of the individuals come to be determined by the genotypes so defined. In this way, there is a loss of plasticity in the phenotypes.
- Cell models constructed to describe the origin of life suggest that within a system characterized by a diverse and unstable state there exists a mechanism through which the state becomes "consolidated" into one exhibiting recursive reproduction controlled by certain chemicals by certain minority chemical species.

Let us consider a system described by a fixed set of variables and its dynamics in the state space of these variables. Usually, if such a system happens to fall into a state characterized by small variations of the variables, it will tend to remain there for a long time. This can be interpreted as directionality in the time evolution of the system toward states of decreasing plasticity.

It is important to carry out theoretical and experimental studies through which we can formulate a quantitative expression of this "law of decreasing plasticity." In addition, it is necessary to elucidate the conditions needed to define a "closed system" characterized by an invariant state space.

## 12.5 The Restoration of Plasticity in an "Open" System

With only the loss of plasticity discussed above, a living system would, from its original "soft" state, gradually approach something resembling a logical machine. In fact, however, this is not what we always see in actual living systems. To resolve this discrepancy, it is necessary to study how and in what cases the plasticity of a system can be restored. In this book, to this point we have considered the following examples.

- *Interference among processes.* There are situations, for example in gastrulation, in which, because of their migration, the arrangement of cells is altered, and cells (or groups of cells) that were previously separated by large distances, and therefore had essentially no interaction, come into immediate contact and begin to interact strongly. Here, there is an interference between two processes representing different aspects of the system, one of a mechanical nature (cell migration) and one of a chemical nature (cell differentiation). Through this interference, a previously stable cellular state is changed, and as a result, a new process of differentiation begins.

- *Interference with a new set of variables.* In Chap. 10, we studied the selection process of a cell through which the fitness, represented by the concentration of some chemical, increases. With this evolutionary process, the plasticity of the phenotype decreases. In this evolutionary process, when we increase the mutation rate, an error catastrophe is encountered. As a result, the distribution that was sharply peaked at the state of highest fitness collapses, and a very broad distribution extending to individuals with very low fitness appears (see Sect. 10.5). Through this process, phenotypic fluctuations and, accordingly, plasticity increase.

  Note that the evolutionary process to increase some phenotypic variable is represented by the increase of a single (scalar) parameter, as discussed in Chap. 10. Recall, however, that a gene in the model there corresponds to a reaction network, which is originally represented by a very high-dimensional space. Only for a given selection process can it be approximately represented by a single parameter, say the Hamming distance from the fittest network. For this reason, when the collapse of the distribution occurs through error catastrophe, the original high dimensionality continues to have an effect on the system behavior, and consequently, the one-dimensional representation of evolution is not valid. Now, the change in network paths influences the phenotype (representing the fitness), and this leads to a restoration of the plasticity.
- *Restoration caused by an external operation.* In somatic clone experiments [Gurodn et al., 1975, Campbell et al., 1996], the restoration of plasticity takes the form of a recovery of the omnipotence of cells that was lost through differentiation. Such experiments have succeeded in restoring the plasticity of differentiated cells by subjecting the nuclei of cells to new conditions. In the field of embryological engineering, such operations capable of restoring plasticity have been sought. In this case, clearly, the operation causing this return to the original plasticity is applied externally. At this time, the question of what kinds of operations can lead to the restoration of plasticity has not been theoretically settled.
- *Symbiosis.* There are situations in which two different organisms exist in a state of strong mutual interaction, and for each, the existence of the other is necessary for its survival and reproduction. Such a state is referred to as "symbiosis."

  In some sense, symbiosis is a miraculous state, as, in general, combining distinct systems requires very delicate matching of functions. For example, consider the joining of two independent logical systems, say two independent programs written separately. If these programs have even a slight degree of complexity, then it would be almost impossible that simply connecting them would yield a combined program that runs without error. However, among biological systems, such joining of separate systems has occurred from time to time, during the course of evolution. For example, it is believed that endosymbiosis (Margulis, 1981) played an essential role in the formation of the eucaryotic cell. It is interesting then to consider how two independent biological systems (e.g., independent biochemical networks) can join to form a new composite system without damaging (or killing) each system.

If two such systems were rigid logical systems programmed by genes through some hard "if–then" rules, such fusion would be quite difficult to realize. However, according to the understanding presented in this book, biological systems are originally dynamically flexible, whereas they may become rigidly determined by genetics later. Also, there is the possibility that when two such rigidly fixed systems are joined under extremely harsh environmental conditions, their rigid genetic rules could be relaxed and they may regain their dynamical flexibility. Recall, for example, the situation regarding context-dependent rules of cell differentiation in Chap. 8. In particular, under certain unusual conditions, such as the removal of all cells of a given type, it is possible for the de-differentiation of determined cells to result and the generated differentiation rules to be lost.

In modeling this kind of situation, among the variables used to describe the internal state of one individual, it is necessary to include variables corresponding to the other (see Fig. 12.1). Thus, this symbiotic relation results in a state space of higher dimension. In such a situation, it is possible that a stable recursive production that would exist for these organisms if they existed independently is now unstable due to the interaction between them, and a new, intrinsically symbiotic state emerges. Indeed, from a theoretical point of view, it is generally the case that a system with many strongly interacting variables is easily destabilized.

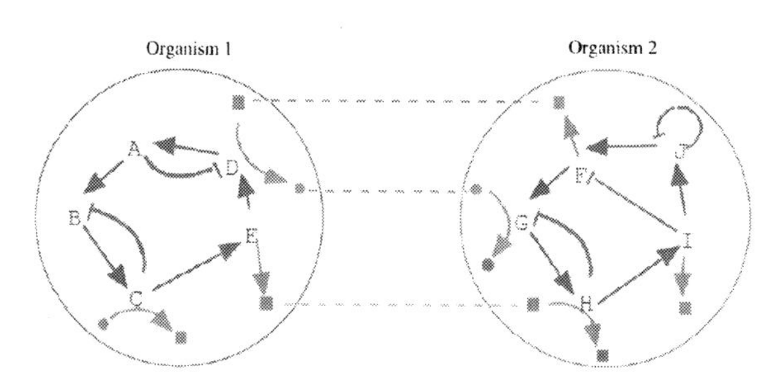


**Fig. 12.1.** Schematic representation of the interference between two organisms represented as a network. Two organisms with different reaction networks influence each other, resulting in the mutual change of their states

The first observation of such symbiosis in laboratory was made by Jeon (1972), somewhat accidentally. Recently, Todoriki et al. (2002a,b) have succeeded in constructing a symbiotic system, using amoeba (slime mold) and bacteria (*E. Coli*). Usually, the latter is eaten by the former, but by creating a situation in which the amoeba could not survive if they completely consumed the bacteria, the two species are found to form a symbiotic relationship. In this case, bacteria are "breeded" by amoeba. In the system studied there, the bacteria and amoeba began to interact strongly, and as a result, the phenotypes that each possessed previously become unstable. Indeed, their gene expressions changed drastically [Matsuyama et al. 2004], and this led to a novel composition of proteins and morphology. This combined system then made a transition to a new stable state that was preserved over a number of generations.

In the experiment described above, the bacteria and amoeba did not form a new individual with recursive production sharing the genes. However, it is believed that in the history of living organisms, there have been cases in which two previously distinct organisms have formed a symbiotic relationship through which they eventually began to share genes and finally merged into a single organism. In such a situation, we can imagine that, while the plasticity of the system is restored when the two individuals first form a symbiotic relationship, after the passage of many generations, the system evolves in a new state space, and through this process it again loses plasticity. This loss of plasticity can be understood as resulting from the "consolidation" of the phenotype into the genotype, as discussed in Chap. 11.

- Environmental change (change in organism population). In the process of the evolution of some species, it will generally be the case that, because of many factors, the environment in which the species exists is constantly changing. Here, that which we regard as the "environment" is not limited to such things as climate. Rather, this will include a much broader range of factors, including the types of other organisms living in the vicinity of the species in question, as well as their populations and distributions. Now, as the result of some change in the environment, we can imagine the states of some individuals of a given species will become unstable, and because of the restored plasticity that this instability represents, evolutionary change can occur. Indeed, we can think of the process of evolution as progressing in this manner.

As discussed in Chap. 10, it is believed that the phenotypic plasticity that emerges in this manner effects change in the genes, which is eventually fixed. In this way, the plasticity of the phenotype is gradually lost.

The restoration of plasticity resulting from environmental changes should be experimentally observable. In fact, this is the case for the experiments studying the differentiation of *E. coli* mentioned in Chap. 8. There, the initial conditions, in which the bacteria are placed in an environment of poor nutritional composition, are such that survival and reproduction become difficult. In those experiments, it was investigated how this change leads to an increase in the plasticity of the phenotype and affects gene expression.

Within the present context, perhaps the most significant environmental change is the variation in population of individuals. Obviously, such changes are important in the study of the evolution of ecosystems. However, they also play an important role in the developmental process of single organisms, where there exist mechanisms for both the increase (reproduction) and decrease (apoptosis) of cell number. Recently, Takagi et al. (2005), studying a model of cell differentiation similar to those discussed in Chaps. 7–9, discovered that at a certain stage in the development process, a large number of cells die spontaneously. They observed that as a result of this change, the states of the remaining cells become unstable and thereby regain plasticity. From these conditions, then, a new process of differentiation occurs, and new cell types appear. Such behavior is clearly related to the processes of metamorphosis exhibited by insects and other organisms, which can be interpreted in the same way: At some stage, the cell structures that existed previously become unstable and plasticity is (locally) restored, with the result that new kinds of cells begin to emerge.

## 12.6 A Theoretical Approach to Plasticity Dynamics

We now consider the question of how to construct a theoretical framework within which we can understand the types of plasticity dynamics discussed above. Of course, here it is necessary to take account of the role of fluctuations, but below we discuss some new points of interest from a dynamical systems point of view. (The treatment given below is intended only for specialists.)

- *Chaotic itinerancy.* As discussed above, itinerant phenomena in which a system makes transitions between stable states that are described by a relatively small number of variables can be interpreted as transition processes characterized by high plasticity between steady states characterized by low plasticity. With this understanding, we see that it is important here to more closely study the nature of chaotic itinerancy.
- *Number of variables (i.e., degrees of freedom).* First, let us address the question of the number of variables allowing for stable states. Of course, this number depends on the relationships among the variables and the strengths of their couplings. (In the case of reaction networks, this is determined by several factors, including the number of reaction paths corresponding to each component chemical, whether or not the structure of the coupling network is random, and the reaction rates.)[4] It is important

[4] In a model describing the variation in populations of an ecosystem, Robert May found that as the number of species increases, it becomes more difficult to realize the coexistence of species as stable fixed points, representing constant popula-

to treat this problem of variable number within the context of some appropriate set of conditions.

Recently, it has been found (Kaneko, 2002d) in chaotic systems in which all elements interact strongly (globally coupled chaotic systems (Kaneko, 1990)) that if the number of variables is at least on the order of 5–10, the system will tend to exhibit chaotic itinerancy. For systems with smaller numbers of variables, it is usually the case that there exist in state space a number of well-separated attractors representing stable states. Then, as the number of variables increases, attractors and the boundaries of their basins of attraction will begin to approach one another (see Sect. 3.2.4) (Kaneko, 1997a, 1998a). When the distances between these become vanishingly small, a system can be caused to jump from one attractor to another by arbitrarily small perturbations. The result is a system that exhibits chaotic itinerancy as it makes transitions among a number of stable states. Thus, the situation seen in the case that the number of variables is less than 5–10 (with attractors and the boundaries of their basins separated by significant distances) gradually changes as the number of variables increases, and itinerant behavior appears.

With regard to the question of why there is a crossover at $N \approx 5$–$10$, it has been asserted that the number representing a sufficiently large system for chaotic itinerancy to occur may be the value $N$ at which $(N-1)!$ becomes larger than $2^N$. The reasoning behind this conjecture is as follows. First, we can think of the 'size' of state space as increasing with the number of elements $N$ as $\sim m^N$, where $m$ represents the number of regions in state space characterized by values of a single variable that we regard as distinct. In the simplest case, we can think of distinguishing between, for example, only negative and positive values of a given variable, in which case we would have $m = 2$. Now, the number of possible ways of forming different clusters of $N$ elements, that is, the number of ways of dividing $N$ elements into different groups, is of order $(N-1)!$. Next, note that a boundary of an attractor basin can be expressed in the form of an equation involving some combination of the variables. In many situations, an equation obtained by replacing the variables appearing in such an equation with any other variables will also represent a boundary (Kaneko, 2002d). If all possible combinations of variables yield such boundaries when substituted into this equation, then the number of boundaries in state space increases as $(N-1)!$. Thus, no matter how finely we wish to divide state space into distinct cells (i.e., no matter how large we choose $m$), as $N$ increases, the number of boundaries grows much more quickly than the size of state space, and therefore eventually they impinge upon one another.

---

tions of individual species (May, 1973). However, in the case that the populations of individual species vary chaotically, in general, there is no such problem of coexistence.

Although the argument given above is quite rough, and indeed to derive a precise number of the kind in which we are interested here would be nearly impossible, the very rapid increase of the quantity $(N-1)!$ strongly suggests that beyond some point, there is a "crowding effect" among the boundaries, and the fact that $(N-1)!$ becomes greater than $2^N$ at approximately $N=6$ indicates that perhaps in the range $N=$ 5–10, the well-separated configuration of boundaries starts to become lost.

The discussion given to this point applies to the simple case in which a set of identical elements all interact in identical manners. However, the argument that the number of combinations of elements, which increases as a factorial, grows more rapidly and will at some point surpass the size of the state space, which increases as an exponential, is both simple and quite generally valid. Thus the idea that the point beyond which the system possesses "many" elements corresponds to the point at which the factorial exceeds the exponential perhaps applies to many types of systems.

For example, Ishihara and Kaneko (2005) considered a system consisting of threshold-type elements that generate an output of $tanh(\beta(x-x_{thr})$ when given an input of $x$. Because each element yields an on–off output depending on whether or not $x > x_{thr}$, this type of system is often used as a model for neural networks and gene networks. Consider a feed-forward network composed of such elements interfering with each other in parallel. When the number of elements acting in parallel, $N$, is larger than 6, it is found that the on–off type behavior is lost, and it is replaced by a fuzzy output. (Recall the discussion given in Sect. 7.1 concerning the collapse of on–off behavior through the influence of noise.) This crossover number again can be understood from the condition $(N-1)! > 2^N$. From this consideration, it is seen why when many elements interfere, "logical" behavior of such an on–off nature is lost.

With a crossover value of order 5–10 in mind, it is quite interesting that, as discussed in Sect. 5.4, stable mutually catalytic states first appear in a system of catalytic chemical components when the number of such strongly interacting components is on the order of 5–10.[5]

The discussion given above regards variables that are mutually interacting. In general, the "magic" number in which we are interested will change with the coupling structure among the variables. Thus, to obtain a better understanding of the situation regarding the number of variables,

[5] The existence of a "magic number" on this order was first pointed out in psychology, where it was found experimentally that the number of items that can be stored in the mind for a short term is $7 \pm 2$. For example, a person can remember a 7-digit number (e.g., a phone number), but certainly not a 20-digit number. It has been argued that this magic number of $7 \pm 2$ is common to all groups of people, noting that, for example, the number of colors that people claim to see in a rainbow varies little among cultures. It is not clear, however, whether the argument given above involving combinatorial divergence offers an understanding of this phenomenon.

it is necessary to study the structure of the coupling network. First, consider the case in which there are several "clusters" of variables that are defined by the condition that variables within each cluster influence each other strongly, while variables in different clusters influence each other only weakly. Here, the situation differs from that described above. Considering that a group of fewer than 5–10 strongly interacting variables is easily stabilized, a system exhibiting hierarchical coupling structure in which there are a few weakly coupled modules consisting of several strongly interacting variables should tend to display stability. In the model of catalytic reaction network in Sect. 5.4, it has been found that modules of this kind will be formed, but as the number of constituents increases, such structure becomes unstable, and itinerant behavior appears. Then, in time, new modular structure is formed, and the itinerant behavior ceases. Thus, we see repeated alternation between the itinerancy discussed in Sect. 12.1.3 and the dynamic consolidation discussed in Sect. 12.1.2.

This behavior can also be interpreted as the repeated loss and restoration of plasticity. If the number of elements coupled strongly is few, there is a deterministic on–off output obtained from inputs, and hence the network has low plasticity, while if the number is larger, the output is fuzzy or chaotic, and the system has high plasticity. For example, in the stem cell dynamics studied in Chap. 8, many strongly interacting chemicals coexist, while the number of such chemicals is smaller for differentiated (committed) cells. The restoration of plasticity discussed in the last subsection often involves the increase of the number of strongly interacting degrees of freedom.

The problem concerning the degrees of freedom considered here is also related to symbiotic behavior. The situation in which distinct organisms exist separately corresponds to the case in which there are separate networks of variables forming closed systems. If those networks begin to interact strongly, the number of coupled variables increases. As a result of this increase, the state of the system becomes unstable, and the phenotype comes to display diversity and plasticity. Then, after some time, the variables again become separated into a number of groups, and fixed, stable states again begin to emerge.

- *Interference of slow and fast processes.* Such properties as reactivity, rate of synthesis and stability vary greatly among molecules. For example, in comparison to DNA, enzymes are synthesized and decomposed quickly. Considering such different rates, we realize that in a system consisting of a number of chemical constituents, the dynamics of variables representing the concentrations of these constituents are characterized by different time scales.

  In physics, in the case that fast and slow variables interact, the method of "adiabatic elimination" has proved useful. In this method, the following approximation is made. As seen from the point of view of the fast variables, the slow variables are nearly static. Then, with the approximation that the behavior of the fast variables can be determined by treating the slow variables as fixed, it is assumed that the fast variables are always in a state of equilibrium with respect to the values taken by the slow variables. In this way, the fast variables can be eliminated from consideration, and the description of the system is made in terms of the slow variables alone. Within this description, the fast variables evolve in an entirely "passive" manner with respect to the slow variables. In connection to this method, it is interesting to reconsider the assumption made in the field of population genetics (see Chap. 10) that the phenotype is determined uniquely by the genotype. Note that because the genotype changes only between generations, whereas the phenotype can change in the process of development during a single generation, we can interpret the genotype as a slow variable and the phenotype as a fast variable. Hence, the assumption made in population genetics is supported by the idea of adiabatic elimination.

Now, it is necessary to consider the realm in which the method of adiabatic elimination is valid, as, in fact, for certain types of time dependence of the fast variables, they cannot be considered as evolving passively. In particular, if the dynamics of the fast variables straddle a transition point (i.e., bifurcation point) at which the system makes a transition from one stable state to another, it is known that there appears a slow time scale associated with these dynamics. The dynamical behavior displayed in speciation (discussed in Chap. 11), which consists of a transition of the phenotype between two states, is precisely of this kind. This bifurcation in the phenotype then becomes fixed in the genotype, and hence, through this transition, the fast variables exercise a "reverse" influence on the slow variables. Also, in the case that the fast variables exhibit chaotic dynamics, because chaos includes motion of arbitrarily long periods, these fast variables actually possess modes much slower than the underlying time scale. Thus in this case too, the fast variables do not change passively with respect to the slow variables but, rather, mutually interact with them. (For an investigation of an actual case of this kind, see Fujimoto & Kaneko, 2003.)

When there exist such transitions or chaos, and the system evolves through the mutual influence of both slow and fast variables, it is no longer sufficient to consider only the slow variables. Here, the effective number of variables (i.e., the number of degrees of freedom) cannot be reduced through the separation of time scales. Because there are more degrees of freedom in this case, there can arise instabilities that do not exist in the case when such a separation of time scales can be carried out. Thus, situations in which there is mutual influence of slow and fast modes represent states of greater plasticity. In connection to problems involving the number of variables necessary to describe a system, it is a very important theoretical problem to obtain a more complete understanding of the interference between dynamics on different time scales.

It is interesting here to reconsider the experiments discussed in Chap. 8. There it was seen that when a cell is subjected to harsh conditions, its metabolic time scale can be increased to the point that it regains plasticity, which results in the ability to create new states. It may be possible to account for this behavior in terms of the interference of slow and fast scales, as here, the slowing down of the metabolic rate can cause its time scale to become of the same order of or slower than that of gene dynamics.

## 12.7 Developmental Phenomenology: Stability, Irreversibility, Operations, Equation of State

In this book, we have sought to lay out a proper phenomenology and ascertain the principles necessary to understand living organisms as dynamical systems. If it happens that the phenomena of life are nothing more than a complicated mixture of many processes, then the search for new principles is of no use, and the study of life would be reduced to an investigation of how these individual processes are delicately combined. Of course, it is important to understand how successful combinations are selected, but ultimately, this is merely Darwin's principle. Contrastingly, in this book, we have attempted to construct a universal phenomenology based on the dynamic relationship between the whole and the parts.

As mentioned in Chap. 1, the field of thermodynamics is an example of an almost miraculously successful phenomenological theory. There, considering appropriate types of systems, and working within certain limitations on their description, the theory has succeeded in extracting universal properties that are independent of the details of individual systems. What we contemplate here is whether such a miraculous phenomenological theory can be developed to describe the phenomena of living systems. Here, let us reconsider the feasibility of such a program as a counterpart to cellular biology (a system consisting of an ensemble of cells).

**Question 1**: The field of thermodynamics defines the concept of equilibrium states and limits its scope to the description of these states and transitions between them. It is this limitation that allowed for the construction of a successful phenomenological theory. Is it possible in the study of life to impose a limitation that can play a similar role?

At the very least, we can say that the properties of internal diversity and the ability to multiply (i.e., create nearly identical copies) are possessed by all life forms. Now, we ask if it is possible, as a counterpart to the key concept of an equilibrium state in thermodynamics, to base the construction of a phenomenological theory of life on the key concept of a *steady growth system*, or in other words, a *reproducing system*.

**Question 2**: Thermodynamics grew out of intuitive knowledge we have based on our own experience with the concepts of heat, temperature, irreversibility, etc. Do we have such knowledge of living systems from which we could create a phenomenological theory?

By comparison to our understanding of thermodynamic phenomena, our comprehension of the phenomena of living systems is more vague, and it probably cannot lead so easily to a mathematically formalized phenomenological theory. However, although it is not as clear as our direct understanding of heat or temperature, we certainly do have some intuitive idea of the qualities possessed by living creatures, and good experimental biologists have a feeling for the characteristic "vigor" or activity of organisms. Of course, we do not wish for this pursuit to degenerate into a simplistic vitalism, but we believe that it should be possible to construct a phenomenological theory of life by supplementing and strengthening our intuitive understanding with phenomenological results derived from appropriate models and developing a theory that can account for these.

**Question 3**: In the case of thermodynamics, the macroscopic state is described by a small number of variables. There, the fact that it is possible to separate the micro and macro scales makes it possible to have a phenomenology expressed in terms of just these few variables. Can something similar be accomplished in the case of living systems?

For living systems, it seems unfeasible to develop a phenomenological theory in which behavior on microscopic levels is completely separated out. If in fact this cannot be done, it would be necessary to take the point of view of complex systems, in which we consider the complementarity of the "whole" and the "parts." However, we conjecture that, to at least some degree, it is possible to describe the universal properties of living systems with a small number of variables. Indeed, when we observe creatures casually, it would seem that the number of variables that we actually apprehend is not large. Therefore, even if we cannot completely separate microscopic phenomena (e.g., the dynamics of individual chemical constituents within a cell) from macroscopic phenomena (e.g., the state of an entire cell), it is reasonable that, properly limiting the scope of our description, we should be able to realize such a separation as a useful approximation.

Of course, from another point of view, the stability that is realized as a consequence of the *inseparability* between micro and macro scales is itself a topic of study in complex systems. In Chaps. 7–9, we considered examples in which the stability of each element (one cell) and the stability of the entire set of elements (a cell complex) are realized cooperatively. In this case, the limitless number of possible microscopic states become segregated into just a few types of cells. Thus, because of this interaction of micro and macro scales, the number of variables needed to describe the system is greatly reduced, at least when we consider one particular property.

**Question 4**: In thermodynamics, stability and irreversibility, which are intimately related properties of a system, are fundamentally important concepts. In fact, the framework of thermodynamics was constructed by taking the point of view that we need only focus on stable properties and can ignore (or perhaps *must ignore*) details. Is an analogous description focusing on stability possible for living systems?

Now, note that irreversibility and stability are also key concepts with regard to living organisms. Of course, in the case of living systems, these properties cannot be expressed simply in terms of ordinary thermodynamic quantities, and it is important to keep in mind that what we consider here is a phenomenological theory that is very different from thermodynamics. However, it is also important to realize that in the mathematical models presented in Chaps. 7–9, stability has been observed from the level of cells to the level of entire cell aggregates, and some degree of understanding of this stability has been obtained from a dynamical systems point of view. In particular, the control of cell-type distributions in the process of cell differentiation (treated in Chap. 8) is such that the system responds to an external perturbation so as to cancel its effect, in a manner analogous to that described by the Le Chatelier principle of thermodynamics. Also, cell differentiation is characterized by irreversibility as it progresses from stem cells to the final, committed cells. These considerations lead us to believe that indeed it may be possible to construct a phenomenological theory of life based on the concepts of stability and irreversibility (Table 12.1).

**Question 5**: Is there some quantity analogous to thermodynamic entropy in terms of which irreversibility can be expressed?

Although it is likely difficult to find a single quantity with which this can be done, we may be able to define some functions of the pattern of gene expressions that can quantify the loss of plasticity. In attempting to construct such functions, clearly, the universal distribution law of gene expression presented in Chap. 6 will certainly be important. In addition, as discussed in Chaps. 7–9, as an organism develops, there is a gradual decrease in both the diversity of chemical constituents within each cell and the instability of the dynamics (or, roughly speaking, the variability of states) displayed by their concentrations. Of course, this directionality characterizes a normal developmental process, and it should be kept in mind that there are exceptional cases. In thermodynamics, for the purpose of clearly exposing the inherent temporal directionality of the phenomena under consideration, the idea of an adiabatic process was introduced. Similarly, to illuminate the irreversibility of cellular processes, we propose to study "idealized developmental processes," in which external operations are regarded as nonexistent.[6]

**Question 6**: In thermodynamics, we define two types of quantities, extensive and intensive. Extensive quantities are those that increase when we

[6] In any case, constructing a theory that demonstrates the impossibility of "eternal youth," in analogy to the impossibility of a perpetual machine of the second kind demonstrated by thermodynamics, would be one worthy goal.

combine two identical systems (e.g., volume and entropy), while intensive quantities are those that remain unchanged (e.g., pressure and temperature). It is not clear, however, whether we can apply similar ideas of additivity to living systems.

Indeed, the existence of an organism that could be scaled to any size seems an utter impossibility, as in most cases, the development process of an organism ceases after it reaches a certain size. In connection to this phenomenon, the question of how each cell "knows" the size of the entire organism is of fundamental importance in the study of living systems. Because there is such a special characteristic size in the case of an organism, the type of scale invariance found in thermodynamics does not hold here. However, we believe that it is possible to account for this "knowledge" possessed by cells as a property of complex systems in which the whole and the parts are mutually determined through their inherent complementarity. Such an understanding could play an important role in the construction of a phenomenological theory of life.

**Question 7**: In the development of thermodynamics, such idealized models as the Carnot cycle played major roles. Can we find models of this kind to play similar roles in the present context?

The Carnot cycle is a repeatable cycle extracted as an idealization from a class of irreversible natural phenomena. Although the Carnot cycle played a crucial role in the development of thermodynamics, once this theory came into being, it could be established independently of the Carnot cycle. Analogously, the models presented in this book were developed as idealizations of the complicated phenomena displayed by living systems. As in the case of the Carnot cycle, if we can formulate a general theory capable of accounting for such fundamental phenomena as stable processes of development and irreversible processes of differentiation, these models too will become unnecessary.[7] Now, it is not unreasonable to speculate that the concept of idealized reversible processes will play an important role here as in thermodynamics. With this in mind, in analogy to the Carnot cycle, perhaps we should consider recent somatic clone experiments [Gurodn et al., 1975, Campbell et al., 1996], as these too present the possibility of repeatable cycles: allowing irreversible differentiation to progress from embryonic stem cells to somatic cells, and then creating embryonic stem cells for the next generation from these somatic cells.[8]

**Question 8**: Although thermodynamics is an independent theory, it is believed to be consistent with the microscopic mechanical description provided by statistical mechanics. However, if the complementarity of micro and macro scales is a necessary aspect of biological systems, then would

[7] In general, we believe that models should act as "scaffolds" to reach conclusions that are independent of the models themselves.

[8] It is important to determine experimentally the degree to which such a process can approximate truly repeatable cycles.

it not be the case that the phenomenological theory we seek cannot be confined to the macro domain?

As stated repeatedly throughout this book, living systems are reproducing systems, and as such they possess the universal properties characterizing such systems. For example, there is the universal statistical law of recursively reproducing systems, presented in Chap. 6. In the theory treated in Chaps. 7–9, the instability of the internal state is important. There, through the instability of the reaction dynamics, a stable developmental process is realized. Similarly, in statistical mechanics, the existence of unstable trajectories (e.g., as in the case of high-dimensional chaos) on the microscopic level precludes a stable description at that level, and this leaves the macroscopic variables considered in thermodynamics as providing the only stable description of the system. The situation regarding the process of development is similar, but in this case, the analogous instability exists in a system in which even the number of degrees of freedom (i.e., the number of cells) is changing. Thus, the underlying dynamics here are not ordinary high-dimensional chaos, but what we might term "open chaos" (Kaneko, 1994b).

There are several other differences between the systems treated within ordinary thermodynamics and the systems we consider. First, because the number of molecules existing within a cell is much smaller than Avogadro's number, a description that treats the stability of trajectories in a phase space of continuous variables is quite likely insufficient. Indeed, to proceed with such an approach, it is probably necessary to first define the very concept of an unstable trajectory representing discrete variables in a noisy system. Second, within a cell, the number of species of molecular constituents is much larger than that in ordinary chemical reactions. Thus, in cellular systems, the limit we consider is not the thermodynamic limit, in which the total number of molecules of each species is large (on the order of or larger than Avogadro's number), but the limit in which the number of species is large (Table 12.1).

## 12.8 Toward an Understanding of the Dynamics of Cognition and Human Society

The keen reader may have realized that some of the phenomena we have treated in this book – differentiation through mutual interaction of elements possessing internal states, formation of rules through system dynamics, control by minority constituents, itinerancy and formation of recursive states – are relevant not only to cellular biology but also to cognition and the dynamics of human society (i.e., history).

With regard to cognition, the point of view we propose differs from that of so-called artificial intelligence, which advocates the description of the

**Table 12.1.** Plan for a systematic comparison to be carried out between thermodynamics and a proposed phenomenological theory of cellular biology

| Field | Complex systems cellular biology | Thermodynamics |
| --- | --- | --- |
| Stability | Cell state, cellular aggregate, organism | Macroscopic |
| Macroscopic stability with respect to perturbation | Regulation of differentiated cell type ratios | Le Chatelier–Braun principle |
| Irreversibility | Creation of committed cells from stem cells, developmental process as a whole | The second law of thermodynamics |
| Quantity defining direction of irreversibility | Degree of variation in pattern of gene expressions (?) | Thermodynamic entropy |
| Cyclic phenomena | Somatic cell clone cycle (?) | Carnot cycle |
| Naive expression on impossibility | Eternal youth (?) | Perpetual machine (of the second kind) |
| Nature of microscopic instability | Open chaos (?) | (Molecular) chaos |
| Large number of relevance | Number of molecular species | Avogadro's number |

cognition process through symbolic logic. In this study, we believe that the important problem is to determine how rules depending on the conditions of the system emerge from the dynamics of its state and how these rules become fixed. From this point of view, we see that properly adapting the theory presented in this book, it can almost certainly be applied to investigating the development of the cognition process as well.

In this book, we have not considered systems in which there exists some fixed set of functional modules with predetermined properties. Rather, we have studied processes of development in which interactions among elements lead to their differentiation and emergence as units possessing context-dependent function. Interestingly, this is precisely the fundamental point at which human cognition differs from processes carried out on computers, as the former does not take place within a system consisting of fixed modules with predefined functions governed by predetermined rules. Also, unlike computers, in the case of the human thought process, a mistake does not result in the system coming to a halt. When humans encounter some such difficulty, they have the ability to change the rules under which they operate and thereby overcome the dilemma and proceed. The differentiation law discussed in Chap. 8 clearly implies a similar property. According to this law, through differentiation, rules come to form in a plastic system. Then, if there is change in the conditions to which the system is subject, its state will become unstable through the interactions among its degrees of freedom. Finally, new rules will form as this unstable state evolves into a new stable state. Phenomena of this kind must be addressed in the study of cognition. Also, the theory treated in the contexts of development and evolution that describes the fixing of patterns and parameter values in a previously plastic state may be applicable here. Specifically, it may be possible to extend this to the process of symbolization that takes place in the brain. In particular, the dynamic consolidation of fast-mode phenomena to slow-mode phenomena should be a key concept in elucidating the dynamics of memory.

Finally, it may also be possible to apply the theoretical ideas presented in this book to the study of human group behavior and the evolution of human society. In this regard, first note that some of the ideas concerning cell differentiation also apply to humans. As the population of a human group increases, eventually there will occur a kind of differentiation in which hierarchical structure is formed. Then (in analogy to the consolidation of the phenotype into the genotype, discussed in Chap. 11), there will be a fixing of this hierarchical structure into societal norms.[9] Expanding these ideas further, we may be able to understand such phenomena as the behavioral norms of individuals in a society, economic and social structure, and forms of civilizations as temporal manifestations characterizing the state of a system that evolves through the circulating interactions of micro and macro scales. Then, history could be interpreted as transitions between such states. In 1979, Braudel, one of the greatest historians of the 20th century, proposed that the development of social systems can be understood as proceeding through the interference of phenomena existing on three different time scales: material civilization, commerce (exchange), and capitalism. We believe that if his work can be reconstructed from the point of view presented in this book, a new theory for the study of history can be developed (Kaneko & Yasutomi, 2002, 2005). Of course, at the present time, introducing this problem as a topic for scientific inquiry would likely receive little support. However, if our approach can first yield a breakthrough in life science, the climate for its acceptance will undoubtedly change, and turning our efforts to the construction of such a theory of history will receive serious attention.

[9] It should be possible to construct a mathematical model that describes the reproduction of a variety of mutually catalyzing commodities, in analogy to the models we have seen describing the reproduction of a variety of mutually catalyzing chemical species in living systems. Here, the emergence of a system that maintains diversity while exhibiting a process of reproduction corresponds to the appearance of a capitalistic society.

# References

Abraham, R. and Shaw, C. (1988) *Dynamics – The Geometry of Behavior*, vols. 1–4 (Aerial Press, Santa Cruz, CA).

Alberts, B., Bray, D., Lewis, J., Raff, M., Roberts, K., and Watson, J.D. (1983, 1989, 1994, 2002) *The Molecular Biology of the Cell.*

Alligood, K., Sauer, T., and Yorke, J.A. (1997) *CHAOS: An Introduction to Dynamical Systems* (Springer).

Almaas, E., Kovacs, B., Vicsek, T., Oltvai, Z.N., and Barabási, A.-L. (2004) Global organization of metabolic fluxes in the bacterium *Escherichia coli. Nature* **427**, 839–843.

Altmeyer, S. and McCaskill, J.S. (2001) Error threshold for spatially resolved evolution in the Quasispecies Model. *Phys. Rev. Lett.* **86**, 5819–5822.

Alon, U., Barkai, N., Notterman, D.A., Gish, K., Ybarra, S., Mack, D., and Levine, A.J. (1999) Broad Patterns of Gene Expression Revealed by Clustering Analysis of Tumor and Normal Colon Tissues Probed by Oligonucleotide Arrays. *Proc. Natl. Acad. Sci. U.S.A.* **96**, 6745–6750.

Ancel, L.W. (2000) Undermining the Baldwin expediting effect: Does phenotypic plasticity accelerate evolution? *Theor Popul Biol.* **58**, 307–319.

Ancel, L.W. and Fontana, W. (2002) Plasticity, evolvability, and modularity in RNA. *J. Exp. Zool.* **288**, 242–283.

Ariizumi, T. and Asashima, M. (2001) In vitro induction systems for analyses of amphibian organogenesis and body patterning. *Int. J. Dev. Biol.* **45**, 273–279.

Bachman, P.A., Luisi, P.L., and Lang, J. (1992) Autocatalytic self-replicating micelles as models for prebiotic structures. *Nature* **357**, 57–59.

Baguna, J., Salo, E., and Auladell, C. (1989) Regeneration and pattern formation in planarians. *Development* **107**, 77–86.

Baldwin, M. (1896) A new factor in evolution. *Am. Nat.* **30**, 441–451; 536–553.

Barabási, A.-L., and Albert, R. (1999) Emergence of scaling in random networks. *Science* **286**, 509–512.

Banerjee, B., Balasubramanian, S., Ananthakrishna, G., Ramakrishnan, T.V., and Shivashankar, G.V. (2004) Tracking operator state fluctuations in gene expression in single cells. *Biophys. J.* **86**, 3052–3059.

Ben-Jacob, E., Schochet, O., Tenenbaum, A., Cohen, I., Czirok A., and Vicsek, T. (1994) Generic modelling of cooperative growth patterns in bacterial colonies. *Nature* **368**, 46–49.

Benner S.A., and Sismour A.M. (2005) Synthetic biology. *Nat. Rev. Genet.* **6**(7), 533–543.

Bignone, F.A. (1993) Cells–gene interactions simulation on a coupled map lattice. *Theo J. Biol.* **161**, 231.

Bjornson, C.R., Rietze, R.L., Reynolds, B.A., Magli, M.C., and Vescovi A.L. (1999) Turning brain into blood: A hematopoietic fate adopted by adult neural stem cells in vivo. *Science* **283**(5401), 534–537.

Boerlijst, M. and Hogeweg, P. (1991) *Physica 48D* **17**; Spiral wave structure in prebiotic evolution – Hypercycles stable against parasites.

Bonner, J.T. (1980) *The Evolution of Culture in Animals* (Princeton University Press).

Bohr, N. (1958) *Atomic Physics and Human Knowledge* (Wiley).

Braudel, F. *Civilisation Materielle, Economie er Capitalisme XV-XVIII Siecle.*

Brillouin, L. (1969) *Science and Information Theory* (Academic Press).

Buss, L.W. (1987) *The Evolution of Individuality* (Princeton University Press).

Cairns, J., Overbaugh, J., and Miller, S. (1988) The origin of mutants. *Nature* **335**, 142–145.

Cairns-Smith, A.G. (1986) *Clay Minerals and the Origin of Life*

Callahan, H.S., Pigliucci, M., and Schlichting, C.D. (1997) Developmental phenotypic plasticity: Where ecology and evolution meet molecular biology. *Bioessays* **19**, 519–525.

Campbell, K.H.S., McWhir, J., Ritchie, W.A., and Wilmut, I. (1996) Sheep cloned by transfer from a cultured cell line. *Nature* **380**, 64–66.

Chow, M., Yao, A., and Rubin, H. (1994) Cellular epigenesis: Topochronology of progressive spontaneous transformation of cells under growth constraint. *Proc. Nat. Acad. Sci. U.S.A.* **91**, 599–603.

Coyne, J.A. and Orr, H.A. (1998) The evolutionary genetics of speciation. *Phil. Trans. R. Soc. Lond.* **B 353**, 287–305.

David, N.D. and MacWilliams, H. (1978) Regulation of the self-renewal probability in Hydra stem cell clones. *Proc. Nat. Acad. Sci. U.S.A.* **75**(2), 886–890.

Darwin, C. (1859) *On the Origin of Species by Means of Natural Selection or the Preservation of Favored Races in the Struggle for Life* (Murray, London).

Dieckmann, U. and Doebeli, M. (1999) On the origin of species by sympatric speciation. *Nature* **400**, 354–357.

Doebeli, M. (1996) A quantitative genetic competition model for sympatric speciation. *J. Evol. Biol.* **9**, 893–909.

Dobzhansky, T. (1937, 1951) *Genetics and the Origin of Species* (Columbia University Press, New York).

Dolmetsch, R.E., Xu, K., and Lewis, R.S. (1998) Calcium oscillations increase the efficiency and specificity of gene expression. *Nature* **392**, 933–936.

Douarin, N.M., and Dupin, E. (1993) Cell lineage analysis in neural crest ontogeny. *J. Neurobiol.* **24**(2), 146–161.

Driever, W. and N.üsselein-Volhard, C. (1988) A gradient of bicoid protein in Drosophila embryos. *Cell* **54**, 83–93.

Dyson, F. (1985) *Origins of Life* (Cambridge University Press).

Edwards, A.W.F. (2000) *Foundations of Mathematical Genetics* (Cambridge University Press).

Eigen, M. (1992) *Steps Towards Life* (Oxford University Press).

Eigen, M. and Schuster, P. (1979) *The Hypercycle* (Springer).

Eigen, M., McCaskill, J., and Schuster, P. (1989) The Molecular Quasi-species. *Adv. Chem. Phys.* **75**, 149–263.

Einstein, A. (1905) Über die von der molekularkinetischen Theorie der Wärme geforderte Bewegung von in ruhe nden Flüssigkeiten suspendierten Teilchen. *Ann. der Physik*, **17**, 549–560.

Einstein, A. (1906) Zur Theorie der Brownschen Bewegung, *Ann. der Physik*, **19**, 371–381.

Elowitz, M. and Shapiro, J., Time-lapse images of *E. coli* cells expressing GFP, figures provided by James Shapiro, under special courtesy (2003).

Elowitz, M.B., and Leibler, S. (2000) A synthetic oscillatory network of transcriptional regulators. *Nature* **403**, 335–338.

Elowitz, M.B., Levine, A.J., Siggia, E.D., and Swain, P.S. (2002) Stochastic gene expression in a single cell. *Science* **297**, 1183–1186.

Farmer, J.D., Kauffman, S.A., and Packard, N.H. (1986) Autocatalytic replication of polymers. *Physica D* **22D**, 50.

Fischer, E.P. and Lipson, C. (1998) *Thinking about Science: Max Delbruck and the Origins of Molecular Biology* (W.W. Norton & Co.).

Fisher, R.A. (1930, 1958) *The Genetical Theory of Natural Selection* (Oxford University Press).

Felsenstein, J. (1981) Skepticism towards Santa Rosalia, or why are there so few kinds of animals? *Evolution* **35**, 124–138.

Fontana, W., and Buss, L.W. (1994). The arrival of the fittest: Toward a theory of biological organization. *Bull. Math. Biol.* **56**, 1–64.

Forgacs, G., and Newman, S.A. (2005) *Biological Physics of the Developing Embryo* (Cambridge University Press).

Fujimoto, K. and Kaneko, K. (2003) How fast elements can affect slow dynamics. *Physica* **180D**, 1–16.

Furusawa, C. and Kaneko, K. (1998a) Emergence of rules in cell society: Differentiation, hierarchy, and stability. *Bull. Math. Biol.* **60**, 659–687.

Furusawa, C. and Kaneko, K. (1998b) Emergence of multicellular organisms with dynamic differentiation and spatial pattern. *Artif. Life* **4**, 78–89.

Furusawa, C. and Kaneko, K. (2000a) Origin of complexity in multicellular organisms. *Phys. Rev. Lett.* **84**, 6130–6133.

Furusawa, C. and Kaneko, K. (2000b) Complex organization in multicullarity as a necessity in evolution. *Artif. Life* **6**, 265–281.

Furusawa, C. and Kaneko, K. (2001) Theory of robustness of irreversible differentiation in a stem cell system: Chaos hypothesis. *J. Theor. Biol.* **209**, 395–416.

Furusawa, C. and Kaneko, K. (2002) Origin of multicellular organisms as an inevitable consequence of dynamical systems. *Anat. Rec.* **268**, 327–342.

Furusawa, C. and Kaneko, K. (2003a) Zipf's Law in gene expression. *Phys. Rev. Lett.* **90**, 088102.

Furusawa, C. and Kaneko, K. (2003b) Robust development as a consequence of generated positional information. *J. Theor. Biol.* **224**, 413–435.

Furusawa, C. and Kaneko, K. (2006a) Evolutionary origin of power-laws in Biochemical Reaction Network; embedding abundance distribution into topology. *Phys. Rev. E* **73**, 011912.

Furusawa, C. and Kaneko, K. (2006b) Morphogenesis, Plasticity, and Irreversibility. *Int. J. Dev. Biol.*, **50**, 223–232.

Furusawa, C., Suzuki, T., Kashiwagi, A. Yomo, T., and Kaneko, K. (2005) Ubiquity of log-normal distributions in intra-cellular reaction dynamics, *Biophysics*, **1**, 25.

Futuyma, D.J. ( 1986) *Evolutionary Biology, 2nd edn.*, (Sinauer Associates Inc., Sunderland, MA).

Freeman, W. and Skarda, C.A. (1985) Spatial EEG patterns, nonlinear dynamics and perception: The Neo-Sherringtonian view. *Brain Res. Rev.* **10**, 147.

Ganti, T. (1975) Organization of chemical reactions into dividing and metabolizing units: the chemotons. *Biosystems* **7**, 189.

Gehring, W. ( 1998) *Master Control Genes in Development and Evolution: The Homeobox Story* (Yale University Press).

Geritz, S.A.H., Kisdi, E., Meszena, G., and Metz, J.A.J. (1998) Evolutionary singular strategies and the adaptive growth and branching of the evolutionary tree. *Evol. Ecol.* **12**, 35–57.

Gilbert, S.F., Opitz, J.M., and Raff, R.A. (1996) Resynthesizing evolutionary and developmental biology. *Dev. Biol.* **173**, 357–372.

Glass, L. and Kauffman, S. (1973) The logical analysis of continuous non!linear biochemical control networks, *J. Theor. Biol.* **39**, 103–129.

Goodsell, D.S. (1998) *The Machinery of Life* (Springer).

Gold, T. (1998) *The Deep Hot Biosphere* (Springer).

Goldbeter, A. (1996) *Biochemical Oscillations and Cellular Rhythms* (Cambridge University Press).

Gould, S.J., and Eldredge, N. (1977) Punctuated equilibria: The tempo and mode of evolution reconsidered, *Paleobiology* **3**, 115–151.

Goodwin, B. (1963) *Temporal Organization in Cells* (Academic Press, London).

Golden, J.W., Robinson, S.J., and Haselkorn, R. (1985) Rearrangement of nitrogen fixation genes during heterocyst differentiation in the cyanobacterium Anabaena, *Nature* **314**, 419–423.

Greenwald, I. and Rubin, G.M. (1992) Making a difference: The role of cell–cell interactions in establishing separate identities for equivalent cells *Cell* **68**, 271–281.

Grey, D., Hutson, V. and Szathmary, E.A. (1995) re-examination of the stochastic corrector model. *Proc. R. Soc. Lond.* **B262**, 29–35.

Goodwin, B. ( 1963) *Temporal Organization in Cells* (Academic Press, London).

Gurdon, J.B., Laskey, R.A., and Reeves, O.R. (1975) The developmetal capacity of nuclei transplanted from keratinized skin cells of adult frogs. *J. Embriol. Exp. Morphol.* **34**, 93–112.

Gurdon, J.B., Lemaire, P., and Kato, K. (1993) Community effects and related phenomena in development *Cell* **75**, 831–834.

Haken H. (1979) *Synergetics* (Springer).

Haldane, J.B.S. (1949) Suggestion as to quantitative measurement of the rate of evolution, *Evolution* **3**, 51–56.

Hanczyc, M., Fujikawa, S.M., and Szostak, J.W. (2003) Experimental models of primitive cellular compartments: Encapsulation, growth, and division science. *Science* **302**, 618–622.

Hasty, J., Pradines, J., Dolnik, M., and Collins, J.J. (2000) Noise-based switches and amplifiers for gene expression. *Proc. Natl. Acad. Sci. U.S.A.* **97**, 2075–2080.

Hess, B. and Boiteux, A. (1971) Oscillatory phenomena in biochemistry. *Ann. Rev. Biochem.* **40**, 237–258.

Hess, B. and Mikhailov, A. (1994) Self-organization in Living Cells, *Science* **264**, 223.

Hofbauer, J. and Sigmund, K. (1988) *Dynamical Systems and the Theory of Evolution* (Cambrige University Press).

Hogeweg, P. (1994) Multilevel evolution: Replicators and the evolution of diversity. *Physica* **75D**, 275–291.

Hotani, H. and Miyamoto, H. (1990) Dynamic features of microtubles as visualized by dark-field microscopy. *Adv. Biophys.* **26**, 135–156.

Holland, J.H., Escaping brittleness: The possibilities of general purpose learning algorithms applied to parallel rule-based systems. In: *Machine Learning II*, ed by R.S. Mishalski, J.G. Carbonell, and T.M. Mitchell, (Kaufman, 1986).

Holman, E. (1987) Recognizability of sexual and asexual species of Rotifers. *Syst. Zool.* **36**, 381–386.

Holmes, L.B. (1979) Penetrance and expressivity of limb malformations *Birth Defects. Orig. Artic. Ser.* **15**, 321–327.

Houchmandzadeh, B., Wieschaus, E., and Leibler, S. (2002) Establishment of developmental precision and proportions in the early Drosophila embryo. *Nature* **415**, 798–802.

Howard, D.J. and Berlocher, S.H. (eds.) (1988) *Endless Form: Species and Speciation* (Oxford University Press).
Hu, M., et al. (1997) Multilineage gene expression precedes commitment in the hemopoietic system. *Genes & Dev.* **11**, 774–785.
Hubbel S.P. (2001) *The Unified Neutral Theory of Biodiversity and Biogeography*, Princeton Univ. Press.
Ishikawa, K., Sato, K., Shima, Y., Urabe, I., and Yomo, T. (2004) Expression of a cascading genetic network within liposomes, *FEBS Lett.* **576** (3), 387–390.
Ishihara, S. and Kaneko, K. (2005) Magic number $7 \pm 2$ in networks of threshold dynamics. *Phys. Rev. Lett.* **94**, 058102.
Ishijima, A., Kojima, H., Funatsu T., Tokunaga M., Higuchi H. and Yanagida, T. (1998) Simultaneous observation of individual ATPase and mechanical events by a single myosin molecule during interaction with actin. *Cell* **92**, 161.
Ikeda, K., Otsuka, K., and Matsumoto, K. (1989) *Prog. Theor. Phys. Suppl.* **99**, 295. Maxell–Bloch turbulence.
Ikegami, T. and Hashimoto, T. (1996) Active mutation in self-reproducing networks of machines and tapes. *Artif. Life* **2**, 305–318.
Ito, Y., Kawama, T., Urabe, I., and Yomo, T. (2004) Evolution of an arbitrary sequence in solubility. *J. Mol. Evol.* **58**, 196–202.
Jablonka, E., Marion, J. (1995) *Epigenetic Inheritance and Evolution: The Lamarckian Dimension* (Oxford University Press).
Jain, S. and Krishna, S. (2002) Large extinctions in an evolutionary model: The role of innovation and keystone species. *Poc Nat. Acad. Sci. U.S.A.* **99**, 2055–2060.
Jeon, K.W. (1972) Development of cellular dependence on infective organisms: Micrurgical studies in amoebas. *Science* **176**, 1122–1123.
Jeong, H., Tombor, B., Albert, R., Oltvai, Z.N., and Barabási, A.-L. (2000) The large-scale organization of metabolic networks. *Nature* **407**, 651–654.
Jeong, H., Mason, S.P., Barabási, A.-L. (2001) *Nature* **411**, 41.
Jones, S.J., et al. (2001) *Genome Res.* **11**(8), 1346. SAGE Data is available from *http://elegans.bcgsc.bc.ca/SAGE/*.
Kaneko, K. (1990) Clustering, coding, switching, hierarchical ordering, and control in network of chaotic elements. *Physica* **41D**, 137–172.
Kaneko, K. (1991) Globally coupled circle maps. *Physica* **54D**, 5–19.
Kaneko, K., ed. (1992) CHAOS focus issue on coupled map lattices. *Chaos* **2**, 279–407.
Kaneko, K., ed. (1993) *Theory and Applications of Coupled Map Lattices* Wiley.
Kaneko, K. (1994a) Relevance of clustering to biological networks. *Physica* **75**D, 55.
Kaneko, K. (1994b) Chaos as a source of complexity and diversity in evolution. *Artif. Life* **1**, 163–177.

Kaneko, K. (1997a) Dominance of milnor attractors and noise-induced selection in a multi-attractor system. *Phys. Rev. Lett.* **78**, 2736–2739.
Kaneko, K. (1997b) Coupled maps with growth and death: An approach to cell differentiation. *Physica* **103D**, 505–527.
Kaneko, K. (1998a) On the strength of attractors in a high-dimensional system: Milnor attractor network, robust global attraction, and noise-induced selection. *Physica* D, **124**, 322–344.
Kaneko, K. (1998b) Diversity, stability, recursivity, hierarchy, and rule generation in a biological system studied as intra-inter dynamics. *Int. J. Mod. Phys. B.* **12**, 285–298.
Kaneko, K. (1998c) Life as complex systems: Viewpoint from intra-inter dynamics. *Complexity* **3**, 53–60.
Kaneko, K. (ed.) (2001) Biophysics of Complex Systems, (Kyoritsu Pub.), in Japanese
Kaneko, K. (2002a) Kinetic origin of heredity in a replicating system with a catalytic network. *J Biol. Phys.* **28**, 781–792.
Kaneko, K. (2002b) From coupled dynamical systems to biological irreversibility. *Adv. Chem. Phys.* **122**, 53–73.
Kaneko, K. (2002c) Symbiotic sympatric speciation: compliance with interaction-driven phenotype differentiation from a single genotype. *Populat. Ecol.* **44**, 71–85.
Kaneko, K. (2002d) Dominance of minlnor attractors in globally coupled dynamical systems with more than $7 \pm 2$ degrees of freedom. *Phys. Rev. E.* **66**, 055201(R).
Kaneko, K. (2003a) *Organization Through Intra–Inter Dynamics in Origination of Organismal Form: Beyond the Gene in Developmental and Evolutionary Biology* (The Vienna Series in Theoretical Biology) (MIT Press).
Kaneko, K. (2003b) Recursiveness, switching, and fluctuations in a replicating catalytic network, *Phys. Rev. E.* **68**, 031909.
Kaneko, K. and Yasutomi, A. (2005) Braudel's viewpoint on hisotory and complex systems in *Perspectives in Economics Theory* (Nippon-Hyoron Pub.) in Japanese.
Kaneko, K. (2005a) On recursive production and evolvabilty of cells: Catalytic reaction network approach. *Adv. Chem. Phys.* **130**, 543–598.
Kaneko, K. (2005b) Inter-intra molecular dynamics as an iterated function system. *J. Phys. Soc. Jpn.* **74**, 2386–2390.
Kaneko, K. and Furusawa, C. (2000) Robust and irreversible development in cell soceity as a general consequence of intra–inter dynamics. *Physica* A **280**, 23–33.
Kaneko, K. and Furusawa, C. (2006) An evolutionary relationship between genetic variation and phenotypic fluctuation. *J. theor. Biol.* **240**, 78–86.
Kaneko, K. and Ikegami, T. (1992) Homeochaos: Dynamics Stability of a symbiotic network with populationdynamics and evolving mutation rates. *Physica* **56**D, 406–429.

Kaneko, K. and Tsuda, I. (1994) Constructive complexity and artificial reality: An introduction. *Physica* **75**D, 1–10.

Kaneko, K. and Tsuda, I. (2000) *Complex Systems: Chaos and Beyond – A Constructive Approach with Applications in Life Sciences* (Springer), pp 1–273.

Kaneko, K. and Tsuda, I., eds. (2003) Chaos focus issue on chaotic itinerancy. *Chaos.*

Kaneko, K. and Yasutomi, A. (2002) *History as Inter-Intra-Dynamics* (in Japanese) (in Perspectives in Social History).

Kaneko, K. and Yomo, T. (1994) Cell division, differentiation, and dynamic clustering, *Physica* **75D**, 89–102.

Kaneko, K. and Yomo, T. (1995) A theory of differentiation with dynamic clustering. In: *Advances in Artificial Life*, ed by E. Moran et al. (Springer) pp 329–340.

Kaneko, K. and Yomo, T. (1997) Isologous diversification: A theory of cell differentiation, *Bull. Math. Biol.* **59**, 139–196.

Kaneko, K. and Yomo, T. (1999) Isologous diversification for robust development of cell society. *J. Theor. Biol.* **199**, 243–256.

Kaneko, K. and Yomo, T. (2000) Symbiotic speciation from a single genotype. *Proc. Roy. Soc.* **B267**, 2367–2373.

Kaneko, K., and Yomo, T. (2002a) On a kinetic origin of heredity: Minority control in replicating molecules. *J. Theor. Biol.* **214**, 563–576.

Kaneko, K. and Yomo, T. (2002b) Symbiotic sympatric speciation through interaction-driven phenotype differentiation. *Evol. Ecol. Res.* **4**, 317–350.

Kaplan, D. and Glass, L. (1995) *Understanding Nonlinear Dynamics* (Springer).

Kashiwagi, A., Kanaya, T., Yomo, T., and Urabe, I. (1998) How small can the difference among competitors be for coexistence to occur, *Res. Populat. Ecol.* **40**, 223.

Kashiwagi, A., Noumachi, W., Katsuno, M., Alam, M.T., Urabe, I., and Yomo, T. (2001) Plasticity of fitness and diversification process during an experimental molecular evolution. *J. Mol. Evol.* **52**, 502–509.

Kashiwagi, A., Urabe, I., Kaneko, K., and Yomo, T. (2005), Adaptive response of a mutually inhibitory gene network to environmental changes by attractor selection. Manuscript submitted for publication.

Kawata M. & Yoshimura J. (2000) *Speciation by sexual selection in hybridizing populations without viability selection. Ev. Ec. Res.* **2**, 897–909.

Kauffman, S.A. (1969) Metabolic stability and epigenesis in randomly constructed genetic nets. *J. Theor. Biol.* **22**, 437.

Kauffman, S.A. (1986) Autocatalytic sets of proteins. *J. Theor. Biol.* **119**, 1–24.

Kauffman, S.A. (1993) *The Origin of Order* (Oxford University Press).

Kenyon, C. (1985) Cell lineage and the control of *Caenorhabditis elegans* development. *Philos. Trans. R. Soc. Lond. (Biol.)* **312**, 21–38.

Kimura, M. (1983) *The Neutral Theory of Molecular Evolution* (Cambridge University Press).

Kirk, D.L. and Harper, J.F. (1986) Genetic, biochemical, and molecular approaches to volvox development and evolution, *Int. Rev. Cytol.*, **99**, 217.

Ko, E., Yomo, and T., Urabe, I. (1994) Dynamic Clustering of bacterial population. *Physica* **75D**, 81–88.

Koch A.L. (1984) Evolution vs the number of gene copies per primitive cell. *J. Mol. Evol.* **20**(1), 71–6 .

Kondrashov, A.S. and Kondrashov, A.F (1999) Interactions among quantitative traits in the course of sympatric speciation, *Nature* **400**, 351–354.

Krishna, S., Banerjee, B., Ramakrishnan, T.V., and Shivashankar, G.V. (2005) Stochastic simulations of the origins and implications of long-tailed distributions in gene expression, *Proc. Nat. Acad. Sci. U.S.A.* **102**, 4771–4776.

Kubo, R., Toda, M., and Hashitsume. N. (1972) *Statistical Physics* in Japanese (English translationi is published from Springer, 1985).

Kuznetsov, V.A., Knott, G.D., and Bonner, R.F. (2002) General statistics of stochastic process of gene expression in eukaryotic cells. *Genetics* **161**, 1321–1332.

Lacalli, T.C. and Harrison, L.G. (1991) From gradient to segments: Models for pattern formation in early Drosophila embryogenesis, *Sems. Dev. Biol.* **2**, 107–117.

Lande, R. (1981) Models of speciation by sexual selection on phylogenic traits, *Proc. Natl. Acad. Sci. U.S.A.* **78**, 3721–3725.

Langer, J.S. (1980) Instabilities and pattern formation in crystal growth. *Rev. Mod. Phys.* **52**, 1–28.

Langton, C., eds. (1989) *Artificial Life* (Adisson Wesley).

Langton, C., eds. (1992) *Artificial Life II* (Adisson Wesley).

Langton, C., eds. (1994) *Artificial Life III* (Adisson Wesley).

Lash, A.E., et al. (2000) *Genome Res.* **10**(7), 1051.

Lee, D.H., Severin K., Yokobayashi Y., and Ghadiri M.R. (1997) Emergence of symbiosis in peptide self-replication through a hypercyclic network. *Nature* **390**, 591–594.

Lev, A.B. and Alexander, N.T. (1994) Biophysical Thermodynamics of Intracellular Processes, (Springer).

Lorenz, E.N. (1963) Deterministic Nonperiodic Flow. *J. Atmos. Sci.* **20**, 130.

Matsuyama, T., and Matsushita, M. (1993) Fractal morphogenesis by a bacterial cell population. *Crit. Rev. Microbiol.* **19**, 117–35.

Matsuyama S., Furusawa C., Todoriki M., Urabe I., and Yomo T. (2004) Global change in Escherichia coli gene expressions in initial stage of symbiosis with Dictyostellium cells. *Biosystems* **73**, 163–171.

Mandelbrot, B.B. (1953) Jeux de communication, Publ. de l'Inst de Statistique de l'Univ de Paris.

Matsuura, T., Yomo, T., Yamaguchi, M., Shibuya, N., Ko-Mitamura, E.P., Shima, Y., and Urabe, I. (2002) Importance of compartment formation for a self-encoding system, *Proc. Nat. Acad. Sci. U.S.A.* **99**, 7514–7517.

Margulis, L. (1981) *Symbiosis in Cell Evolution* (W.H. Freemand and Company).
May, R. (1973) *Stability and Complexity in Model Ecosystems* (Princeton University Press).
May R.M. (1999) Unanswered questions in ecology, Philosophical Transactions of the Royal Society B: Biological Sciences **354**, 1951–1959.
Maynard-Smith, J. (1979) Hypercycles and the origin of life, *Nature* **280**, 445–446.
Maynard -Smith, J. (1989) *Evolutionary Genetics* (Oxford University Press).
Maynard-Smith, J. (1966) Sympatric Speciation, *The American Naturalist* **100**, 637–650.
Maynard-Smith, J. and Szathmary, E. (1995) *The Major Transitions in Evolution* (W.H. Freeman).
Maynard-Smith, J., Burian, R., Kauffman, S., Alberch, P., Campbell, J., Goodwin, B., Lande, R., Raup, D., and Wolpert, L. (1985) Developmental constraints and evolution. *Q. Rev. Biol.* **60**, 265–287.
Meinhardt, H. and Gierer, A. (2000) Pattern formation by local self-activation and lateral inhibition. *Bioessays.* **22**, 753–60.
Mezard, M., Parisi, G., and Virasoro, M.A. eds. (1987) *Spin Glass Theory and Beyond* (World Science Publication).
Mikhailov, A. and Hess, B. (1995) Fluctuations in living cells and itracellular traffic. *J. Theor. Biol.* **176**, 185–192.
Mikhailov, A.S. and Calenbuhr, V. (2002) *From Cells to Societies* (Springer).
Miller, M.B. and Bassler, B.L. (1973) Quorum sensing in bacteria. *Annu. Rev. Microbiol.* **55**, 165–199.
Miller, G.A. (1975) *The Psychology of Communication* Basic Books, New York,.
Mills, D.R., Kramer F.R., and Spiegelman, S. (1973) Complete nucleotide sequence of a replicating RNA molecule, *Science* **180**, 916.
Mills, D.R., Peterson R.L., and Spiegelman, S. (1967) An extracellular Darwinian experiment with a self-duplicating nucleic acid molecule, *Proc. nat. Acad. Sci. U.S.A.* **58**, 217.
Milnor, J. (1985) On definition of an attractor. *Comm. Math. Phys.* **99**, 177 (1985); **102**, 517.
Mjolsness, E., Sharp, D.H., and Reinitz, J. (1991) A connectionist model of development. *J. Theor. Biol.* **152**, 429–453.
Muller, H.J. (1964) Mutat Res. 1; 1–9. The Relation of Recombination to Mutational Advance.
Nakagawa, N. and Kaneko, K. (2004) Autonomous energy transducer: Proposition, example, basic characteristics. *Physica* A **338**, 511–536.
Nakahata, T., Gross, A.J., and Ogawa, M. (1982) A stochastic model of self-renewal and commitment to differentiation of the primitive hemopoietic stem cells in culture. *J. Cell. Phy.* **113**, 455.
von Neumann, J. (1966) *Theory of Self-Reproducing Automata*, ed by A.W. Burks (University of Illinois Press).

Newman, S.A. (1994) Generic physical mechanisms of tissue morphogenesis: A common basis for development and evolution. *J. Evol. Biol.* **7**, 467–488.

Newman, S.A. (2003) From physics to development: The evolution of morphogenetic mechanism to appear in *Origins of Organismal Form* ed by G.B. Müller and S.A. Newman (MIT Press, Cambridge).

Newman, S.A. and Comper, W.D. (1990) Generic physical mechanisms of morphogenesis and pattern formation. *Development* **110**, 1–18.

Nicolis, G. and Prigogine, I. (1977) Self-organization in Nonequilibrium Systems. (Wiley).

Noireaux V. and Libchaber A., A vesicle bioreactor as a step toward an artificial cell assembly. *Proc. Natl. Acad. Sci. U.S.A.* **101**, 17669–17674

Ogawa, M. (1993) Differentiation and proliferation of hematopoietic stem cells. *Blood* **81**, 2844.

Oparin A.I. (1967) The origin of life. In: *Origin of Life*, ed by J.D. Bernal (Wiesenfeld and Nicholson).

Oosawa, F. (2000) The loose coupling mechanism in molecular machines in living cells, *Genes to Cells* **5**, 9–16.

Oosawa, F. (2001) Autonomy, Spontanienty, and individuality. in Biophysics of Complex Systems, (Kyoritsu Pub.) ed. K. Kaneko, in Japanese.

Oosawa, F. Biophysics, Maruzen Pub. (1998), in Japanese.

Oosawa, F. and Hayashi, S. (1986) The loose coupling mechanism in molecular machines in living cells, *Adv. Biophys.* **22**, 151–183 .

Opitz, J.M. (1981) Some comments on penetrance and related subjects, *Am. J. Med. Genet.* **8**, 265–274.

Osawa, M., et al. (1996) Long-term lymphohematopoietic reconstitution by a single CD34-low/negative hematopoietic stem cell. *Science* **273**, 242–245.

Preston F.W. (1962) The Canonical Distribution of Commonness and Rarity: Part II, Ecology, **43**, 410–432.

Roberts, M.S., and Cohan, F.M. (1995) Recombination and migration rates in natural populations of *Bacillus subtilis* and *Bacillus mojavensis*, *Evolution* **49**, 1081–1094.

Rosen, R. *Dynamical System Theory in Biology*, (Wiley and Sons, 1970).

Ronen, M., Rosenberg, R., Shraiman, B., and Alon, U. (2002) Assigning numbers to the arrows: Parameterizing a gene regulation network by using accurate expression kinetics, *Proc Natl. Acad. Sci. U.S.A.* **99**, 10555–10560.

Rubin, H. (1990) The significance of biological heterogeneity. *Cancer Metastasis Rev.* **9**, 1–20

Rubin, H. (1994a) Cellular epigenetics: Control of the size, shape, and spatial distribution of transformed foci by interactions between the transformed and nontransformed cells. *Proc. Nat. Acad. Sci. U.S.A.* **91**, 1039–1043.

Rubin, H. (1994b) Experimental control of neoplastic progression in cell populations; Fould's rules revisited. *Proc. Nat. Acad. Sci. U.S.A.* **91**, 6619–6623.

Salzar-Ciudad, I., Garcia-Fernandez, J., and Sole, R.V. (2000) Gene networks capable of pattern formation: from induction to reaction-diffusion. *J. Theor. Biol.* **205**, 587–603.

Sato K., Ito Y., Yomo, T., and Kaneko, K. (2003) On the Relation between fluctuation and response in biological systems. *Proc. Natl. Acad. Sci. U.S.A.* **100**, 14086–14090.

Sato, K., Obinata, K., Sugawara, T., Urabe, I., and Yomo, T. (2005). Quantification of structural properties of cell-sized individual liposomes by flow cytometry.

Schrödinger, E. (1946) "*What Is Life* (Cambridge University Press).

Segré, D., Ben-Eli, D., and Lancet, D. (2000) Compositional genomes: prebiotic information transfer in mutually catalytic noncovalent assemblies, *Proc. Natl. Acad. Sci. U.S.A.* **97**, 4112–4117.

Shannon, C. and Weaver, W. (1949) The Mathematical Theory of Communication, (Univ. of llinois Press).

Shapiro, J.A. (1995) Adaptive mutation: Who's really in the garden?, *Science* **268**, 373–374.

Shapiro, J.A., Dworkin M., ed. (1997) *Bacteria As Multicellular Organisms*. Oxford. Oxford University Press.

Shnerb, N.M., Louzoun, Y., Bettelheim, E., and Solomon, S. (2000) The importance of being discrete: Life always wins on the surface. *Proc. Natl. Acad. Sci. U.S.A.* **97**, 10 322.

Simon, H.A. (1955) On a class of skew distribution functions Biometrika **42**, 425–440.

Solari, F. and Ahringer J. (2000) NURD-complex genes antagonise Ras-induced vulval development in *Caenorhabditis elegans*, *Curr. Biol.* **10**, 223–226.

Solé, R.V., Bascompte, J., and Vallis, J. (1992) Nonequilibrium dynamics in lattice ecosystems *Chaos* **2**, 387.

Sprinzak D. and Elowitz M.B. (2005) Reconstruction of genetic circuits. *Nature* **438**, 443–448.

Spitze, K. and Sadler, T.D. (1996) Evolution of a generalist genotype: Multivariate analysis of the adaptiveness of phenotypic plasticity, *Am. Nat.* **148**, 108–123.

Spudich, J.L. and Koshland D.E., Jr. (1976) Non-genetic individuality: Chance in the single cell. *Nature* **262**, 467–471.

Stadler, P.F. and Schuster, P. (1990) Dynamics of small autocatalytic reaction networks–I. Bifurcations, permanence and exclusion. *Bull. Math Biol.* **52**, 485–508.

Stadler, P.F., Fontana, W., and Miller, J.H. (1993) Random catalytic reaction networks, *Physica* **63**D, 378.

Stange, P., Mikhailov, A.S., and Hess, B. (1998) Mutual synchronization of molecular turnover cycles in allosteric enzymes, *J. Phys. Chem.* B **102**, 6273.

Sternberg, P.W. and Han, M. (1988) Genetics of RAS signalling in *C-elegans*, Trends Genet. **14**, 466–472.

Steward, F.C., Mapes, M.O., and Mears, K. (1958) Growth and organized develoment of cultured cells. II. Organization in cultures from freely suspended cells, *Am. J. Bot.* **45**, 705–708.

Strogatz, S. (2001) *Nonlinear Dynamics and Chaos: With Applications to Physics, Biology, Chemistry, and Engineering*, (Perseus Books).

Suel, G.M., Garcia-Ojalvo, J., Liberman, L.M., and Elowitz, M.B. (2006) "An excitable gene regulatory circuit induces transient cellular differentiation." *Nature*. **440**, 545–50.

Sunami, T., Sato, K., Tsukada, K., Matsuura, T., Urabe, I., and Yomo, T. (2005) Population analysis of liposomes with protein synthesis and a cascading genetic network. In: *Protocells*, ed by Pacakrd et al. (MIT Press).

Szostak, J.W., Bartel, D., and Luisi, P.L. (2001) Synthesizing life, *Nature* **409**, 387–390.

Szathmary, E., and Demeter, L. (1987) Group selection of early replicators and the origin of life. *J. Theor. Biol.* **128**, 463–86.

Szathmary, E. and Maynard-Smith, J. (1995) The major evolutionary transitions. *Nature* **374**, 227.

Szathmary E. and Maynard-Smith J. (1997) From replicators to reproducers: The first major transitions leading to life, *J. Theor. Biol.* **187**, 555–571.

Takagi, H., and Kaneko, K. (2001) Differentiation and replication of spots in a reaction diffusion system with many chemicals. *Europhys. Lett.* **56**, 145–151.

Takagi, H., and Kaneko, K. (2002) Dynamic relationship between diversity and plasticity of cell types in multi-cellular state. In: *Proceedings of the Eighth International Conference of Artificial Life.*

Takagi, H. and Kaneko, K. (2005) Dynamical systems basis of metamorphosis: Diversity and plasticity of cellular states in reaction diffusion network. *J. Theor. Biol.* **234**, 173–186.

Takagi, H., Kaneko, K., and Yomo, T. (2000) Evolution of genetic code through isologous diversification of cellular states, *Artif. Life* **6**, 283–305.

Takakura K. and Sugawara, T. (2004) Membrane dynamics of a myelin-like giant multiamellar vesicle applicable to a self-reproducing system. *Langmuir* **20**, 3832–3834.

Takakura, K., Toyota, T., and Sugawara, T. (2003) A novel system of self-reproducing giant vesicles. *J. Am. Chem. Soc.* **125**, 8134.

Tam, L. and Kirk, D.L. (1991) The program for cellular differentiation in Volvox carteri as revealed by molecular analysis of development in a gonidialess/somatic regenerator mutant, *Development*, **112**, 571.

Till, J.E., McCulloch, E.A., and Siminovitch, L. (1964) A stochastic model of stem cell proliferation, based on the growth of spleen colony-forming cells. *Proc. Natl. Acad. Sci. U.S.A.* **51**, 29.

Tilman, D. (1976) Ecological competition between algae: Experimental confirmation of resource-based competition theory, *Science* **192**, 463–465.

Tilman, D. (1981) Test of resource competition theory using four species of lake Michigan algae, *Ecology* **62**, 802–815.

Todoriki, M., Oki, S., Matsuyama, S.-I., Urabe, I., and Yomo, T. (2002a) Uniqe colony housing the coexisting *Escherichia coli* and *Dictyostelium discoideum*. *J. Biol. Phys.* **28**(4), 793–797.

Todoriki, M., Oki, S., Matsuyama, S., Ko-Mitamura, E.P., Urabe, I., and Yomo, T. (2002b) An observation of the initial stage towards a symbiotic relationship. *BioSystems.* **65**, 105–112

Togashi, Y. and Kaneko, K. (2001) Transitions induced by the discreteness of molecules in a small autocatalytic system. *Phys. Rev. Lett.* **86**, 2459.

Togashi, Y. and Kaneko, K. (2003) Alteration of chemical concentrations through discreteness-induced transitions in small autocatalytic systems. *J. Phys. Soc. Jpn.* **72**, 62–68.

Togashi, Y. and Kaneko, K. (2004). Discreteness of molecules in reaction-diffusion systems can induce a novel steady state, *Phys. Rev. E* **70**, 020901(R).

Togashi,Y. and Kaneko, K. (2005) Discreteness-induced stochastic steady state in reaction diffusion systems: Self-consistent analysis and stochastic simulations. *Physica* D **205**, 87–99.

Tsuda, I. (1991a) In: *Neurocomputers and Attention* (Manchester university press) pp 430.

Tsuda, I. (1991b) Chaotic itinerancy as a dynamical basis of Hermeneutics in brain and mind. *World Futures* **32**, 167.

Tsuda, I. (1992) Dynamic link of memory–chaotic memory map in nonequilibrium neural networks. *Neural Networks* **5**, 313.

Turner, G.F. and Burrows M.T. (1995) A model for sympatric speciation by sexual selection. *Proc. R. Soc. Lond.* **B 260**, 287–292.

Tyson, J.J., Novak B., Odell G.M., Chen K., and Thron C.D. (1996) Chemical kinetic theory: Understanding cell-cycle regulation. *Trends Biochem. Sci.* **21**(3), 89–96.

Turing, A.M. (1952) The chemical basis of morphogenesis *Phil. Trans. Roy. Soc.* **B 237**, 37–72.

Ueda, H.R., et al. (2004) Universality and flexibility in gene expression from bacteria to human. *Proc. Natl. Acad. Sci. U.S.A.* **101**(11), 3765–3769.

Ueda, M., Sako, Y., Tanaka, T., Devreotes, P., and Yanagida, T. (2001) Single-molecule analysis of chemotactic signaling in Dictyostelium cells. *Science* **294**, 864–867.

Uochi, T. and Asashima, M. (1996) Sequential gene expression during pronephric tubule formation in vitro in Xenopus ectoderm. *Dev. Growth Diff.* **38**, 625–634.

de Visser, J.A.G.M., et al. (2003) Evolution and detection of genetic robustness. *Evolution* **57**, 1959–1972.

Veening, J.W., Hamoen, L.W., and Kuipers, O.P. (2005) Phosphatases modulate the bistable sporulation gene expression pattern in Bacillus subtilis. *Mol. Microbiol.* **56**, 1481–1494.

Velculescu V.E., Zhang L., Zhou W., Vogelstein J., Basrai M.A., Bassett D.E. Jr., Hieter P., Vogelstein B., Kinzler K.W. (1997) Characterization

of the yeast transcriptome. *Cell* **88**, 243: SAGE Data is available from http://www.sagenet.org/.
Velculescu, V.E., Zhang L., Vogelstein B., Kinzler K.W. (1995) Serial analysis of gene expression. *Science* **270**, 484.
Volkov, E.L., Stolyarov M.N., and Brooks, R.F. (1992) The modelling of heterogeneity in proliferative capacity during clonal growth. In: *Proceedings of the Lebededv Physics Institute Biophysical Approach to Complex Biological Phenomena*, Vol. **194**, ed. by E. Volkov (Nova Publishers).
Wačhtershäuser, G. (1990) Evolution of the first metabolic cycles. *Proc. Natl. Acad. Sci. U.S.A.* **87**, 200–204.
Watson, J.D. (1965) *Molecular Biology of the Gene* Benjamin (Revised 1970, 1975).
Waddington, C.H. (1957) *The Strategy of the Genes* (George Allen and Unwin L.D., Bristol).
Weinig, C. (2000) Plasticity versus canalization: Population differences in the timing of shade-avoidance responses., *Evolution* **54**, 441–451.
Weismann, A. (1893) *The Germ-plasm; A Theory of Heredity* (Charles Scribner's Sons, New York).
West-Eberhard, M.J. (2003) *Developmental Plasticity and Evolution* (Oxford University Press).
Wiener, N. (1948) *Cybernetics, or Control and Communication in the Animal and the Machine* (John Wiley).
Wilson, A.B., Noack-Kunnmann, K., and Meyer, A. (2000) Incipient speciation in sympatric Nicaraguan crater lake cichlid fishes: Sexual selection versus ecological diversification *Proc. Roy. Soc. Lond.* **B 267**, 2133–2141.
Wilson, E.O. (1992) *The Diversity of Life* (W.W. Norton and Company Inc.).
Wolpert, L. (1969) Positional information and the spatial pattern of cellular formation. *J. Theor. Biol.* **25**, 1–47.
Xu, W.Z., Kashiwagi, A.T. Yomo, and Urabe, I. (1996) Fate of a mutant emerging at the initial stage of evolution, *Res. Popul. Ecol.* **38**, 231–237.
Yamauchi, A., Nakashima, T., Tokuriki, N., Hosokawa, M., Nogami, H., Arioka S., Urabe I., and Yomo, T. (2002) Evolvability of random polypeptides through functional selection within a small library, *Protein Eng.* **15**, 619–26.
Yanagida, T., Arata, T., and Oosawa, F. (1985) Sliding distance of actin filament induced by a myosin cross-bridge during one ATP hydrolysis cycle. *Nature* **316**, 366–369.
Yoshida, H., Furusawa, C., and Kaneko, K. (2005) Selection of initial condition for recursive production of multicellular organisms *J. Theor. Biol.* **233**, 501–514.
Yoshimura, J., and Shields, W.M. (1987) Probablistic optimization of phenotypic distributions: A general solution for the effects of uncertainty on natural selection? *Evol. Ecol.* **1**, 125–138.
Yoon, H.-S. and Golden, J.W. (1998) Hereocyst pattern formation controlled by a diffusible peptide, *Science* **282**, 935–938.

You, L., Cox, R.S., Weiss, R., and Arnold, F.H. (2004) Ppopulation control by cell–cell communication and regulated killing. *Nature* **42**, 8868–8871.

Yates, F.E. (1980) Physical Causality and brain theories. *Am. J. Physiol.* **238**, 277.

Yu,W., Sato, K., Wakabayashi, M., Nakaishi, T., Ko-Mitamura, E.P., Shima, Y., Urabe, I., and Yomo, T. (2001) Synthesis of functional protein in liposome. *J. Biosci. Bioeng.* **92**(6), 590–593.

Zandstra, P. and Nagy, A. (2001) Stem cell bioengineering, *Annu. Rev. Biomed. Eng.* **3**, 275–305.

Zaug, A.J. and Cech, T.R. (1986) The intervening sequence RNA of *Tetrahymena* is an enzyme. *Science* **231**, 470–475.

Zipf, G.K. (1949) *Human Behavior and the Principle of Least Effort* (Addison-Wesley, Cambridge.)

# Index

# Understanding Complex Systems

## Edited by J.A. Scott Kelso

Jirsa, V.K.; Kelso, J.A.S. (Eds.)
Coordination Dynamics: Issues and Trends
XIV, 272 p. 2004 [3-540-20323-0]

Kerner, B.S.
The Physics of Traffic:
Empirical Freeway Pattern Features,
Engineering Applications, and Theory
XXIII, 682 p. 2004 [3-540-20716-3]

Kleidon, A.; Lorenz, R.D. (Eds.)
Non-equilibrium Thermodynamics
and the Production of Entropy
XIX, 260 p. 2005 [3-540-22495-5]

Kocarev, L.; Vattay, G. (Eds.)
Complex Dynamics in Communication Networks
X, 361 p. 2005 [3-540-24305-4]

McDaniel, R.R.Jr.; Driebe, D.J. (Eds.)
Uncertainty and Surprise in Complex Systems:
Questions on Working with the Unexpected
X, 200 p. 2005 [3-540-23773-9]

Ausloos, M.; Dirickx, M. (Eds.)
The Logistic Map and the Route to Chaos:
From The Beginnings to Modern Applications
XVI, 411 p. 2006 [3-540-28366-8]

Kaneko, K.
Life: An Introduction to Complex Systems Biology
XIV, 371 p. 2006 [3-540-32666-9]

Braha, D.; Minai, A.A.; Bar-Yam Y. (Eds.)
Complex Engineered Systems: Science Meets Technology
VIII, 394 p. 2006 [3-540-32831-9]

Printed in the United States
146357LV00003BA/19/P

9 783540 326663